Lecture Notes in Mathematics

Edited by A. Dold, B. Eckmann and F. Takens

1464

D.L. Burkholder
E. Pardoux
A. Sznitman

Ecole d'Eté de Probabilités de
Saint-Flour XIX – 1989

Editor: P.I Henneguin

Springer-Verlag

Berlin Heidelberg N w York London Paris
Tokyo Hong Kong Barcelona Budapest

Authors

Donald L. Burkholder
University of Illinois at Urbana-Champaign
Department of Mathematics
273 Altgeld Hall
1409 West Green Street
Urbana, IL 61801, USA

Etienne Pardoux
Université de Provence
Département de Mathématiques
Place Victor Hugo
13001 Marseille, France

Alain-Sol Sznitman
Département de Mathématiques
ETH-Zentrum
CH-8092 Zürich, Switzerland

Editor

Paul-Louis Hennequin
Université Blaise Pascal, Clermont-Ferrand
Mathématiques Appliquées
63177 Aubière Cedex, France

Mathematics Subject Classification (1980): 60-02, 31A05, 35R60, 60G35, 60G46, 60J65, 60K35, 60M15, 62M20, 70-02, 82-02, 82A31

ISBN 3-540-53841-0 Springer-Verlag Berlin Heidelberg New York
ISBN 0-387-53841-0 Springer-Verlag New York Berlin Heidelberg

© Springer-Verlag Berlin Heidelberg 1991
Printed in Germany

Printing and binding: Druckhaus Beltz, Hemsbach/Bergstr.
2146/3140-543210 – Printed on acid-free paper

INTRODUCTION

Ce volume rassemble les trois cours donnés à l'Ecole d'Eté de Calcul des Probabilités de Saint-Flour du 16 Août au 2 Septembre 1989. Nous rattrapons ainsi peu à peu le retard accumulé depuis cinq ans dans la publication.

Nous remercions les auteurs qui ont effectué un gros travail de rédaction définitive qui fait de leurs cours un texte de référence.

L'Ecole a rassemblé près de quatre vingt participants dont 28 ont présenté, dans un exposé, leur travail de recherche.

On trouvera ci-dessous la liste des participants et de ces exposés dont un résumé pourra être obtenu sur demande.

Afin de faciliter les recherches concernant les écoles antérieures, nous redonnons ici le numéro du volume des "Lecture Notes" qui leur est consacré :

Lecture Notes in Mathematics
1971 : n° 307 - 1973 : n° 390 - 1974 : n° 480 - 1975 : n° 539 -
1976 : n° 598 - 1977 : n° 678 - 1978 : n° 774 - 1979 : n° 876 -
1980 : n° 929 - 1981 : n° 976 - 1982 : n° 1097 - 1983 : n° 1117 -
1984 : n° 1180 - 1985 - 1986 et 1987 : n° 1362 - 1988 : n° 1427

Lecture Notes in Statistics
1986 : n° 50

TABLE DES MATIERES

D.L. BURKHOLDER : "EXPLORATIONS IN MARTINGALE THEORY AND ITS APPLICATIONS"(*)

(*) It is a pleasure to thank my fellow students at the Ecole d'Eté de Probabilités de Saint-Flour (1989) and especially the organizer, P.L. Hennequin, for an enjoyable meeting in a beautiful milieu.

E. PARDOUX : "FILTRAGE NON LINEAIRE ET EQUATIONS AUX DERIVEES PARTIELLES STOCHASTIQUES ASSOCIEES"

A. S. SZNITMAN : "TOPICS IN PROPAGATION OF CHAOS"

EXPLORATIONS IN MARTINGALE THEORY
AND ITS APPLICATIONS

Donald L. BURKHOLDER

EXPLORATIONS IN MARTINGALE THEORY
AND ITS APPLICATIONS[1]

Donald L. Burkholder
Department of Mathematics
University of Illinois
Urbana, Illinois 61801

[1]It is a pleasure to thank my fellow students at the École d'Été de Probabilités de Saint-Flour (1989) and especially the organizer, P. L. Hennequin, for an enjoyable meeting in a beautiful milieu.

1. INTRODUCTION

These lectures center on new and other recent work giving sharp inequalities for martingales and stochastic integrals with some applications to harmonic analysis and the geometry of Banach spaces. But to introduce some of the ideas and notation, we shall begin with an example from earlier work. At the end of the introduction is a summary of the remaining chapters.

A ±1-inequality. Let $(\Omega, \mathcal{F}_\infty, P)$ be a probability space and $\mathcal{F} = (\mathcal{F}_n)_{n \geq 0}$ a non-decreasing sequence of sub-σ-fields of $\mathcal{F}_\infty$. Suppose $d = (d_n)_{n \geq 0}$ is a real martingale difference sequence relative to $\mathcal{F}$ and $\varepsilon = (\varepsilon_n)_{n \geq 0}$ is a sequence of numbers in $\{1, -1\}$. Then [Bur 66, 84], for $1 < p < \infty$,

$$(1.1) \qquad \left\| \sum_{k=0}^{n} \varepsilon_k d_k \right\|_p \leq (p^* - 1) \left\| \sum_{k=0}^{n} d_k \right\|_p$$

where $p^* = p \vee q$ and $1/p + 1/q = 1$. Observe that $p^* - 1$ is the maximum of $p - 1$ and $1/(p - 1)$. Let f and g be the martingales defined by

$$f_n = \sum_{k=0}^{n} d_k,$$

$$g_n = \sum_{k=0}^{n} \varepsilon_k d_k.$$

Set $\|f\|_p = \sup_{n \geq 0} \|f_n\|_p$. Because $(d_0, \ldots, d_n, 0, 0, 0, \ldots)$ is also a martingale difference sequence and $\|f_0\|_p \leq \cdots \leq \|f_n\|_p$, the inequality above is equivalent to

$$(1.2) \qquad \|g\|_p \leq (p^* - 1)\|f\|_p.$$

As with any inequality (cf. [Har-Lit-Pól 34]), a number of questions can be asked:
- Is the inequality interesting?
- Is it sharp?
- When does equality hold?
- Is it important to know the best constant?
- If so, how is it possible to discover it?

Two of these questions can be answered at once. Inequality (1.2) is sharp: if $\beta < p^* - 1$, then there is a probability space and martingales f and g as above such that $\|g\|_p > \beta\|f\|_p$. Moreover, strict inequality holds if and only if $0 < \|f\|_p < \infty$ and $p \neq 2$. See [Bur 84] or the discussion below. The other questions are less precise, but we can throw some light on them.

An inequality for stochastic integrals. Suppose that $v = (v_n)_{n \geq 0}$ is a real predictable sequence relative to $\mathcal{F}$: v_n is measurable with respect to $\mathcal{F}_{0 \vee (n-1)}$. If v is uniformly bounded in absolute value by 1, then (1.2) holds with the ±1-transform g replaced by the transform g of f by v: $g_n = \sum_{k=0}^n v_k d_k$. This is an immediate consequence of (1.2) and the decomposition lemma in the Appendix. In turn, this leads to the analogous inequality for stochastic integrals where now the probability space $(\Omega, \mathcal{F}_\infty, P)$ is complete and $\mathcal{F} = (\mathcal{F}_t)_{t\geq 0}$ is a nondecreasing right-continuous family of sub-σ-fields of $\mathcal{F}_\infty$ where $\mathcal{F}_0$ contains all $A \in \mathcal{F}_\infty$ with $P(A) = 0$. Let $M = (M_t)_{t\geq 0}$ be a real martingale adapted to $\mathcal{F}$ such that the paths of M are right-continuous on $[0, \infty)$ and have left limits on $(0, \infty)$. Let $V = (V_t)_{t\geq 0}$ be a predictable process with values in $[-1, 1]$ and $N = V \cdot M$ the stochastic integral of V with respect to M: N is a right-continuous martingale with left limits such that

$$N_t = \int_{[0,t]} V_s \, dM_s \quad \text{a.s.}$$

Set $\|N\|_p = \sup_{t\geq 0} \|N_t\|_p$. Then [Bur 84, 87], for $1 < p < \infty$,

(1.3) $$\|N\|_p \leq (p^* - 1)\|M\|_p$$

and $p^* - 1$ is the best constant. Suppose that $0 < \|M\|_p < \infty$. Then equality holds if and only if $p = 2$ and $P(\int(1 - V_t^2) \, d[M, M]_t > 0) = 0$. Here $[M, M]$ is the square bracket [Del-Mey 80] of M. Note that (1.3) contains (1.2). On the other hand, (1.2) provides the key to (1.3) and has closer connections to many applications.

Unconditional constants. Consider any sequence $e = (e_n)_{n\geq 0}$ in the real Lebesgue space $L^p = L^p[0, 1)$ where, again, $1 < p < \infty$. The *unconditional constant* $\beta_p(e)$ of e is the least $\beta \in [1, \infty]$ with the property that if n is a nonnegative integer and $a_0, \ldots, a_n$ are real numbers such that $\|\sum_{k=0}^n a_k e_k\|_p = 1$, then

$$\left\| \sum_{k=0}^n \varepsilon_k a_k e_k \right\|_p \leq \beta$$

for all choices of signs $\varepsilon_k \in \{1, -1\}$. The sequence e is a *basis* of L^p if, for every $f \in L^p$, there is a unique sequence a satisfying $\|f - \sum_{k=0}^n a_k e_k\|_p \to 0$ as $n \to \infty$.

To recall the definition of the Haar system, a particularly interesting basis of L^p, we shall use the same notation for an interval and its indicator function:

$$h_0 = [0, 1), \qquad h_1 = [0, \tfrac{1}{2}) - [\tfrac{1}{2}, 1),$$
$$h_2 = [0, \tfrac{1}{4}) - [\tfrac{1}{4}, \tfrac{1}{2}), \qquad h_3 = [\tfrac{1}{2}, \tfrac{3}{4}) - [\tfrac{3}{4}, 1),$$
$$h_4 = [0, \tfrac{1}{8}) - [\tfrac{1}{8}, \tfrac{1}{4}), \qquad h_5 = [\tfrac{1}{4}, \tfrac{3}{8}) - [\tfrac{3}{8}, \tfrac{1}{2}), \ldots.$$

Using an inequality of Paley (1932) for the Walsh system, Marcinkiewicz (1937) proved that the Haar system $h = (h_n)_{n\geq 0}$ is an unconditional basis of L^p:

(1.4) $$\beta_p(h) < \infty.$$

This is also a consequence of (1.1): set $d_k = a_k h_k$. To see that d is a martingale difference sequence, use $E h_{n+1} = 0$ and the fact that any function of $h_0, \ldots, h_n$ is constant on the set where $h_{n+1} \neq 0$.

Moreover, the Haar system satisfies the following double inequality. If d is any martingale difference sequence and e is any basis of L^p, then

$$(1.5) \qquad \beta_p(d) \leq \beta_p(h) \leq \beta_p(e).$$

For the right-hand side, see [Ole 67, 75] or [Lin-Pel 71]; a convenient reference is [Lin-Tza 79]. For the left-hand side, see [Mau 75] or, for a proof resting on the simple fact that a nonnegative midpoint concave function is concave, see [Bur 86b]. Inequality (1.5) shows that h is extremal both as a basis of L^p and as a martingale difference sequence so it is natural to ask for the value of $\beta_p(h)$. But the fact that $p^* - 1$ is the best constant in (1.1) now yields

$$(1.6) \qquad \beta_p(h) = p^* - 1.$$

Indeed, $p^* - 1$ is the unconditional constant of any monotone basis of L^p. Recall that a basis e of L^p is monotone if for all real sequences a and nonnegative integers n,

$$\left\| \sum_{k=0}^{n} a_k e_k \right\|_p \leq \left\| \sum_{k=0}^{n+1} a_k e_k \right\|_p.$$

Using a theorem of Ando (1966) and our nonsharp original version of (1.1), Dor and Odell (1975) and Pełczyński and Rosenthal (1975) proved that $\beta_p(e)$ is finite for every monotone basis e. The sharp version yields $\beta_p(e) = p^* - 1$. See [Bur 84].

Contractive projections. The following inequality is closely related. It is the sharp version of an inequality of Dor and Odell (1975). Let $1 < p < \infty$. Suppose that $0 = P_0, P_1, P_2, \ldots$ is a nondecreasing sequence of contractive projections in L^p of a positive measure space: $P_k : L^p \to L^p$ is linear with a norm not exceeding 1 and $P_j P_k = P_k P_j = P_j$ if $0 \leq j \leq k$. If $a_k \in [-1, 1]$ for all $k \geq 1$, and $f \in L^p$, then [Bur 84]

$$\left\| \sum_{k=1}^{\infty} a_k (P_k - P_{k-1}) f \right\|_p \leq (p^* - 1) \|f\|_p.$$

Strict inequality holds if $p \neq 2$ and $\|f\|_p > 0$.

This inequality can be used to study the spectral representation of operators on L^p; see Doust (1989) for some recent work in this direction.

The inequality in a Banach space setting. Let $\mathbf{B}$ be a real or complex Banach space. Inequality (1.1), with or without the best constant, leads to the question of whether or not a similar inequality is valid for $\mathbf{B}$-valued martingale difference sequences. Let $\beta_p(\mathbf{B})$ be the least $\beta \in [1, \infty]$ such that if d is a $\mathbf{B}$-valued martingale difference sequence, ε is a sequence in $\{1, -1\}$, and n is a nonnegative integer, then

$$(1.7) \qquad \left\| \sum_{k=0}^{n} \varepsilon_k d_k \right\|_p \leq \beta \left\| \sum_{k=0}^{n} d_k \right\|_p.$$

Note that $\beta_p(\mathbf{B}) \geq \beta_p(\mathbf{R}) = p^* - 1$ provided $\mathbf{B}$ is nondegenerate, as we shall always assume. Another simple consequence of (1.1) and its sharpness is that $\beta_p(\ell^p) = p^* - 1$: integrate term by term. Similarly, $\beta_p(L^p) = p^* - 1$.

A Banach space $\mathbf{B}$ is *unconditional for martingale difference sequences*, which we write as $\mathbf{B} \in$ UMD, if $\beta_p(\mathbf{B})$ is finite.

Some early work on such spaces was carried out by Maurey and Pisier. For example, the class UMD does not depend on p for $1 < p < \infty$ and $\mathbf{B} \in$ UMD implies that $\mathbf{B}$ is superreflexive [Mau 75]. On the other hand, there are superreflexive spaces that are not UMD [Pis 75a].

We shall see some other properties of UMD spaces later—as well as recall (in Chapter 4) the definition of superreflexivity. Here we focus on how it is possible, at least in principle, to find the value of $\beta_p(\mathbf{B})$.

Let L be the greatest function $u \colon \mathbf{B} \times \mathbf{B} \times [0, \infty) \to \mathbf{R}$ such that the mappings

$$(x, t) \mapsto u(x, y, t) \quad \text{and} \quad (y, t) \mapsto u(x, y, t)$$

are convex on $\mathbf{B} \times [0, \infty)$, and

$$u(x, y, t) \leq \left| \frac{x + y}{2} \right|^p \quad \text{if} \quad \left| \frac{x - y}{2} \right|^p \geq t,$$

where $|\cdot|$ denotes the norm of $\mathbf{B}$. Then [Bur 81a, 84]

$$(1.8) \qquad \beta_p^p(\mathbf{B}) = 1/L(0, 0, 1) \quad \text{if} \quad L(0, 0, 1) > 0,$$

$$(1.9) \qquad \qquad = \infty \qquad \quad \text{if} \quad L(0, 0, 1) = 0.$$

It is possible to find L in the case $\mathbf{B} = \mathbf{R}$ and thus find the best constant in (1.1). This was the original method.

Here is another characterization of $\beta_p(\mathbf{B})$. Let $\beta \in [1, \infty)$. Then [Bur 86b]

$$(1.10) \qquad \beta_p(\mathbf{B}) \leq \beta$$

if and only if there is a biconcave function $u \colon \mathbf{B} \times \mathbf{B} \to \mathbf{R}$ such that, for $(x, y) \in \mathbf{B} \times \mathbf{B}$,

$$\left| \frac{x + y}{2} \right|^p - \beta^p \left| \frac{x - y}{2} \right|^p \leq u(x, y).$$

We shall recall in the next chapter the ideas behind the proof of this characterization and show how it also leads to the value of $\beta_p(\mathbf{R})$.

ζ-convex spaces. A Banach space $\mathbf{B}$ is *ζ-convex* if there is a biconvex function $\zeta \colon \mathbf{B} \times \mathbf{B} \to \mathbf{R}$ such that $\zeta(0, 0) > 0$ and

$$(1.11) \qquad \zeta(x, y) \leq |x + y| \quad \text{if} \quad |x| = |y| = 1.$$

Each of the two characterizations of $\beta_p(\mathbf{B})$ given above yields a characterization of UMD spaces. Here is a simpler characterization [Bur 81a, 86b]:

$$(1.12) \qquad \mathbf{B} \in \text{UMD} \iff \mathbf{B} \text{ is } \zeta\text{-convex}.$$

The condition of ζ-convexity has a simple martingale interpretation. Let g^*, defined by $g^*(\omega) = \sup_{n \geq 0} |g_n(\omega)|$, be the maximal function of a ± 1-transform g of a $\mathbf{B}$-valued martingale f. Then

$$(1.13) \qquad\qquad g^* \geq 1 \text{ a.s.} \implies 2\|f\|_1 \geq \zeta(0,0)$$

where ζ is any biconvex function on $\mathbf{B} \times \mathbf{B}$ satisfying (1.11). The inequality on the right is sharp for the largest possible value of $\zeta(0,0)$, which we denote by $\zeta_{\mathbf{B}}(0,0)$. Furthermore, for $1 < p < \infty$,

$$(1.14) \qquad\qquad \zeta_{\mathbf{B}}(0,0) \geq 1/\beta_p(\mathbf{B}).$$

The M. Riesz inequality. Inequality (1.2) compares the size of a martingale f with the size of a ± 1-transform of f. The M. Riesz inequality compares the size of a harmonic function with the size of a properly normalized conjugate. Let u and v be $\mathbf{B}$-valued functions harmonic in the open unit disk of the complex plane with $v(0) = 0$. Suppose that the Cauchy-Riemann equations, $u_x = v_y$ and $u_y = -v_x$, are satisfied and set

$$\|u\|_p^p = \sup_{0 < r < 1} \int_0^{2\pi} |u(re^{i\theta})|^p \, d\theta.$$

Define $\alpha_p(\mathbf{B})$ to be the least $\alpha \in [1, \infty]$ such that

$$\|v\|_p \leq \alpha \|u\|_p$$

for all pairs (u, v) as above. If $1 < p < \infty$, then $\alpha_p(\mathbf{R})$ is finite. This is the classical result of M. Riesz (1927) who also recognized its implications for the Hilbert transform. If $1 \leq p < \infty$ and $f \in L_{\mathbf{B}}^p(\mathbf{R})$, then the Hilbert transform of f is defined by

$$Hf(x) = \lim_{\varepsilon \downarrow 0} \frac{1}{\pi} \int_{|y| > \varepsilon} \frac{f(x-y)}{y} \, dy$$

provided this limit exists for almost all $x \in \mathbf{R}$, in which case

$$\|Hf\|_p \leq \alpha_p(\mathbf{B})\|f\|_p, \qquad 1 < p < \infty,$$

and, in fact, if $\alpha_p(\mathbf{B})$ is finite, it is the norm of H as an operator on $L_{\mathbf{B}}^p(\mathbf{R})$. For the case $\mathbf{B} = \mathbf{R}$, see [Rie 27] and [Zyg 59].

For convenience, we shall write $\mathbf{B} \in \text{HT}$ if $\alpha_p(\mathbf{B})$ is finite. The Hilbert transform has special significance. If $\mathbf{B} \in \text{HT}$, then, as in the scalar-valued case, the norm inequality for the Hilbert transform together with the method of rotation introduced by Calderón and Zygmund (1956) can be used to establish norm inequalities for a large class of singular integral operators on $L_{\mathbf{B}}^p(\mathbf{R}^n)$.

But under what conditions does $\mathbf{B} \in \text{HT}$? Inequality (1.1) and its $\mathbf{B}$-space version provide the key: $\mathbf{B} \in \text{UMD} \implies \mathbf{B} \in \text{HT}$. This result, due to McConnell and the author, is in [Bur 83] where it is also suggested that the converse might be true. This was verified later by Bourgain (1983). Accordingly,

$$(1.15) \qquad\qquad \mathbf{B} \text{ is } \zeta\text{-convex} \iff \mathbf{B} \in \text{UMD} \iff \mathbf{B} \in \text{HT}.$$

Differential subordination. The above is just one example of the analogy—and more—between conjugate harmonic functions and martingale transforms. There is an idea that underlies both of these concepts: differential subordination. In some of the later chapters, we shall study the consequences of this less restrictive condition.

A summary of what is to follow. Chapter 2 contains a proof of the second characterization of $\beta_p(\mathbf{B})$ given above. The method of the proof will be used again in the later chapters. This characterization is already contained in [Bur 86b] where it is used for theoretical work and not for calculation. Recently, McConnell (1989) and Hitczenko (unpublished manuscript) have used it in a similar way. They show that an inequality of Zinn (1985) for real tangent martingales carries over to tangent martingales with values in a UMD space. We shall show how this characterization can be used to calculate $\beta_p(\mathbf{R})$ without having to construct examples. Examples that are extremal or nearly extremal can be constructed more easily at a later stage. To illustrate some of the ideas further, we shall give a new proof of Doob's L^p-inequality for the maximal function of a nonnegative submartingale.

Chapter 3 contains an extension of the inequality (1.2) to differentially subordinate martingales with values in a real or complex Hilbert space $\mathbf{H}$. Among other things, it makes possible a simple proof of the square-function inequality, leads to a sharp inequality for the integral of a complex predictable process with respect to a complex martingale M, and implies that the complex unconditional constant of a monotone basis of complex $L^p[0,1)$ is p^*-1, the same as in the real case. Moreover, the inequality mentioned above for a nondecreasing sequence of contractive projections carries over to the complex case with no change in the best constant.

The results of Chapter 3 yield a sharp inequality (Theorem 4.1) for the number of ε-escapes of a differentially subordinate martingale. Chapter 4 also contains escape-inequality characterizations of Hilbert space, spaces isomorphic to a Hilbert space, UMD spaces, and superreflexive spaces.

Chapter 5 contains weak-type and L^p inequalities for differentially subordinate harmonic functions on domains of $\mathbf{R}^n$. Among other things, these extend the classical results of Kolmogorov and Riesz for conjugate harmonic functions on the open unit disk.

The number of ε-escapes of a real continuous local martingale M is studied further in Chapter 6 and, for each positive integer j, a sharp upper bound on $P(C_\varepsilon(M) \geq j)$ is obtained.

In Chapter 7, we give probability bounds on the number of ε-escapes of a harmonic function along a Brownian path to the boundary.

Chapters 8 and 9 contain sharp exponential inequalities for differentially subordinate martingales and stochastic integrals. Here is a simple consequence. If f is a real martingale, $\|f\|_\infty \leq 1, \alpha = e^2/4$, and $\lambda > 2$, then there does not exist a predictable sequence v uniformly bounded in absolute value by 1 such that

$$P\left(\sum_{k=0}^n v_k d_k \geq \lambda \text{ for some } n \geq 0\right) \geq \alpha e^{-\lambda}.$$

But if $\beta < \alpha e^{-\lambda}$, then there is a probability space, a martingale f as above, and a positive integer n such that

$$P\left(\sum_{k=0}^n (-1)^k d_k \geq \lambda\right) > \beta.$$

Chapter 10 contains a new characterization of superreflexivity. It is similar to the characterization of UMD spaces in terms of the function ζ of (1.11) above. There are no martingales in this characterization but the proof rests partly on Pisier's (1975b) identification of superreflexive spaces with those having martingale cotype.

In Chapter 11, we describe some of the recent work of Wang (1989) in which he sharpens some square-function inequalities for conditionally symmetric martingales.

Sharp martingale inequalities, particularly those for martingale transforms and stochastic integrals, are relevant to problems in the optimal control of martingales. Chapter 12 contains a recent such result of Choi (1988).

Most of the work presented in the following chapters dates from 1986 to 1990. Among the new or newly-sharpened results are Theorems 3.2, 3.4, 3.9, 4.4, 4.5, 5.2, 6.1, 7.1, 7.2, 7.3, 7.4, 8.1, 8.2, 9.1, 9.2, 10.1, and 12.1. Some of the other results, which are included here to give a more complete picture, have appeared recently in [Bur 88b, 89a, 89b]. For a more thorough treatment of some of the topics discussed as background above, see [Bur 86b] and the references given there. Since 1986, there has been a substantial amount of additional work showing that UMD spaces are the right spaces for many problems in analysis. Much of this work is not described here but we have included many of the relevant papers in the list of references.

2. SEARCHING FOR THE BEST: AN EXAMPLE

Suppose that we are interested in proving an inequality of the form

$$(2.1) \qquad \|g\|_p \le \beta \|f\|_p$$

where g is a ± 1-transform of a B-valued martingale f. The first step is to notice that it is enough to prove (2.1) for simple martingales. A martingale f is *simple* if f_n is a simple function for all nonnegative integers n and there is some nonnegative integer n such that $f_n = f_{n+1} = \cdots$. This reduction is straightforward; see [Bur 81a]. So assume that f is a simple martingale. Then g is also a simple martingale and the expectation of any function of f and g is finite

The next step is to focus on a new martingale, one that is easier to work with in this context. Let $Z_n = (X_n, Y_n)$ where

$$X_n = g_n + f_n = \sum_{k=0}^n (\varepsilon_k + 1) d_k,$$

$$Y_n = g_n - f_n = \sum_{k=0}^n (\varepsilon_k - 1) d_k.$$

The simple martingale $Z = (Z_n)_{n \ge 0}$ has the *zigzag property*: if $n \ge 1$, then either

$$(2.2) \qquad X_n - X_{n-1} \equiv 0 \quad \text{or} \quad Y_n - Y_{n-1} \equiv 0.$$

This implies that if u is a biconcave function on $\mathbf{B} \times \mathbf{B}$, then

$$(2.3) \qquad Eu(Z_n) \leq Eu(Z_{n-1}) \leq \cdots \leq Eu(Z_0).$$

To see this, notice that $Z_n = (X_n, Y_{n-1})$ or $Z_n = (X_{n-1}, Y_n)$. For example, assume that the former possibility holds. Then, by Jensen's inequality for conditional expectations,

$$E[u(Z_n)|\mathcal{F}_{n-1}] = E[u(X_n, Y_{n-1})|\mathcal{F}_{n-1}] \leq u(E[X_n|\mathcal{F}_{n-1}], Y_{n-1}) = u(Z_{n-1}).$$

Now take expectations to obtain $Eu(Z_n) \leq Eu(Z_{n-1})$ for all $n \geq 1$.

Let F be the function defined on $\mathbf{B} \times \mathbf{B}$ by

$$(2.4) \qquad F(x, y) = \left| \frac{x+y}{2} \right|^p - \beta^p \left| \frac{x-y}{2} \right|^p$$

and suppose the biconcave function u majorizes F on $\mathbf{B} \times \mathbf{B}$. Then

$$(2.5) \qquad \|g_n\|_p^p - \beta^p \|f_n\|_p^p = EF(Z_n) \leq Eu(Z_n).$$

We can assume that u satisfies the homogeneity condition

$$(2.6) \qquad u(\alpha x, \alpha y) = |\alpha|^p u(x, y), \qquad \alpha \in \mathbf{R}.$$

If u does not already satisfy (2.6), replace $u(x, y)$ by $\inf_{\lambda \neq 0} u(\lambda x, \lambda y)/|\lambda|^p$ to obtain a function that does satisfy (2.6) as well as the conditions placed on the original function. If $Z_0 = (X_0, 0)$, then

$$u(Z_0) = [u(X_0, 0) + u(-X_0, 0)]/2 \leq u(0, 0) = 0$$

so $Eu(Z_0) \leq 0$ with a similar conclusion holding for the other possibility $Z_0 = (0, Y_0)$. Therefore, by (2.5) and (2.3),

$$\|g_n\|_p^p - \beta^p \|f_n\|_p^p \leq 0,$$

which is equivalent to (2.1).

So we have proved one half of the following characterization of $\beta_p(\mathbf{B})$, already mentioned in Chapter 1:

THEOREM 2.1. *Suppose that* $\beta \in [1, \infty)$. *Then*

$$(2.7) \qquad \beta_p(\mathbf{B}) \leq \beta$$

if and only if there is a biconcave function $u: \mathbf{B} \times \mathbf{B} \to \mathbf{R}$ *such that, for* $(x, y) \in \mathbf{B} \times \mathbf{B}$,

$$(2.8) \qquad \left| \frac{x+y}{2} \right|^p - \beta^p \left| \frac{x-y}{2} \right|^p \leq u(x, y).$$

To find the value of $\beta_p(\mathbf{B})$, we can of course try to guess its value β, use the half of the theorem that we have already proved to verify that β is an upper bound as in (2.7), and then show that it is the least upper bound by constructing examples of f and g. However, it is useful to know that this method has the possibility of succeeding—that the other half of the theorem is true. The full theorem also makes it possible, at least in principle, to find the value of $\beta_p(\mathbf{B})$ before, rather than after, the construction of examples.

To prove the other half, assume that (2.7) holds. If $(x,y) \in \mathbf{B} \times \mathbf{B}$, let $\mathbf{Z}(x,y)$ be the set of all simple zigzag martingales Z on the Lebesgue unit interval $[0,1)$ that start at (x,y) and have their values in $\mathbf{B} \times \mathbf{B}$. The filtration may vary with Z. Since Z is simple, its pointwise limit Z_∞ exists. Let F be given by (2.4) as before and define $U: \mathbf{B} \times \mathbf{B} \to (-\infty, \infty]$ by

$$(2.9) \qquad U(x,y) = \sup\{EF(Z_\infty): Z \in \mathbf{Z}(x,y)\}.$$

Note that $U(x,y) \geq F(x,y)$ so (2.8) holds: choose $Z_n \equiv (x,y)$.

Let $x_1, x_2 \in \mathbf{B}$, $m_1, m_2 \in \mathbf{R}$, $U(x_i, y) > m_i$, $i = 1, 2$, and $x = \alpha x_1 + (1-\alpha)x_2$ where $0 < \alpha < 1$. Then $U(x,y) > \alpha m_1 + (1-\alpha)m_2$ so

$$(2.10) \qquad U(x,y) \geq \alpha U(x_1, y) + (1-\alpha)U(x_2, y).$$

To prove this, choose $Z^i \in \mathbf{Z}(x_i, y)$ so that $EF(Z_\infty^i) > m_i$. We can assume that Z^1 and Z^2 satisfy

$$Y_{2n+1}^i - Y_{2n}^i \equiv 0 \quad \text{and} \quad X_{2n+2}^i - X_{2n+1}^i \equiv 0.$$

Let Z be defined by $Z_0 \equiv (x,y)$ and

$$\begin{aligned} Z_{n+1}(\omega) &= Z_n^1(\omega/\alpha) & \text{if} \quad \omega \in [0,\alpha), \\ &= Z_n^2((\omega - \alpha)/(1-\alpha)) & \text{if} \quad \omega \in [\alpha, 1). \end{aligned}$$

The martingale Z is *the splice of Z^1 and Z^2 with weight α*. It is easy to check that $Z \in \mathbf{Z}(x,y)$ and

$$U(x,y) \geq EF(Z_\infty) = \alpha EF(Z_\infty^1) + (1-\alpha)EF(Z_\infty^2).$$

This implies (2.10). By symmetry, a similar inequality holds for $U(x, \cdot)$.

The next step is to show that U has its values in $\mathbf{R}$. Since it majorizes F, it is locally bounded from below. The assumption (2.7) implies that $U(0,0) \leq 0$: any $Z \in \mathbf{Z}(0,0)$ generates a pair f and g satisfying (2.1), so, by the first part of (2.5), $EF(Z_\infty) \leq 0$. But this implies that $U(x,0)$ is finite:

$$U(x,0) + U(-x,0) \leq 2U(0,0) \leq 0.$$

This implies in a similar way that $U(x,y)$ is finite. Therefore, U is a biconcave function majorizing F and the theorem is proved.

The theorem for dyadic martingales. Let $\mathcal{F}$ be the dyadic filtration of the Lebesgue unit interval: $\mathcal{F}_n$ is generated by $[0, 2^{-n}), \ldots, [1-2^{-n}, 1)$, $n \geq 0$. Let $\beta_p^0(\mathbf{B})$ be the analogue of $\beta_p(\mathbf{B})$ for martingales with respect to this dyadic filtration. Then

$$(2.11) \qquad \beta_p^0(\mathbf{B}) \leq \beta$$

if and only if there is a midpoint biconcave function u satisfying (2.8). The proof is the same as the one given above except that Z is always a simple zigzag martingale relative to the dyadic filtration and the splicing is with weight $1/2$. Note that the midpoint concave functions $u(\cdot, y)$ and $u(x, \cdot)$ are locally bounded from below and therefore are concave. So Theorem 2.1 holds with (2.7) replaced by (2.11). This gives a conceptually simple proof of Maurey's (1975) observation that

$$\beta_p^0(\mathbf{B}) = \beta_p(\mathbf{B}).$$

The left-hand side of (1.5) also follows since $\beta_p(h) = \beta_p^0(\mathbf{R})$: it is easy to see that if d is a martingale difference sequence relative to the dyadic filtration, then there is a sequence a such that $d_0 = a_0 h_0$, $d_1 = a_1 h_1$, $d_2 = a_2 h_2 + a_3 h_3$, $d_3 = a_4 h_4 + \cdots + a_7 h_7, \ldots$.

A new derivation of the best constant in the real case. Can the search for a biconcave function on $\mathbf{B} \times \mathbf{B}$ satisfying the conditions of Theorem 2.1 be replaced by a search for a suitable function on $\mathbf{B}$? Here is the answer for $\mathbf{B} = \mathbf{R}$. Let F be as before.

LEMMA 2.1. *Suppose that $\beta \in [1, \infty)$. Then*

$$\beta_p(\mathbf{R}) \le \beta$$

if and only if there is a concave function $w : \mathbf{R} \to \mathbf{R}$ majorizing $F(\cdot, 1)$ such that

$$(2.12) \qquad\qquad w(x) = |x|^p w(1/x), \qquad x \ne 0.$$

PROOF: Suppose there is such a function w. Its concavity implies that it is continuous and satisfies

$$(2.13) \qquad\qquad w(0) \le 0 :$$

by (2.12), $w(0) > 0$ would imply that $w(x) \to \infty$ as $x \to \infty$ and also as $x \to -\infty$, a contradiction. Define u on $\mathbf{R}^2$ by

$$\begin{aligned} u(x, y) &= |y|^p w(x/y) & \text{if } \quad y \ne 0, \\ &= |x|^p w(0) & \text{if } \quad y = 0. \end{aligned}$$

It is easy to check that u majorizes F and that $u(x, y) = u(y, x)$. For example, if $x, y \ne 0$, then, by (2.12),

$$u(y, x) = |x|^p w(y/x) = |y|^p w(x/y) = u(x, y)$$

and the other cases are equally easy. Using (2.13) and the concavity of w, we see that $u(\cdot, y)$ is concave so, by symmetry, u is biconcave. As a consequence, Theorem 2.1 gives $\beta_p(\mathbf{R}) \le \beta$.

In the other direction, $\beta_p(\mathbf{R}) \le \beta$ implies the existence of a biconcave majorant u of F satisfying (2.6) so $w = u(\cdot, 1)$ is a concave majorant of $F(\cdot, 1)$ satisfying (2.12).

LEMMA 2.2. *Let $\beta \in [1, \infty)$. There exists a concave majorant w of $F(\cdot, 1)$ satisfying (2.12) if and only if $p^* - 1 \leq \beta$.*

Lemmas 2.1 and 2.2 imply that $\beta_p(\mathbf{R}) = p^* - 1$.

PROOF: Let w be such a function and set $x_0 = (\beta - 1)/(\beta + 1)$. Then $0 \leq x_0 < 1$ and $w(x_0) \geq F(x_0, 1) = 0$. Similarly, $w(1) \geq F(1, 1) = 1$.

Let L be the limit of $(w(1) - w(x))/(1 - x)$ as $x \uparrow 1$ and R the limit of $(w(x) - w(1))/(x - 1)$ as $x \downarrow 1$. Because w is concave, both of these limits exist. If $0 < x < 1$, then

$$\frac{w(1) - w(x)}{1 - x} = \frac{w(1) - x^p w(1/x)}{1 - x} = \frac{1 - x^p}{1 - x} w(1) - x^{p-1} \frac{w(1/x) - w(1)}{(1/x) - 1}.$$

Letting $x \uparrow 1$, we see that $L = pw(1) - R$. Using the nonnegativity of $w(x_0)$ and the concavity of w, we obtain

$$(2.14) \qquad pw(1) = L + R \leq 2L \leq 2\frac{w(1) - w(x_0)}{1 - x_0} \leq (\beta + 1)w(1)$$

and the strict positivity of $w(1)$ implies that $\beta \geq p - 1$.

The proof of the inequality $\beta \geq p^* - 1$ can be completed by showing in a similar way that $\beta \geq q - 1$. If $-1 < x < 0$, then

$$\frac{w(x) - w(-1)}{x - (-1)} = -w(-1)\frac{1 - |x|^p}{1 - |x|} - |x|^{p-1}\frac{w(1/x) - w(-1)}{(1/x) - (-1)}$$

and this implies that

$$-pw(-1) \geq 2\frac{w(x_0) - w(-1)}{x_0 - (-1)} \geq -w(-1)\frac{\beta + 1}{\beta}.$$

Since $w(-1) + w(1) \leq 2w(0)$, we see that $w(-1)$ is strictly negative. Therefore, $p \geq (\beta + 1)/\beta$ or, equivalently, $\beta \geq 1/(p - 1) = q - 1$.

To prove the converse, we can assume that $\beta = p^* - 1$.

Let $p = 2$. In this case we can take $w(x) = F(x, 1) = x$.

Now suppose that $p > 2$. Then $F(x, 1)$ is concave on the interval $(-\infty, x_0]$ and also on the interval $[1/x_0, \infty)$. Furthermore, $\beta = p^* - 1 = p - 1$ and equality must hold throughout (2.14) for the desired function w. So w must satisfy $L = R = w(1)/(1 - x_0)$ with $w(x_0) = 0$. This implies that w must be of the form $w(x) = ax - b$ on the interval $[x_0, 1]$. So, by (2.12), $w(x) = ax^{p-1} - bx^p$ on the interval $[1, 1/x_0]$. Accordingly, $L = R$ implies that $b = a(1 - 2/p)$. Since $w(x_0) = F(x_0, 1)$, in order for w to be concave and majorize $F(\cdot, 1)$ it must satisfy

$$w'(x_0+) = \partial F(x, 1)/\partial x|_{x=x_0}$$

and this gives the equation $a = \frac{1}{2}p^2(1 - 1/p)^{p-1}$. These equations for a and b determine w uniquely on the interval $[x_0, 1/x_0]$. Outside of this interval, w can be defined by $w(x) = F(x, 1)$ but there is no uniqueness. In fact, let

$$(2.15) \qquad w(x) = \alpha_p \left\{ \left|\frac{x+1}{2}\right| - (p^* - 1)\left|\frac{x-1}{2}\right| \right\} \left\{ \left|\frac{x+1}{2}\right| + \left|\frac{x-1}{2}\right| \right\}^{p-1}$$

for all $x \in \mathbf{R}$ where $\alpha_p = p(1 - 1/p^*)^{p-1}$. Then, as can be checked, w is also a concave majorant of $F(\cdot, 1)$ on $\mathbf{R}$ satisfying (2.12). This function of course agrees with the earlier one on the interval $[x_0, 1/x_0]$ and was suggested by it.

Finally suppose that $1 < p < 2$. Similar reasoning leads to the function w defined in (2.15) This completes the proof of the lemma.

On the construction of examples. The point of this proof of $\beta_p(\mathbf{R}) = p^* - 1$ is that $p^* - 1$ is discovered in the course of the proof. Our original approach was to construct examples of f and g and these led to the inequality $\beta_p(\mathbf{R}) \geq p^* - 1$. Guessing that $p^* - 1$ might be the correct value, we were able to show that the reverse inequality does hold; see [Bur 84] and, for a shorter proof, [Bur 85]. The derivation of $\beta_p(\mathbf{R})$ given above does not require knowing any examples of f and g such that the ratio $\|g\|_p / \|f\|_p$ is near $p^* - 1$. Such examples (one is contained in [Bur 82]) can be constructed easily by first constructing appropriate zigzag martingales. The least biconcave majorants that appear above and in similar problems provide the guide. Consult [Bur 84] or Chapter 8 below.

A new proof of Doob's maximal inequality. To illustrate further some of the ideas used above, we shall give a new proof of Doob's L^p-inequality for the maximal function of a nonnegative submartingale. Let $1 < p < \infty$ and $q = p/(p - 1)$. If $f = (f_n)_{n \geq 0}$ is a nonnegative submartingale and f^* is its maximal function ($f_n^* = |f_0| \vee \cdots \vee |f_n| \uparrow f^*$), then, as Doob (1953) has proved,

$$(2.16) \qquad \|f^*\|_p \leq q\|f\|_p.$$

To see this from a new point of view, let u and v be the functions defined on $S = \{(x, y) \in \mathbf{R}^2 : 0 \leq x \leq y\}$ by

$$u(x, y) = py^{p-1}(y - qx),$$
$$v(x, y) = y^p - q^p x^p.$$

If $y > 0$, then $u(\cdot, y)$ is linear and $v(\cdot, y)$ is concave on $[0, y]$ with u and its partial derivative u_x agreeing with v and v_x, respectively, at the point $(y/q, y)$. Therefore, $u(\cdot, y)$ majorizes $v(\cdot, y)$ on $[0, y]$ and, since both u and v vanish at the origin, u majorizes v on S. If $(x, y) \in S$ and $x + h \geq 0$, then

$$u(x + h, (x + h) \vee y) \leq u(x, y) - pqy^{p-1}h.$$

Equality holds if $x + h \leq y$; strict inequality holds if $x + h > y$. Therefore, for $n \geq 1$,

$$u(f_n, f_n^*) = u(f_{n-1} + d_n, (f_{n-1} + d_n) \vee f_{n-1}^*)$$
$$(2.17) \qquad \leq u(f_{n-1}, f_{n-1}^*) - pq(f_{n-1}^*)^{p-1}d_n$$

with strict inequality holding on the set where $f_n > f_{n-1}^*$.

To prove (2.16), we can assume that $\|f\|_p$ is finite. Then d_n, f_n, and f_n^* belong to L^p, and, by Hölder's inequality, all the terms in (2.17) are integrable. By the submartingale property, the expectation of the last term is nonnegative. Accordingly,

$$\|f_n^*\|_p^p - q^p\|f_n\|_p^p = Ev(f_n, f_n^*) \leq Eu(f_n, f_n^*)$$
$$\leq Eu(f_{n-1}, f_{n-1}^*) \leq \cdots \leq Eu(f_0, f_0^*)$$

where $Eu(f_0, f_0^*) = Eu(f_0, f_0) = p(1 - q)\|f_0\|_p^p \leq 0$, and this implies (2.16).

The inequality (2.16) also holds if f is a B-valued martingale in which case $(|f_n|)_{n \geq 0}$ is a nonnegative submartingale.

Strictness. The proof of (2.16) in [Doo 53] can be used to show that strict inequality holds if $0 < \|f\|_p < \infty$. The above method yields strictness as follows: Let m be the least nonnegative integer n such that $\|f_n\|_p > 0$. Then

$$\|f^*\|_p^p - q^p\|f\|_p^p \leq \lim_{n \to \infty} Eu(f_n, f_n^*) \leq Eu(f_m, f_m^*)$$
$$= Eu(f_m, f_m) = p(1-q)\|f_m\|_p^p < 0.$$

Sharpness—even in the dyadic case. The constant q is best possible since it is already best possible for Hardy's inequality, a classical inequality that follows easily from (2.16); for example, see [Cha 73]. By constructing suitable martingales, Wang (to appear) has proved that q is already best possible for the family of real dyadic martingales. This answers a question of Gundy and Journé (personal communication). Here is a different proof, one that rests on the fact that a nonnegative midpoint concave function is concave.

Suppose that β is a positive number such that $\|f^*\|_p \leq \beta\|f\|_p$ for all real dyadic martingales f. Then $\beta \geq q$ as can be seen as follows.

Define F and U on $\mathbf{R} \times [0, \infty)$ by $F(x, y) = y^p - \beta^p|x|^p$ and

$$U(x, y) = \sup\{EF(f_\infty, f^* \vee y)\}$$

where the supremum is taken over all simple dyadic martingales f on the Lebesgue unit interval with $f_0 \equiv x$. Then $U(0, 0) \leq 0$,

$$F(x, |x| \vee y) \leq U(x, y) = U(x, |x| \vee y),$$

and $U(\alpha x, \alpha y) = \alpha^p U(x, y)$ for all $\alpha \in \mathbf{R}$. A splicing argument similar to the one used in proving Theorem 2.1 but with weight $1/2$ shows that $U(\cdot, y)$ is midpoint concave on $\mathbf{R}$ for each $y \geq 0$. Since $U(\cdot, y)$ is locally bounded from below, it is concave.

So the function $G : \mathbf{R} \to \mathbf{R}$ defined by $G(x) = U(x, 1)$ is concave and $G(x) \geq F(x, |x| \vee 1)$. In particular, $G(0) \geq F(0, 1) = 1$. If $|x| \geq 1$, then $G(x) = U(x, |x|) = |x|^p G(1)$. All of this implies that $G(1) < 0$; otherwise, G would not be concave. Also, $\beta > 1$ since $1 - \beta^p = F(1, 1) \leq G(1) < 0$. Therefore,

$$pG(1) = G'(1+) \leq \frac{G(1) - G(1/\beta)}{1 - (1/\beta)} \leq \frac{G(1) - F(1/\beta, 1)}{1 - (1/\beta)} = \frac{G(1)}{1 - (1/\beta)}$$

so $p \geq 1/(1 - (1/\beta))$. This implies that $\beta \geq q$.

3. DIFFERENTIAL SUBORDINATION OF MARTINGALES

If v is a real predictable sequence uniformly bounded by 1 in absolute value, then (1.1) remains valid with the same best constant if the numbers ε_k are replaced by the functions v_k. A similar result holds for the analogous inequality in which the d_k are

B-valued: the constant $\beta_p(\mathbf{B})$ is unchanged if in its definition the ε_k are replaced by the real v_k. This can be seen in several different ways, for example, by using Lemma A.1. But is this also true if the v_k are complex-valued? For example, consider a mobile of the kind that Alexander Calder created in his early experiments with the concept, beginning about 1931. To each configuration of such a mobile, there corresponds a martingale with a similar arrangement of successive centers of gravity. This martingale can be expressed as the transform of a martingale corresponding to the initial configuration with both the d_k and the v_k complex-valued. In fact, $v_k(\omega)$ is a unimodular complex number. How can we compare the size of two different configurations, equivalently, the size of the two martingales that they determine? If both configurations are flat, the problem can be reduced to the comparison of real martingales as in [Bur 84]. The complex case seems to be much more difficult—and it is if the methods of the real case are tried. But there is another way.

Suppose that f and g are B-valued martingales with respect to the same filtration. Let d be the difference sequence of f and e the difference sequence of g: for all $n \geq 0$,

$$f_n = \sum_{k=0}^{n} d_k \quad \text{and} \quad g_n = \sum_{k=0}^{n} e_n.$$

Then g *is differentially subordinate to* f if, for all $\omega \in \Omega$ and $k \geq 0$,

$$(3.1) \qquad |e_k(\omega)| \leq |d_k(\omega)|.$$

Certainly, (3.1) is satisfied for transforms of the kind discussed above. It is also satisfied, as we shall see, for martingales that arise naturally in other quite different settings.

The following theorem shows that (1.2) holds for differentially subordinate martingales taking values in $\mathbf{R}$, in $\mathbf{C}$, or in any real or complex Hilbert space $\mathbf{H}$. There is no increase in the value of the best constant.

THEOREM 3.1. *Let* $1 < p < \infty$ *and suppose that* f *and* g *are* H-*valued martingales with respect to the same filtration. If* g *is differentially subordinate to* f, *then*

$$(3.2) \qquad \|g\|_p \leq (p^* - 1)\|f\|_p$$

and the constant $p^* - 1$ *is best possible. In the nontrivial case* $0 < \|f\|_p < \infty$, *there is equality in (3.2) if and only if* $p = 2$ *and equality holds in (3.1) for almost all* ω *and all* $k \geq 0$.

Let $U, V \colon \mathbf{H} \times \mathbf{H} \to \mathbf{R}$ be defined by

$$(3.3) \qquad U(x, y) = \alpha_p(|y| - (p^* - 1)|x|)(|x| + |y|)^{p-1}$$

and

$$(3.4) \qquad V(x, y) = |y|^p - (p^* - 1)^p |x|^p$$

where $\alpha_p = p(1 - 1/p^*)^{p-1}$. Note that if $u \colon \mathbf{R}^2 \to \mathbf{R}$ is defined as in the proof of Lemma 2.1 using the function w of (2.15), then

$$U(x, y) = u(|y| + |x|, |y| - |x|),$$

and if F is the function of Chapter 2 with $\beta = p^* - 1$, then

$$V(x, y) = F(|y| + |x|, |y| - |x|).$$

Therefore, U majorizes V, which can also be seen directly.

To prove (3.2), we can assume that $\|f\|_p$ is finite. Then

$$\|g_n\|_p^p - (p^* - 1)^p \|f_n\|_p^p = EV(f_n, g_n) \le EU(f_n, g_n).$$

The next step, see [Bur 88b] for the details, is to show that

$$EU(f_n, g_n) \le \cdots \le EU(f_0, g_0).$$

The property of U that can be used to establish this is the following: if x, y, h, k are in H and $|k| \le |h|$, then the mapping

$$t \mapsto U(x + ht, y + kt)$$

is concave on $\mathbf{R}$. Inequality (3.2) then follows from

$$EU(f_0, g_0) \le \alpha_p (2 - p^*) \|f_0\|_p^p \le 0.$$

Since the constant is already best possible in the special case (1.2), it is best possible in (3.2). Note also that we can assume in the case $0 < \|f\|_p < \infty$ that $\|f_0\|_p$ is strictly positive and so, for $p \neq 2$, we obtain

$$EU(f_0, g_0) < 0$$

and this implies that strict inequality holds in (3.2). If $p = 2$, we can use the fact that $\|f\|_2^2 = \sum_{k=0}^{\infty} \|d_k\|_2^2$.

The complex unconditional constant of the Haar basis. Let $1 < p < \infty$. Pełczyński (1985) conjectured that the complex unconditional constant of the Haar basis of $L_{\mathbf{C}}^p[0, 1]$ is the same as the unconditional constant for the real case, equivalently, $\beta_p(h, \mathbf{C}) = \beta_p(h)$ where $\beta_p(h)$ is defined as in Chapter 1 and $\beta_p(h, \mathbf{C})$ is defined to be the least β such that if n is a nonnegative integer, $c_0, \ldots, c_n$ are complex numbers, and $\theta_0, \ldots, \theta_n$ are real numbers, then

$$\left\| \sum_{k=0}^{n} e^{i\theta_k} c_k h_k \right\|_p \le \beta \left\| \sum_{k=0}^{n} c_k h_k \right\|_p.$$

By Theorem 3.1, we have that $\beta_p(h, \mathbf{C}) \le p^* - 1$ and it is clear that $\beta_p(h, \mathbf{C}) \ge \beta_p(h)$. Therefore, by (1.6), the complex unconditional constant of the Haar basis is the same as the unconditional constant for the real case. See [Bur 88a] for a slightly different proof and for related results.

Square-function inequalities. Theorem 3.1 leads to the following sharp inequality.

THEOREM 3.2. *Suppose that f and g are H-valued martingales with respect to the filtration $\mathcal{F}$. Let u_{jk} and v_{jk} be scalar-valued $\mathcal{F}_{0 \vee (k-1)}$-measurable functions such that*

$$\sum_{j=0}^{\infty} |v_{jk}(\omega)|^2 \leq \sum_{j=0}^{\infty} |u_{jk}(\omega)|^2$$

for all $\omega \in \Omega$ and $k \geq 0$. If g is differentially subordinate to f, then, for $1 < p < \infty$ and $n \geq 0$,

(3.5)
$$\left\| \left(\sum_{j=0}^{\infty} \left| \sum_{k=0}^{n} v_{jk} e_k \right|^2 \right)^{\frac{1}{2}} \right\|_p \leq (p^* - 1) \left\| \left(\sum_{j=0}^{\infty} \left| \sum_{k=0}^{n} u_{jk} d_k \right|^2 \right)^{\frac{1}{2}} \right\|_p$$

and the constant $p^* - 1$ is best possible. If the right-hand side is finite and strictly positive, then there is equality if and only if $p = 2$ and

$$\sum_{j=0}^{\infty} |v_{jk}(\omega)|^2 |e_k(\omega)|^2 = \sum_{j=0}^{\infty} |u_{jk}(\omega)|^2 |d_k(\omega)|^2$$

for almost all ω and all k satisfying $0 \leq k \leq n$.

PROOF: Let $\mathbf{K} = \ell_{\mathbf{H}}^2$, the Hilbert space of sequences $x = (x_j)_{j \geq 0}$ with $x_j \in \mathbf{H}$ and norm

$$|x|_{\mathbf{K}} = \left(\sum_{j=0}^{\infty} |x_j|^2 \right)^{\frac{1}{2}}.$$

To prove (3.5), we can assume that the right-hand side is finite for all $n \geq 0$. Let

$$D_k = (u_{jk} d_k)_{j \geq 0} \quad \text{and} \quad E_k = (v_{jk} e_k)_{j \geq 0}.$$

Then

$$|E_k|_{\mathbf{K}}^2 = |e_k|^2 \sum_{j=0}^{\infty} |v_{jk}|^2 \leq |d_k|^2 \sum_{j=0}^{\infty} |u_{jk}|^2 = |D_k|_{\mathbf{K}}^2.$$

Set $F_n = (\sum_{k=0}^{n} u_{jk} d_k)_{j \geq 0}$ and $G_n = (\sum_{k=0}^{n} v_{jk} e_k)_{j \geq 0}$ so

$$|F_n|_{\mathbf{K}} = \left(\sum_{j=0}^{\infty} \left| \sum_{k=0}^{n} u_{jk} d_k \right|^2 \right)^{\frac{1}{2}}$$

with a similar expression for $|G_n|_{\mathbf{K}}$. It is now clear that F and G are $\mathbf{K}$-valued martingales, that G is differentially subordinate to F, and that (3.5) and the statement about equality follow from Theorem 3.1 applied to such F and G.

Suppose that $u_{jk} = v_{jk}$ where $v_{0k} \equiv 1$ and $v_{jk} \equiv 0$ if $j \geq 1$. Then (3.5) becomes $\|g_n\|_p \leq (p^* - 1) \|f_n\|_p$. So in (3.5) no constant smaller than $p^* - 1$ suffices. This completes the proof of the theorem.

Recall that if f is a **B**-valued martingale, its square function $S(f)$ is defined by

$$S(f) = \left(\sum_{k=0}^{\infty} |d_k|^2 \right)^{\frac{1}{2}}.$$

THEOREM 3.3. *Let* $1 < p < \infty$. *If* f *is an* **H**-valued martingale, then

$$(3.6) \qquad (p^* - 1)^{-1} \|S(f)\|_p \leq \|f\|_p \leq (p^* - 1) \|S(f)\|_p.$$

In particular,

$$(3.7) \qquad \|f\|_p \geq (p - 1) \|S(f)\|_p \quad \text{if} \quad 1 < p \leq 2,$$
$$(3.8) \qquad \|f\|_p \leq (p - 1) \|S(f)\|_p \quad \text{if} \quad 2 \leq p < \infty,$$

and the constant $p - 1$ *is best possible. If* $0 < \|f\|_p < \infty$, *then equality holds if and only if* $p = 2$.

This improves one of the inequalities of [Bur 66] but further improvement is possible since the best constants in the cases not covered by (3.7) and (3.8) are not yet known.

Notice that (3.6) is an immediate consequence of (3.5). For example, to prove the left-hand side of (3.6), set $e_k = d_k$, $v_{jk} \equiv 1$ if $j = k$, $v_{jk} \equiv 0$ if $j \neq k$, $u_{jk} \equiv 1$ if $j = 0$, $u_{jk} \equiv 0$ if $j \geq 1$, and then let $n \to \infty$ in (3.5). The proof of the right-hand side of (3.6) is similar. Inequalities (3.7) and (3.8) follow at once. For the remainder of the proof of Theorem 3.3, see [Bur 88b].

The maximal function $f^* = \sup_{n \geq 0} |f_n|$ satisfies

$$(3.9) \qquad \|f^*\|_p \leq p \|S(f)\|_p \quad \text{if} \quad 2 \leq p \leq \infty,$$

and strict inequality holds if $0 < \|f\|_p < \infty$. This follows from (3.8) and Doob's inequality $\|f^*\|_p \leq q \|f\|_p$, which is a strict inequality when $0 < \|f\|_p < \infty$ as can be seen from his proof [Doo 53]. In (3.9) the constant p is best possible.

Klincsek (1977) proved (3.9) for $p = 3, 4, 5, \ldots$ and conjectured that it holds for all $p \geq 2$. Pittenger (1979) proved part of (3.8), namely, the case $p \geq 3$. Both work with real martingales but their proofs can be carried over to **H**. Our approach is quite different and yields (3.8) and (3.9) for the full interval $2 \leq p < \infty$ and, in addition, (3.7) for $1 < p \leq 2$.

Let $\sigma = (\sigma_n)_{n \geq 0}$ be a nondecreasing sequence of stopping times with values in $\{0, 1 \ldots \infty\}$. For convenience, let $\sigma_{-1} \equiv -1$ and $\sigma_\infty = \lim_{n \to \infty} \sigma_n$. Let u_{jk} be the indicator function of the set $\{\sigma_{j-1} < k \leq \sigma_j\}$. Define $S_n(f, \sigma)$ by

$$(3.10) \qquad S_n(f, \sigma) = \left(\sum_{j=0}^{\infty} \left| \sum_{k=0}^{n} u_{jk} d_k \right|^2 \right)^{\frac{1}{2}}.$$

Define $S_n(g, \tau)$ similarly. Inequality (3.11) below is an immediate consequence of (3.5) and, for some purposes, is a more useful inequality than (3.6), which it contains.

THEOREM 3.4. *Suppose that relative to the same filtration f and g are $\mathbf{H}$-valued martingales and σ and τ are nondecreasing sequences of stopping times such that $\tau_\infty \leq \sigma_\infty$. If g is differentially subordinate to f, then, for $1 < p < \infty$ and $n \geq 0$,*

$$(3.11) \qquad \|S_n(g, \tau)\|_p \leq (p^* - 1)\|S_n(f, \sigma)\|_p$$

and the constant $p^* - 1$ is best possible. If the right-hand side is finite and strictly positive, then there is equality if and only if $p = 2$ and

$$J_k(\omega)|e_k(\omega)| = I_k(\omega)|d_k(\omega)|$$

for almost all ω and all k satisfying $0 \leq k \leq n$ where I_k is the indicator function of $\{\sigma_\infty \geq k\}$ and J_k is the indicator function of $\{\tau_\infty \geq k\}$.

Notice that the choice $\sigma_j \equiv \infty$ for all $j \geq 0$ yields

$$(3.12) \qquad \|S_n(g, \tau)\|_p \leq (p^* - 1)\|f_n\|_p$$

and $p^* - 1$ is best possible: choose $\tau_j \equiv \infty$, $j \geq 0$, to show that (3.12) contains (3.2). Therefore, the constant in (3.11) and (3.12) cannot be replaced by a smaller constant.

Weak-type inequalities. If f and g are $\mathbf{H}$-valued martingales with respect to the same filtration and g is differentially subordinate to f, then [Bur 89a]

$$(3.13) \qquad P(|g_n| \geq \lambda) \leq \frac{2}{\lambda}\|f_n\|_1.$$

The constant 2 is best possible since it is already best possible in the special case that g is a ± 1-transform of f. If $\|f\|_1$ is finite, this inequality leads to an easy proof, as we shall see, of the almost everywhere convergence of g. The following is also true.

THEOREM 3.5. *If f and g are $\mathbf{H}$-valued martingales with respect to the same filtration and g is differentially subordinate to f, then, for $0 < \lambda < \infty$,*

$$(3.14) \qquad P\left(\sup_{n \geq 0}\{|f_n| + |g_n|\} \geq \lambda\right) \leq \frac{2}{\lambda}\|f\|_1.$$

The key to the proof is the existence of a function $L : \mathbf{H} \times \mathbf{H} \to \mathbf{R}$ satisfying

$$(3.15) \qquad L(x, y) \leq 2|x| \quad \text{if} \quad |x| + |y| \geq 1,$$
$$(3.16) \qquad L(x, y) \leq 2|x| + 1,$$
$$(3.17) \qquad L(x, y) \geq 1 \quad \text{if} \quad |y| \leq |x|,$$

and the further property that

$$(3.18) \qquad EL(f_n, g_n) \geq \cdots \geq EL(f_0, g_0).$$

Then, by (3.15) and (3.16),

$$\begin{aligned}
P(|f_n| + |g_n| \geq 1) &\leq P(2|f_n| \geq L(f_n, g_n)) \\
&= P(2|f_n| - L(f_n, g_n) + 1 \geq 1) \\
&\leq E\{2|f_n| - L(f_n, g_n) + 1\}.
\end{aligned}$$

By (3.17) and (3.18), $EL(f_n, g_n) \geq EL(f_0, g_0) \geq 1$ so

$$P(|f_n| + |g_n| \geq 1) \leq 2\|f_n\|_1.$$

Homogeneity and a stopping time argument now give the inequality

$$P\left(\sup_{n \geq 0}\{|f_n| + |g_n|\} > \lambda\right) \leq \frac{2}{\lambda}\|f\|_1.$$

Replace λ by λ' in this inequality and let $\lambda' \uparrow \lambda$ to obtain (3.14).

Suppose that $L: \mathbf{H} \times \mathbf{H} \to \mathbf{R}$ is defined by

$$(3.19) \qquad \begin{aligned} L(x, y) &= 1 + |x|^2 - |y|^2 & \text{if} \quad |x| + |y| < 1, \\ &= 2|x| & \text{if} \quad |x| + |y| \geq 1. \end{aligned}$$

As is shown in [Bur 89a], this is a function that satisfies (3.15)–(3.18). The property of L that can be used to prove (3.18) is the following: if x, y, h, k are in $\mathbf{H}$ and $|k| \leq |h|$, then the mapping

$$t \mapsto L(x + ht, y + kt)$$

is convex on $\mathbf{R}$.

There is a weak-type version of Theorem 3.2, which, with the use of the martingales F and G that are defined in the proof of that theorem, follows at once from (3.13). This, in turn, gives weak-type versions of Theorems 3.3 and 3.4. For example, a consequence of (3.13) is the inequality

$$(3.20) \qquad P(S_n(g, \tau) \geq \lambda) \leq \frac{2}{\lambda}\|f_n\|_1.$$

In fact, letting $S^*(g, \tau) = \sup_{n \geq 0} S_n(g, \tau)$, we see that Theorem 3.5 yields a stronger result.

THEOREM 3.6. *Let f and g be $\mathbf{H}$-valued martingales relative to the same filtration and τ a nondecreasing sequence of stopping times. If g is differentially subordinate to f, then*

$$(3.21) \qquad P(S^*(g, \tau) \geq \lambda) \leq \frac{2}{\lambda}\|f\|_1.$$

The constant 2 is best possible in (3.20), which contains (3.13), and so is best possible in (3.21).

Theorem 3.5 yields a new proof of the following theorem from [Bur 86a], which also contains some related results for stochastic integrals and $\mathbf{B}$-valued martingales.

THEOREM 3.7. *Let a and b be real numbers satisfying $a < b$ and $a \leq 0 \leq b$. If g is the transform of an $\mathbf{H}$-valued martingale f by a predictable sequence v satisfying $a \leq v_k(\omega) \leq b$ for all k and ω, then*

$$(3.22) \qquad P(g^* \geq \lambda) \leq \frac{b - a}{\lambda}\|f\|_1$$

and the constant $b - a$ is best possible.

PROOF: Let $\alpha = (b - a)/2$ and $\beta = (b + a)/2$. Define the predictable sequence w by

$$v_k = \alpha w_k + \beta$$

and note that w_k has its values in the interval $[-1, 1]$. Let h be the transform of f by w. Then f and h are martingales relative to the same filtration, h is differentially subordinate to f, and, since $|\beta| \leq \alpha$,

$$|g_n| = |\alpha h_n + \beta f_n| \leq \alpha \{|h_n| + |f_n|\}.$$

Now apply Theorem 3.5 to αf and αh to obtain (3.22). See [Bur 86a] for an example showing that the constant $b - a$ is best possible.

An inequality for stochastic integrals. Let M be a right-continuous martingale with limits from the left, and let U and V be predictable processes. The probability space and filtration are the same as for (1.3), but here we consider (i) the case in which U and V are H-valued and M is scalar-valued, and (ii) the case in which U and V are scalar-valued and M is H-valued. We assume that H is separable and that $\int_{[0,\infty)} |U_t|^2 \, d[M, M]_t$ is finite a.s.

THEOREM 3.8. *Let $1 < p < \infty$. If, for all $\omega \in \Omega$ and $t \geq 0$,*

$$(3.23) \qquad |V_t(\omega)| \leq |U_t(\omega)|,$$

then, for both cases (i) and (ii),

$$(3.24) \qquad \|V \cdot M\|_p \leq (p^* - 1)\|U \cdot M\|_p$$

and the constant is best possible.

So complex-valued M, U, and V, for example, do not require an increase in the constant needed for the real case.

The key to this theorem is (3.2). See [Bur 88b] for the proof.

Nondecreasing sequences of contractive projections: The complex case. The inequality for the real case (see Chapter 1) carries over to the complex case with no change in the constant.

THEOREM 3.9. *Let $1 < p < \infty$. Suppose that $0 = P_0, P_1, P_2, \ldots$ is a nondecreasing sequence of contractive projections in the complex L^p of a positive measure space. If $c_k \in \mathbf{C}$ where $|c_k| \leq 1$ for all $k \geq 1$, and $f \in L^p$, then*

$$(3.25) \qquad \left\| \sum_{k=1}^{\infty} c_k (P_k - P_{k-1})f \right\|_p \leq (p^* - 1)\|f\|_p.$$

Strict inequality holds if $p \neq 2$ and $\|f\|_p > 0$.

PROOF: Let $\alpha_p = p(1 - 1/p^*)^{p-1}$ as in the proof of Theorem 3.1. Then

$$(3.26) \qquad \left\| \sum_{k=1}^{\infty} c_k (P_k - P_{k-1})f \right\|_p^p + \alpha_p(p^* - 2)\|P_1 f\|_p^p \leq (p^* - 1)^p \|f\|_p^p.$$

This not only implies (3.25) but also the assertion about strict inequality. Suppose that $\|f\|_p > 0$ and $p \neq 2$ so that $p^* > 2$. If the left-hand side of (3.25) vanishes then strict inequality holds. On the other hand, if the left-hand side is positive, let m be the least integer n such that $\|P_n f\|_p > 0$. We can assume that $m = 1$ and this gives the strict inequality in (3.25).

To prove (3.26), we can assume that the positive measure space is finite (cf. [Tza 69]). Then the results of Ando (1966), and Dor and Odell (1975), on contractive projections in complex L^p can be used to reduce the proof of (3.26) to the case in which $(P_n)_{n \geq 1}$ is a nondecreasing sequence of conditional expectation operators. In this case, (3.26) is an immediate consequence of Theorem 3.1 and its proof.

For a recent discussion of how inequalities of the above kind can be used to study the spectral representation of operators, see Doust (1988, 1989).

4. ON THE NUMBER OF ESCAPES OF A MARTINGALE

Probability bounds on the number of their escapes illuminate both the behavior of martingales and the geometry of the Banach spaces in which they take their values.

Consider a sequence $x = (x_n)_{n \geq 0}$ in a real or complex Banach space $\mathbf{B}$. By Cauchy's criterion,

$$x \text{ converges} \iff C_\varepsilon(x) < \infty \text{ for all } \varepsilon > 0$$

where $C_\varepsilon(x)$ is *the number of ε-escapes of x*. The counting function $C_\varepsilon(\cdot)$, the notation is in honor of Cauchy, is defined as follows: $C_\varepsilon(x) = 0$ and $\nu_0(x) = \infty$ if the set $\{n \geq 0 : |x_n| \geq \varepsilon\}$ is empty. If the set is nonempty, let

$$\nu_0(x) = \inf\{n \geq 0 : |x_n| \geq \varepsilon\}.$$

In this case $C_\varepsilon(x) = 1$ and $\nu_1(x) = \infty$ if $\{n > \nu_0(x) : |x_n - x_{\nu_0(x)}| \geq \varepsilon\}$ is empty. If nonempty, continue as above. If $\nu_j(x)$ is not defined by this induction, equivalently, if there is a nonnegative integer $i < j$ such that $\nu_i(x) = \infty$, set $\nu_j(x) = \infty$. Then $C_\varepsilon(x) \leq j$ if and only if $\nu_j(x) = \infty$.

The upcrossing method of Doob (1953) provides a convenient counting function for real-valued martingales. For nonnegative martingales, there is also the method of rises introduced by Dubins (1962). The counting function $C_\varepsilon(\cdot)$, related to one used by Davis (1969) to study real-valued martingales f, is dimension free and can be used to study vector-valued f and g [Bur 89b]. Even in the real case it leads to new understanding.

THEOREM 4.1. *Let f and g be $\mathbf{H}$-valued martingales with respect to the same filtration. If g is differentially subordinate to f, then, for all $j \geq 1$,*

$$(4.1) \qquad P(C_\varepsilon(g) \geq j) \leq 2\|f\|_1 / \varepsilon j^{1/2}.$$

Both the constant 2 and the exponent 1/2 are best possible.

The function $C_\varepsilon(g)$ is defined by $C_\varepsilon(g)(\omega) = C_\varepsilon(g(\omega))$. In (4.1), g can of course be equal to f.

An immediate consequence of (4.1) is that the finiteness of $\|f\|_1$ implies that $C_e(g)$ is finite and g converges almost everywhere.

PROOF: To prove (4.1), we can assume that the martingale f is of finite length. So there is a positive integer m such that $f_n = f_m$ and $g_n = g_m$ for all $n \geq m$. Let $\tau_j = \nu_j(g) \wedge m$ where $\nu_j(g)(\omega) = \nu_j(g(\omega))$. Then $\tau = (\tau_j)_{j \geq 0}$ is a nondecreasing sequence of stopping times and, by the definition of $S_n(g, \tau)$ in Chapter 3,

$$\varepsilon^2 C_e(g) \leq S_m^2(g, \tau).$$

Therefore, by (3.20),

$$P(C_e(g) \geq j) \leq P(S_m^2(g, \tau) \geq \varepsilon^2 j) \leq 2\|f\|_1 / \varepsilon j^{1/2}.$$

The proof of (4.1) rests on the inequality $P(|g_n| \geq \lambda) \leq 2\|f_n\|_1/\lambda$. But the order can be reversed:

$$P(g^* \geq \lambda) \leq \lim_{\varepsilon \uparrow \lambda} P(C_\varepsilon(g) \geq 1) \leq \frac{2}{\lambda}\|f\|_1.$$

This implies that 2 is the best constant in (4.1). To see that the exponent $1/2$ is best possible, consider simple random walk stopped at time j.

Escape-inequality characterizations of Hilbert space, spaces isomorphic to a Hilbert space, UMD spaces, and superreflexive spaces. Inequality (4.1) characterizes the class of Hilbert spaces. If **B** is a Banach space and (4.1) holds for all **B**-valued martingales f and g as above, then **B** is a Hilbert space. On the other hand, if in (4.1), the constant 2 were replaced by a real number $\beta = \beta(\mathbf{B})$, then the inequality would characterize the class of spaces **B** that are isomorphic to a Hilbert space. Recall that g converges almost everywhere for all g differentially subordinate to an L^1-bounded martingale f as above if and only if **B** is isomorphic to a Hilbert space [Bur 81a].

So for g differentially subordinate to f, the good Banach spaces are those isomorphic to a Hilbert space. There is a much larger class of good spaces for martingale transforms.

THEOREM 4.2. *If **B** is a Banach space, then **B** $\in$ UMD if and only if there are strictly positive real numbers α and β such that if f is a **B**-valued martingale and g is the transform of f by a scalar-valued predictable sequence v uniformly bounded in absolute value by 1, then, for all $j \geq 1$,*

$$(4.2) \qquad P(C_e(g) \geq j) \leq \beta\|f\|_1 / \varepsilon j^\alpha.$$

If we set $g = f$, then we have an even larger class of good spaces.

THEOREM 4.3. *If **B** is a Banach space, then **B** is superreflexive if and only if there are strictly positive real numbers α and β such that if f is a **B**-valued martingale, then, for all $j \geq 1$,*

$$(4.3) \qquad P(C_e(f) \geq j) \leq \beta\|f\|_1 / \varepsilon j^\alpha.$$

For the proofs of these two theorems, see [Bur 89b].

A note on superreflexivity. Suppose that X and Y are Banach spaces. Then X is *finitely representable* in Y if, for all finite-dimensional subspaces E of X and all $\lambda > 1$, there is a linear map $T : E \to Y$ such that, for all $x \in E$,

$$\lambda^{-1}|x|_X \leq |Tx|_Y \leq \lambda|x|_X.$$

A Banach space $\mathbf{B}$ is *superreflexive* if it is reflexive and every Banach space that is finitely representable in $\mathbf{B}$ is also reflexive. This concept was introduced and studied by James. In particular, see his papers of 1972. Since a uniformly convex Banach space is reflexive, a Banach space that can be given an equivalent uniformly convex norm is superreflexive. Enflo (1972) proved the converse. Pisier (1975b) extended Enflo's result in several directions and, among other things, found a condition equivalent to superreflexivity based on a modification of the left-hand side of the L^p-inequality for the martingale square function. If $1 < p, q < \infty$, let $\gamma_{p,q}(\mathbf{B})$ be the least $\gamma \leq \infty$ such that $\|S(f,q)\|_p \leq \gamma\|f\|_p$ for all $\mathbf{B}$-valued martingales f where

$$S(f,q) = \left(\sum_{k=0}^{\infty} |d_k|^q \right)^{1/q}.$$

The filtration may vary with f. The probability space must also be allowed to vary unless it is assumed to be nonatomic. If $q < 2$, then $\gamma_{p,q}(\mathbf{R}) = \infty$. If $q \geq 2$, then $S(f,q) \leq S(f)$ and, by Theorem 3.3, $\gamma_{p,q}(\mathbf{H}) \leq p^* - 1$. Pisier (1975b) showed that $\mathbf{B}$ is superreflexive if and only if there is a $q \in [2, \infty)$ such that $\gamma_{p,q}(\mathbf{B})$ is finite. By the extrapolation (good λ) method of [Bur-Gun 70], the finiteness of $\gamma_{p,q}(\mathbf{B})$ for some $p \in (1, \infty)$ implies its finiteness for all such p.

On the constants in Theorems 4.2 and 4.3. Inequality (4.3) holds with $\alpha = 1/q$ and $\beta = 2\gamma_{2,q}(\mathbf{B})$. In the case $\mathbf{B} = \mathbf{H}$ and $q = 2$, the constants are best possible. If the predictable sequence v is real-valued, then the inequality (4.2) holds with $\alpha = 1/q$ and $\beta = 2\beta_2(\mathbf{B})\gamma_{2,q}(\mathbf{B})$ where $\beta_p(\mathbf{B})$ is defined in Chapter 1. Again, the constants are best possible if $\mathbf{B} = \mathbf{H}$ and $q = 2$. If v is complex-valued, then $\beta_2(\mathbf{B})$ is replaced by $\beta_2(\mathbf{B}, \mathbf{C})$ where $\|g\|_2 \leq \beta_2(\mathbf{B}, \mathbf{C})\|f\|_2$ and the inequality is to hold for all complex-valued predictable sequences uniformly bounded in modulus by 1. See [Bur 89b].

On the number of escapes of a right-continuous martingale. Here the probability space $(\Omega, \mathcal{F}_\infty, P)$ is complete, $\mathcal{F}$ is a nondecreasing right-continuous family $(\mathcal{F}_t)_{t \geq 0}$ of sub-σ-fields of $\mathcal{F}_\infty$, and $\mathcal{F}_0$ contains all $A \in \mathcal{F}_\infty$ with $P(A) = 0$.

THEOREM 4.4. *Let N be an $\mathbf{H}$-valued right-continuous martingale. If there is an $\mathbf{H}$-valued continuous martingale M such that, for all bounded stopping times T,*

$$(4.4) \qquad E\left(|N_T|^2|\mathcal{F}_0\right) \leq E\left(|M_T|^2|\mathcal{F}_0\right),$$

then, for all $j \geq 1$,

$$(4.5) \qquad P(C_\varepsilon(N) \geq j) \leq 2\|M\|_1/\varepsilon j^{1/2}.$$

Both the constant 2 and the exponent $1/2$ are best possible.

The counting function $C_\varepsilon(N)$ is defined as in the discrete-time case.

PROOF: We can assume in the proof that the real or complex Hilbert space is separable and that M_0 is a constant function. Let $\tau = (\tau_j)_{j \geq 0}$ be a nondecreasing sequence of bounded stopping times of $\mathcal{F}$ and set

$$S(N, \tau) = \left(|N_{\tau_0}|^2 + \sum_{j=1}^{\infty} |N_{\tau_j} - N_{\tau_{j-1}}|^2 \right)^{\frac{1}{2}}.$$

Then, as we shall show,

(4.6) $$P(S(N, \tau) \geq \lambda) \leq \frac{2}{\lambda} \|M\|_1.$$

This implies (4.5) just as (3.20) implies (4.1).

If the constant function $|M_0|$ exceeds $\lambda/2$, then (4.6) holds: the right-hand side is greater than 1. So suppose that $|M_0| \leq \lambda/2$, define the stopping time

$$\sigma = \inf\{t > 0 : |M_t| > \lambda\},$$

and notice that the martingale $M^\sigma = (M_{\sigma \wedge t})_{t \geq 0}$ is uniformly bounded in norm by λ. We have that

$$
\begin{aligned}
\lambda^2 P(S(N, \tau) \geq \lambda, \ M^* \leq \lambda) &\leq \lambda^2 P(S(N^\sigma, \tau) \geq \lambda) \\
&\leq \|S(N^\sigma, \tau)\|_2^2 \\
&= \lim_{j \to \infty} E|N_{\sigma \wedge \tau_j}|^2 \\
&\leq \lim_{j \to \infty} E|M_{\sigma \wedge \tau_j}|^2 \\
&\leq \lambda \|M\|_1.
\end{aligned}
$$

This inequality and the inequality $\lambda P(M^* \geq \lambda) \leq \|M\|_1$ imply (4.6).

That 2 and 1/2 are best possible follows from discrete-time examples and Skorohod embedding.

Remarks. (i) The condition that M is continuous cannot be eliminated entirely. Inequality (4.1) does not hold if the condition of differential subordination is replaced by the discrete-time analogue of (4.4). (ii) Instead of assuming the martingale condition on M in Theorem 4.4, we can assume that M is a nonnegative continuous submartingale. There is no change in the proof.

THEOREM 4.5. *Let* **B** *be a superreflexive space and* N *a* **B**-*valued right-continuous martingale. If there is a* **B**-*valued continuous martingale* M *such that, for all bounded stopping times* T,

(4.7) $$E\left(|N_T|^2 | \mathcal{F}_0 \right) \leq E\left(|M_T|^2 | \mathcal{F}_0 \right),$$

then, for all $j \geq 1$,

(4.8) $$P(C_\varepsilon(N) \geq j) \leq 2\gamma_{2,q}(\mathbf{B}) \|M\|_1 / \varepsilon j^{1/q}.$$

PROOF: The main step is to prove that

(4.9) $$P(S(N, q, \tau) \geq \lambda) \leq \frac{2\gamma_{2,q}(\mathbf{B})}{\lambda} \|M\|_1.$$

where

$$S(N, q, \tau) = \left(|N_{\tau_0}|^q + \sum_{j=1}^{\infty} |N_{\tau_j} - N_{\tau_{j-1}}|^q \right)^{1/q}.$$

This is similar to the corresponding step in the proof of Theorem 4.4 with the stopping time σ defined here by $\sigma = \inf\{t > 0 : |M_t| > \lambda/\gamma\}$ where $\gamma = \gamma_{2,q}(\mathbf{B})$.

5. DIFFERENTIAL SUBORDINATION OF HARMONIC FUNCTIONS

If u is harmonic and v is conjugate to u on a domain D of $\mathbf{C}$, then $|\nabla v| = |\nabla u|$ on D. This follows at once from the Cauchy-Riemann equations. The martingale analogue is $|\pm d_k| = |d_k|$. As we have seen, a number of sharp inequalities for martingale transforms hold also for differentially subordinate $\mathbf{H}$-valued martingales and in this new setting lead to new applications. There are analogous results for harmonic functions.

Let D be an open connected set of points $x = (x_1, \ldots, x_n) \in \mathbf{R}^n$ and $\mathbf{H}$ a real or complex Hilbert space with norm $|\cdot|$. Suppose that u is harmonic on D with values in $\mathbf{H}$: the partial derivatives $u_k = \partial u / \partial x_k$ and $u_{jk} = \partial^2 u / \partial x_j \partial x_k$ exist and are continuous, and $\Delta u = \sum_{k=1}^{n} u_{kk} = 0$, the origin of $\mathbf{H}$. Note that $u_k(x) \in \mathbf{H}$ and write

$$|\nabla u(x)| = \left(\sum_{k=1}^{n} |u_k(x)|^2 \right)^{\frac{1}{2}}.$$

Suppose that $v \colon D \to \mathbf{H}$ is also harmonic. Then v is *differentially subordinate to* u if, for all $x \in D$,

$$|\nabla v(x)| \leq |\nabla u(x)|.$$

Fix a point $\xi \in D$ and let D_0 be a bounded subdomain satisfying

$$\xi \in D_0 \subset D_0 \cup \partial D_0 \subset D.$$

Denote by μ the harmonic measure on ∂D_0 with respect to ξ. If $1 \leq p < \infty$, let

$$\|u\|_p = \sup_{D_0} \left[\int_{\partial D_0} |u|^p \, d\mu \right]^{1/p}$$

where the supremum is taken over all such D_0.

THEOREM 5.1. *Suppose that u and v are harmonic in the domain $D \subset \mathbf{R}^n$ with values in $\mathbf{H}$. If $|v(\xi)| \leq |u(\xi)|$ and v is differentially subordinate to u, then, for all measures μ as above and all $\lambda > 0$,*

$$(5.1) \qquad \mu(|u| + |v| \geq \lambda) \leq \frac{2}{\lambda} \|u\|_1$$

and the constant 2 is best possible. Furthermore, for $1 < p < \infty$,

$$(5.2) \qquad \|v\|_p \leq (p^* - 1)\|u\|_p.$$

This theorem is of interest even for real-valued u and v on the open unit disk: the usual conjugacy condition on v in this classical setting is replaced by the condition of differential subordination, a condition that makes sense on domains of $\mathbf{R}^n$.

PROOF: To prove inequality (5.1), we can use the same argument as in the proof of Theorem 3.5 to obtain

$$\mu(|u| + |v| \geq 1) \leq \int_{\partial D_0} [2|u| - L(u,v) + 1]\, d\mu.$$

The next step is to notice that the function $L\colon \mathbf{H} \times \mathbf{H} \to \mathbf{R}$ defined in (3.19) has the property that $L(u,v)$ is subharmonic [Bur 89a]. Therefore, using the condition that $|v(\xi)| \leq |u(\xi)|$, we have that

$$\int_{\partial D_0} L(u,v)\, d\mu \geq L(u(\xi), v(\xi)) \geq 1$$

and this gives (5.1). The constant 2 is the best possible: fix $a \in \mathbf{H}$ satisfying $2|a| = \lambda$ and consider $u = v \equiv a$.

The proof of the martingale inequality (3.2) rests on the function $U\colon \mathbf{H} \times \mathbf{H} \to \mathbf{R}$ defined in (3.3). This same function is used in a similar way in the proof of (5.2). It has the property that $U(u,v)$ is superharmonic on D as is proved in [Bur 89a].

The best constant for (5.2) is not yet known. It is less than or equal to $p^* - 1$ but greater than or equal to $\cot(\pi/2p^*)$, the best constant for the classical case in which v is conjugate to u on the open unit disk of $\mathbf{C}$ (see [Pic 72] and [Gam 78]).

Remark. If u and v are $\mathbf{B}$-valued where $\mathbf{B}$ is isomorphic to a Hilbert space then, apart from the choice of the constants, (5.1) and (5.2) continue to hold. This is no longer true for other spaces.

THEOREM 5.2. *Let $\mathbf{B}$ be a Banach space and β a positive real number. Suppose that, for all positive integers n and harmonic functions $u, v\colon D_n \to \mathbf{B}$ with $v(0) = u(0) = 0$ and v differentially subordinate to u on the open unit ball D_n of $\mathbf{R}^n$, the following inequality holds:*

$$(5.3) \qquad \|v\|_2 \leq \beta \|u\|_2.$$

Then **B** *is isomorphic to a real or complex Hilbert space.*

PROOF: Let $1 \leq m \leq n$ and $a_1, \ldots a_m, b_1, \ldots b_m \in \mathbf{B}$ where $|b_k| \leq |a_k|$ for all $k \leq m$. Define u and v on D_n by $u(x) = a_1 x_1 + \cdots a_m x_m$ and $v(x) = b_1 x_1 + \cdots b_m x_m$. Then u and v are harmonic on D_n, vanish at the origin, and satisfy

$$|\nabla v|^2 = \sum_{k=1}^m |b_k|^2 \leq \sum_{k=1}^m |a_k|^2 = |\nabla u|^2.$$

So inequality (5.3) holds for this choice of u and v. Consider the equivalent inequality for a Brownian motion $X = (X_t)_{t \geq 0}$ in $\mathbf{R}^n$ starting at the origin: with τ_n the least t such that the Euclidean norm of $X_t = (X_{1,t}, \ldots, X_{n,t})$ in $\mathbf{R}^n$ is equal to 1, we have the inequality

(5.4)
$$\left\| \sum_{k=1}^m b_k X_{k,\tau_n} \right\|_2 \leq \beta \left\| \sum_{k=1}^m a_k X_{k,\tau_n} \right\|_2.$$

By the rotational invariance of Brownian motion, X_{τ_n} is uniformly distributed on the unit sphere of $\mathbf{R}^n$ so, by a classical result of Poincaré (for example, see [McK 73]), the m random variables $\sqrt{n} X_{1,\tau_n}, \ldots, \sqrt{n} X_{m,\tau_n}$ are asymptotically independent and each has the standard normal distribution in the limit as $n \to \infty$. Since $\|\sqrt{\tau_n}\|_4 \sim 1/\sqrt{n}$ (cf. (2.4) in [Bur 77]) and $\|X_{k,\tau_n}\|_4 \leq 3\|\sqrt{\tau_n}\|_4$, we have that

$$\sup_n \left\| \sum_{k=1}^m a_k \sqrt{n} X_{k,\tau_n} \right\|_4 \leq \sup_n \sum_{k=1}^m |a_k| \sqrt{n} \|X_{k,\tau_n}\|_4 < \infty$$

with a similar result for the b_k. Therefore, if we multiply each side of (5.4) by $\sqrt{n}$ and take the limit as $n \to \infty$, we obtain

(5.5)
$$\left\| \sum_{k=1}^m b_k Y_k \right\|_2 \leq \beta \left\| \sum_{k=1}^m a_k Y_k \right\|_2$$

where the Y_k are independent and each has the standard normal distribution.

If $a \in \mathbf{B}$ with $|a| = 1$ and $a_k = |b_k| a$ for all $k \leq m$, then (5.5) yields

$$\left\| \sum_{k=1}^m b_k Y_k \right\|_2 \leq \beta \left\| \sum_{k=1}^m |b_k| Y_k \right\|_2 = \beta \left(\sum_{k=1}^m |b_k|^2 \right)^{\frac{1}{2}}.$$

Similarly,

$$\left\| \sum_{k=1}^m b_k Y_k \right\|_2 \geq \beta^{-1} \left\| \sum_{k=1}^m |b_k| Y_k \right\|_2 = \beta^{-1} \left(\sum_{k=1}^m |b_k|^2 \right)^{\frac{1}{2}}.$$

Therefore, by a theorem of Kwapień (1972), **B** is isomorphic to a Hilbert space.

An application of Theorem 5.1 to Riesz systems of harmonic functions. Here it is convenient to let D be a domain of points $x = (x_0, \ldots, x_n)$ in $\mathbf{R}^{n+1}$. Let $u : D \to$

H be harmonic. We write $u_k = \partial u/\partial x_k$ and $u_{jk} = \partial^2 u/\partial x_j \partial x_k$ as above. Then $F = (u_0, \ldots, u_n)$ is a Riesz system: the u_j are harmonic and satisfy the generalized Cauchy-Riemann equations

$$(5.6) \qquad \sum_{k=0}^{n} u_{kk} = 0 \quad \text{and} \quad u_{jk} = u_{kj}.$$

Let $F^{00} = (u_0, 0, \ldots, 0)$ and $F^0 = (0, u_1, \ldots, u_n)$. Then F, F^0, and F^{00} are harmonic from D to $\mathbf{K} = \mathbf{H}^{n+1}$, F^{00} is differentially subordinate to $n^{1/2} F^0$, and F is differentially subordinate to $(n+1)^{1/2} F^0$. To see this, use (5.6) or consult [Bur 89a]. Suppose there is a point $\xi \in D$ such that $|u_0(\xi)| = |F^{00}(\xi)|_{\mathbf{K}} \le n^{1/2} |F^0(\xi)|_{\mathbf{K}}$ and μ has the same meaning as above. Then, by Theorem 5.1,

$$(5.7) \qquad \mu\left(\left|\frac{\partial u}{\partial x_0}\right| \ge \lambda\right) \le \frac{2n^{1/2}}{\lambda} \left\| \left(\sum_{j=1}^{n} \left|\frac{\partial u}{\partial x_j}\right|^2\right)^{1/2} \right\|_1$$

and, for $1 < p < \infty$,

$$(5.8) \qquad \left\|\frac{\partial u}{\partial x_0}\right\|_p \le n^{1/2}(p^* - 1) \left\| \left(\sum_{j=1}^{n} \left|\frac{\partial u}{\partial x_j}\right|^2\right)^{1/2} \right\|_p.$$

The bounds on the right, which improve earlier bounds (see Stein (1970) and Essén (1984)) are of the right order of magnitude in n: consider $u(x) = nx_0^2 - x_1^2 - \cdots - x_n^2$ on a small ball of $\mathbf{R}^{n+1}$ centered at $(1, \ldots, 1)$. Similar inequalities follow from the differential subordination of F to $(n+1)^{1/2} F^0$.

6. SHARP PROBABILITY BOUNDS FOR REAL MARTINGALES

By setting $N = M$ in Theorem 4.4, we obtain the escape inequality

$$(6.1) \qquad P(C_\varepsilon(M) \ge j) \le 2\|M\|_1/\varepsilon\sqrt{j}$$

but there is no longer any positive integer j for which the bound $2/\sqrt{j}$ is sharp. Here we let $\mathbf{H} = \mathbf{R}$ and determine the sharp bound for each j.

THEOREM 6.1. *Let M be a real continuous local martingale with $M_0 \equiv 0$. If $\varepsilon > 0$ and j is a positive integer, then*

$$(6.2) \qquad P(C_\varepsilon(M) \ge j) \le \frac{\|M\|_1}{\varepsilon\|r_1 + \cdots + r_j\|_1}.$$

The inequality is sharp and equality can hold. If $\theta \in [0, 1]$, there is an M such that each side of (6.2) is equal to θ.

Here $\|r_1 + \cdots + r_j\|_1$ is the integral over the Lebesgue unit interval $[0, 1)$ of the absolute value of the sum of $r_1, \ldots, r_j$ where r_k is the kth Rademacher function, the function on $[0, 1)$ taking the values $+1$ and -1 alternately on $[0, 1/2^{k+1})$, $[1/2^{k+1}, 2/2^{k+1})$, ... It is known and easy to check that if f is a real martingale indexed by the set of nonnegative integers, then, for $n \geq 1$,

$$(6.3) \qquad \|f_n\|_1 - \|f_{n-1}\|_1 = 2 \int_{f_{n-1} \leq 0} f_n^+ + 2 \int_{f_{n-1} > 0} f_n^-$$

and, for $j = 2m$ where m is a positive integer, this gives

$$(6.4) \qquad \|r_1 + \cdots + r_{j-1}\|_1 = \|r_1 + \cdots + r_j\|_1 = \sum_{k=0}^{m-1} \binom{2k}{k} 2^{-2k}.$$

If j is any positive integer, then (cf. [Sza 76])

$$\|r_1 + \cdots + r_j\|_1 \geq \sqrt{j/2}$$

so the right-hand sides of (6.1) and (6.2) can be replaced by $\sqrt{2}\|M\|_1/\varepsilon\sqrt{j}$. Also, $1/\|r_1 + \cdots + r_j\|_1 \sim \sqrt{\pi/2j} = 1.25\ldots/\sqrt{j}$ as $j \to \infty$.

To prove Theorem 6.1 and the other escape inequalities here, we may set $\varepsilon = 1$: $C_\varepsilon(M) = C_1(\varepsilon^{-1}M)$. We shall write C for C_1.

LEMMA 6.1. *Let $x \in \mathbf{R}$ and suppose that Y is a real-valued random variable with $EY = 0$ and $P(|Y| \geq 1) = 1$. Then*

$$(6.5) \qquad E|x + Y| \geq \|x + r_k\|_1.$$

PROOF: The mapping $x \mapsto E|x + Y|$ is affine on the interval $[-1, 1]$ and, for all real x, $E|x+Y| \geq |x + EY| = |x|$. Therefore, $E|x+Y| \geq |x| \vee 1$, which is the value of $\|x + r_k\|_1$.

If x is a real number, i is a nonnegative integer, and j is a positive integer, let

$$(6.6) \qquad u_j(x, i) = \left\| x + \sum_{k=i+1}^{j} r_k \right\|_1$$

where the sum is taken to be 0 if $i \geq j$. Then

$$(6.7) \qquad u_j(x, i) = |x| \quad \text{if} \quad i \geq j,$$

and, by the triangle inequality, $u_j(x, i) \leq |x| + u_j(0, i)$ so

$$(6.8) \qquad u_j(x, i) \leq |x| + u_j(0, 0) \quad \text{if} \quad i \geq 0.$$

The function u_j also satisfies a kind of convexity property:

LEMMA 6.2. *Let Y be a real-valued random variable with $EY = 0$ and such that $|Y(\omega)| < 1$ implies that $Y(\omega) = 0$. Then*

(6.9) $$Eu_j(x + Y, i + 1(|Y| \geq 1)) \geq u_j(x, i).$$

PROOF: The left-hand side of (6.9) is equal to

$$\int_0^1 \int_{|Y| \geq 1} |x + Y + \sum_{k=i+2}^j r_k(t)| \, dP \, dt + \int_0^1 \int_{|Y| < 1} |x + \sum_{k=i+1}^j r_k(t)| \, dP \, dt.$$

Using Lemma 6.1 and the fact that $E(Y | |Y| \geq 1) = 0$, the first double integral is greater than or equal to

$$P(|Y| \geq 1) \int_0^1 \int_0^1 |x + r_{i+1}(s) + \sum_{k=i+2}^j r_k(t)| \, ds \, dt$$

which is equal to $P(|Y| \geq 1)u_j(x, i)$. The second is equal to $P(|Y| < 1)u_j(x, i)$ so the sum of the two integrals is equal to the right-hand side of (6.9).

LEMMA 6.3. *If f is a real martingale with difference sequence d such that, for each nonnegative integer k, the inequality $|d_k(\omega)| < 1$ implies that $d_k(\omega) = 0$, then*

(6.10) $$P(C(f) \geq j) \leq \frac{\|f\|_1}{\|r_1 + \cdots + r_j\|_1}.$$

PROOF: We can assume in the proof that $f_0 \equiv 0$. If this does not already hold, replace f by the splice of f and $-f$ with weight $1/2$ (see Chapter 2).

Let $I_n = \sum_{k=0}^n 1(|d_k| \geq 1)$. Then

$$Eu_j(f_n, I_n) \geq Eu_j(f_{n-1}, I_{n-1}) \geq \cdots \geq Eu_j(f_0, I_0) = u_j(0, 0).$$

To prove the first inequality, use Lemma 6.2 to obtain

$$E[u_j(f_n, I_n)|\mathcal{F}_{n-1}] = E[u_j(f_{n-1} + d_n, I_{n-1} + 1(|d_n| \geq 1))|\mathcal{F}_{n-1}]$$
$$\geq u_j(f_{n-1}, I_{n-1}),$$

and then take expectations. Then, keeping in mind (6.7), (6.8), Chebyshev's inequality for nonnegative functions, and $u_j(0, 0) = \|r_1 + \cdots + r_j\|_1$, observe that

$$P(I_n \geq j) \leq P(|f_n| - u_j(f_n, I_n) \geq 0)$$
$$= P(|f_n| - u_j(f_n, I_n) + u_j(0, 0) \geq u_j(0, 0))$$
$$\leq E[|f_n| - u_j(f_n, I_n) + u_j(0, 0)]/u_j(0, 0)$$
$$\leq \|f_n\|_1/\|r_1 + \cdots + r_j\|_1.$$

Finally use the condition on f to see that $I_n \uparrow C(f)$ as $n \uparrow \infty$.

PROOF OF THEOREM 6.1: We can assume in the proof that M is a martingale. To prove (6.2) under this assumption, fix a positive number t and let $M^t = (M_{s \wedge t})_{s \geq 0}$. It is then enough to show that the inequality holds with $\varepsilon = 1$ for the martingale M^t.

By enlarging the probability space if necessary, we can assume there is a random variable $Y : \Omega \to [0,1)$ with $P(Y \leq y) = y$ for $y \in [0,1)$ such that Y is independent of M. Let N be defined by $N_s = M_{s \wedge t}$ if $s < t + 1$, and by

$$N_s = [M_t] + 1(Y \leq M_t - [M_t])$$

if $s \geq t + 1$. Here, for x a real number, $[x]$ is the largest integer less than or equal to x. It is easy to see that N is a right-continuous martingale. Let $\tau_n = \nu_n(N) \wedge (t+1)$ and $f = (N_{\tau_n})_{n \geq 0}$ where the ν_n are the times of escape of N. Then f is an integer-valued martingale satisfying the conditions of Lemma 6.3. Therefore, (6.10) holds for f. Since $C(M^t) \leq C(N) = C(f)$ and $\|f\|_1 \leq \|N\|_1 = \|M^t\|_1$, the inequality (6.2) holds for M^t.

To finish the proof of Theorem 6.1, we shall construct a continuous martingale M so that each side of (6.2) is equal to 1. If $\theta \in [0,1]$, a slight modification of the example (stop at 0 with probability $1 - \theta$) shows that each side can be equal to θ.

Let $\tau_0 \equiv 0$ and, if τ_n has been defined, let

$$\tau_{n+1} = \inf\{t > \tau_n : |B_t - B_{\tau_n}| = \varepsilon\}$$

where B is a real Brownian motion starting at the origin. Then the martingale $M = B^{\tau_j} = (B_{\tau_j \wedge t})_{t \geq 0}$ has the desired properties since the distribution of $(B_{\tau_n} - B_{\tau_{n-1}})_{1 \leq n \leq j}$ has the same distribution as $(\varepsilon r_n)_{1 \leq n \leq j}$.

Some questions. Does (6.2) hold for all real right-continuous local martingales? In particular, does (6.10) hold for all real martingales f? If so, simple random walk started at 0 and stopped at j is an extremal martingale for (6.10) among all real martingales, not just among those satisfying the conditions of Lemma 6.3.

Under the conditions of Lemma 6.3, the proof of (6.10) gives

$$(6.11) \qquad P\left(\sum_{k=0}^{\infty} 1(|d_k| \geq 1) \geq j\right) \leq \frac{\|f\|_1}{\|r_1 + \cdots + r_j\|_1}.$$

This inequality does *not* hold for all real martingales f as the following example shows:

$$P(d_0 = \pm 1/2, \, d_1 = 1/2, \, d_2 = \pm 1) = 1/6,$$
$$P(d_0 = \pm 1/2, \, d_1 = -2d_0, \, d_2 = 0) = 1/6,$$

and $d_n \equiv 0$, $n \geq 3$. Then, for $j = 1$, the left-hand side of (6.11) is 1 but the right-hand side is 5/6.

Does (6.10) hold for all complex martingales f? No. Consider the example

$$P(d_0 = 1, \, d_1 = 1, \, d_2 = \pm 1) = 1/6,$$
$$P(d_0 = 1, \, d_1 = e^{2\pi i/3}, \, d_2 = \pm f_1) = 1/6,$$
$$P(d_0 = 1, \, d_1 = e^{4\pi i/3}, \, d_2 = \pm f_1) = 1/6.$$

Here also, let $d_n \equiv 0$, $n \geq 3$. Then $P(C(f) \geq 3) = 1$ but the right-hand side of (6.10) is 8/9.

7. ON THE NUMBER OF ESCAPES OF A HARMONIC FUNCTION

Using Theorem 4.4, we can obtain probability bounds on the number of ε-escapes of the harmonic functions u and v along a Brownian path from ξ to the boundary of the domain $D \subset \mathbf{R}^n$. Let X be a Brownian motion in $\mathbf{R}^n$ with $X_0 \equiv \xi$ and let $C_\varepsilon(u)$ be the number of ε-escapes of $(u(X_t), 0 \leq t < \tau_D)$ where

$$\tau_D = \inf\{t > 0 \colon X_t \notin D\}.$$

Equivalently, $C_\varepsilon(u) = \sup C_\varepsilon(M)$ where the supremum is taken over all martingales M with $M_t = u(X_{\sigma \wedge t})$ and $\sigma = \tau_{D_0}$. Here D_0 is a bounded subdomain of D satisfying

$$\xi \in D_0 \subset D_0 \cup \partial D_0 \subset D,$$

μ is the harmonic measure on ∂D_0 with respect to ξ, and $\|u\|_1$ is also as in Chapter 5.

THEOREM 7.1. *Suppose that u and v are harmonic in the domain $D \subset \mathbf{R}^n$ with values in $\mathbf{H}$. If $|v(\xi)| \leq |u(\xi)|$ and v is differentially subordinate to u, then, for all $j \geq 1$,*

$$(7.1) \qquad P(C_\varepsilon(v) \geq j) \leq 2\|u\|_1/\varepsilon j^{1/2}.$$

PROOF: We can assume as before that $\mathbf{H}$ is separable. Let $\sigma = \tau_{D_0}$ and define M and N by $M_t = u(X_{\sigma \wedge t})$ and $N_t = v(X_{\sigma \wedge t})$. Then, for all bounded stopping times T,

$$\begin{aligned}
E|N_T|^2 &= |v(\xi)|^2 + E \int_0^{\sigma \wedge T} |\nabla v(X_t)|^2 \, dt \\
&\leq |u(\xi)|^2 + E \int_0^{\sigma \wedge T} |\nabla u(X_t)|^2 \, dt \\
&= E|M_T|^2.
\end{aligned}$$

The filtration $\mathcal{F}$ can be chosen so that $\mathcal{F}_0$ contains only the null sets and their complements. Inequality (4.4) then follows for the continuous martingales M and N. Therefore, (4.5) and the inequality

$$\|M\|_1 = \int_{\partial D_0} |u| \, d\mu \leq \|u\|_1$$

lead at once to (7.1).

THEOREM 7.2. *Suppose that u is real-valued and harmonic in the domain $D \subset \mathbf{R}^n$ and $u(\xi) = 0$. Then, for all $\varepsilon > 0$ and $j \geq 1$,*

$$(7.2) \qquad P(C_\varepsilon(u) \geq j) \leq \frac{\|u\|_1}{\varepsilon\|r_1 + \cdots + r_j\|_1}.$$

PROOF: Apply Theorem 6.1 to the martingale M with $M_t = u(X_{\sigma \wedge t})$ where $\sigma = \tau_{D_0}$ and then take the supremum over the subdomains D_0.

THEOREM 7.3. *Let* $\mathbf{B} \in UMD$. *Then there are positive real numbers* α *and* β *such that if* u *and* v *are harmonic in the domain* $D \subset \mathbf{R}^2$ *with values in* $\mathbf{B}$ *and* v *is conjugate to* u *with* $|v(\xi)| \leq |u(\xi)|$, *then, for all* $j \geq 1$,

$$(7.3) \qquad\qquad P(C_\varepsilon(v) \geq j) \leq \beta \|u\|_1 / \varepsilon j^\alpha.$$

PROOF: Let $\sigma = \tau_{D_0}$ as above, and define the continuous martingales M and N here by $M_t = \rho u(X_{\sigma \wedge t})$ and $N_t = v(X_{\sigma \wedge t})$ where ρ is a positive constant such that

$$E|N_T|^2 \leq E|M_T|^2$$

for all bounded stopping times T. The constant ρ exists by the work of McConnell and the author ([Bur 83], [McC 84]) on conjugate harmonic functions in UMD spaces. Also see [Gar 86] and [Bur 86b]. Since a UMD space is superreflexive ([Mau 75], [Ald 79]), Theorem 4.5 yields (7.3) with $\alpha = 1/q$ and $\beta = 2\rho \gamma_{2,q}(\mathbf{B})$.

THEOREM 7.4. *Let* $\mathbf{B}$ *be superreflexive. Then there are positive real numbers* α *and* β *such that if* u *is harmonic in the domain* $D \subset \mathbf{R}^n$ *with values in* $\mathbf{B}$, *then, for all* $j \geq 1$,

$$(7.4) \qquad\qquad P(C_\varepsilon(u) \geq j) \leq \beta \|u\|_1 / \varepsilon j^\alpha.$$

PROOF: Apply Theorem 4.5 with $N = M$.

Inequality (7.4) holds with $\alpha = 1/q$ and $\beta = 2\gamma_{2,q}(\mathbf{B})$.

8. A SHARP EXPONENTIAL INEQUALITY

Let $\mathbf{H}$ be a real or complex Hilbert space with norm $|\cdot|$ as above. Recall that g is differentially subordinate to f if its difference sequence e satisfies $|e_n(\omega)| \leq |d_n(\omega)|$ for all $\omega \in \Omega$ and $n \geq 0$. Several sharp exponential inequalities for such $\mathbf{H}$-valued martingales f and g relative to the same filtration are contained in [Bur 88b] and follow from the following sharp inequality proved there: $\|f\|_\infty \leq 1$ implies that

$$\sup_n E\Phi(|g_n|) < \frac{1}{2} \int_0^\infty \Phi(t) e^{-t} \, dt$$

where Φ is any increasing convex function on $[0, \infty)$ such that $\Phi(0) = 0$, the integral $\int_0^\infty \Phi(t) e^{-t} \, dt$ is finite, and Φ is twice differentiable on $(0, \infty)$ with a strictly convex first derivative satisfying $\Phi'(0+) = 0$.

The following theorem gives a sharp probability bound on the maximal function of g. It is new even for g a ± 1-transform of a real martingale.

THEOREM 8.1. *Let f and g be* **H**-*valued martingales with respect to the same filtration. If g is differentially subordinate to f and $\|f\|_\infty \le 1$, then*

$$
\begin{aligned}
P(g^* \ge \lambda) &\le 1 && \text{if } 0 < \lambda \le 1, \\
&\le 1/\lambda^2 && \text{if } 1 < \lambda \le 2, \\
&< \alpha e^{-\lambda} && \text{if } \lambda > 2,
\end{aligned}
$$

where $\alpha = e^2/4$. For each λ, the bound on the right is sharp. If $0 < \lambda \le 2$, then equality can hold.

This theorem implies an analogous inequality for stochastic integrals but with one difference: if λ is any positive number, then equality can hold.

PROOF: If $0 < \lambda \le 1$, the inequality is trivial. To see that equality can hold, choose $x \in \mathbf{H}$ so that $|x| = 1$ and let $f_n(\omega) = g_n(\omega) = x$ for all ω and n.

Observe next that $(|g_n|)_{n \ge 0}$ is a nonnegative submartingale so, by Doob's weak-L^2 inequality for the maximal function,

$$
P(g^* \ge \lambda) \le \|g\|_2^2/\lambda^2, \quad \lambda > 0.
$$

This would be true if the martingale g had its values in any Banach space and the same for $\|f\|_2 \le \|f\|_\infty \le 1$. Here f and g are **H**-valued martingales and g is differentially subordinate to f so we have the further result that

$$
\|g\|_2^2 = \sum_{k=0}^\infty \|e_k\|_2^2 \le \sum_{k=0}^\infty \|d_k\|_2^2 = \|f\|_2^2.
$$

These inequalities imply that

$$
P(g^* \ge \lambda) \le 1/\lambda^2, \quad \lambda > 0.
$$

What is not so transparent is that equality can hold if λ is a number satisfying $1 < \lambda \le 2$. We shall prove this below after it will become clear how an example can be constructed.

A key inequality. Now we come to the main step, the proof of

(8.1) $$ P(|g_n| \ge \lambda) \le \alpha e^{-\lambda} \quad \text{if } \lambda > 2. $$

A stopping time argument will then yield the desired inequality for g^* but without the strictness, which will require an additional argument.

A special case. The proof of (8.1) has several steps. We shall begin with the special case where $\mathbf{H} = \mathbf{R}$ and g is a ± 1-transform of the martingale f. Accordingly, for some sequence of numbers $\varepsilon_k \in \{1, -1\}$,

$$
f_n = \sum_{k=0}^n d_k,
$$

$$g_n = \sum_{k=0}^n \varepsilon_k d_k.$$

There is a related martingale. Let $Z_n = (X_n, Y_n)$ where

$$X_n = g_n + f_n = \sum_{k=0}^n (\varepsilon_k + 1)d_k,$$

$$Y_n = g_n - f_n = \sum_{k=0}^n (\varepsilon_k - 1)d_k.$$

The martingale $Z = (Z_n)_{n\geq 0}$ has the property that if $n \geq 1$, then either

(8.2) $$X_n - X_{n-1} \equiv 0 \quad \text{or} \quad Y_n - Y_{n-1} \equiv 0.$$

It also has the property that almost all of its values belong to the set

$$S = \{(x,y) \in \mathbf{R}^2 : |x - y| \leq 2\}.$$

This follows from $X_n - Y_n = 2f_n$ and the assumption that $\|f\|_\infty \leq 1$. In fact, we can assume that $Z_n(\omega) \in S$ for all ω and n.

Fix $\lambda > 0$ and suppose that u is a biconcave function on S, that is, for all $y \in \mathbf{R}$, the function $u(\cdot, y)$ is concave on the interval $[y - 2, y + 2]$ and, for all $x \in \mathbf{R}$, the function $u(x, \cdot)$ is concave on $[x - 2, x + 2]$. Also, suppose that u is nonnegative on S and satisfies $u(x, y) \geq 1$ if $|x + y| \geq 2\lambda$. Then, by Chebyshev's inequality,

$$P(|g_n| \geq \lambda) = P(|X_n + Y_n| \geq 2\lambda) \leq P(u(Z_n) \geq 1) \leq Eu(Z_n).$$

Moreover,

$$Eu(Z_n) \leq Eu(Z_{n-1}) \leq \cdots \leq Eu(Z_0).$$

To see this, notice that $Z_n = (X_n, Y_{n-1})$ or $Z_n = (X_{n-1}, Y_n)$. For example, assume that the former possibility holds. Then, by Jensen's inequality for conditional expectations,

$$E[u(Z_n)|\mathcal{F}_{n-1}] = E[u(X_n, Y_{n-1})|\mathcal{F}_{n-1}] \leq u(E[X_n|\mathcal{F}_{n-1}], Y_{n-1}) = u(Z_{n-1}).$$

Now take expectations to obtain $Eu(Z_n) \leq Eu(Z_{n-1})$ for all $n \geq 1$. Therefore,

$$P(|g_n| \geq \lambda) \leq Eu(Z_0).$$

We can assume that u satisfies the symmetry conditions

(8.3) $$u(x, y) = u(y, x) = u(-x, -y) = u(-y, -x).$$

(If u does not already satisfy these symmetry conditions, replace $u(x, y)$ by the minimum of the four numbers in (8.3) to obtain a function that does satisfy (8.3) as well as the conditions placed on the original function u.) If $Z_0 = (X_0, 0)$, then

$$u(Z_0) = [u(X_0, 0) + u(-X_0, 0)]/2 \leq u(0, 0)$$

and $Eu(Z_0) \leq u(0,0)$ with a similar conclusion holding for the other possibility $Z_0 = (0, Y_0)$. Therefore,

$$(8.4) \qquad\qquad P(|g_n| \geq \lambda) \leq u(0,0).$$

So the problem is to find a biconcave function u on S that majorizes the indicator function of the subset of S where $|x + y| \geq 2\lambda$. Any such biconcave function will lead to the inequality (8.4). The problem is to find the least such function, or perhaps any biconcave majorant with $u(0,0)$ as small as possible.

Let $\lambda > 2$. Consider the function u on S satisfying (8.3) and the following additional conditions: $u(x, y) = 0$ if $|x - y| = 2$ and $|x + y| < 2\lambda$, $u(x, y) = 1$ if $|x - y| = 2$ and $|x + y| \geq 2\lambda$, the restriction of u to the interior of S is continuous, and

$$
\begin{aligned}
u(x, y) &= \alpha(1 + xy)e^{-\lambda} && \text{if } (x, y) \in D_0, \\
&= \alpha(y - x + 2)e^{x - \lambda - 1} && \text{if } (x, y) \in D_1, \\
&= \frac{(y - x + 2)(x - \lambda + 3)}{4(\lambda - x + 1)} && \text{if } (x, y) \in D_2, \\
&= 1 - \frac{\lambda}{2} + \frac{x + y}{4} && \text{if } (x, y) \in D_3, \\
&= 1 && \text{if } (x, y) \in D_4,
\end{aligned}
$$

where

$$
\begin{aligned}
D_0 &= \{(x, y): |x| \vee |y| < 1\}, \\
D_1 &= \{(x, y): 1 < x < \lambda - 1 \text{ and } x - 2 < y < x\}, \\
D_2 &= \{(x, y): \lambda - 1 < x < \lambda + 1 \text{ and } x - 2 < y < \lambda - 1\}, \\
D_3 &= \{(x, y): \lambda - 1 < x < \lambda + 1 \text{ and } \lambda - 1 < y < 2\lambda - x\}, \\
D_4 &= \{(x, y): x + y > 2\lambda \text{ and } x - 2 < y < x + 2\}.
\end{aligned}
$$

Note that u is continous on S except at $(\lambda + 1, \lambda - 1)$, $(\lambda - 1, \lambda + 1)$, $(-\lambda - 1, -\lambda + 1)$, and $(-\lambda + 1, -\lambda - 1)$.

Elementary calculations show that u is biconcave on S. For $0 \leq k \leq 4$, the partial derivatives u_{xx} and u_{yy} are nonpositive on D_k so u is biconcave on D_k. The next step is to check that u has the correct behavior at the boundary of each D_k: if $-1 < y \leq 1$, then $u_x(1-, y) = u_x(1+, y)$; if $\lambda - 1 < x < \lambda + 1$, then

$$u_y(x, \lambda - 1-) > u_y(x, \lambda - 1+);$$

and so forth.

Because u is biconcave and has the desired majorization property, the inequality (8.4) holds with $u(0,0) = \alpha e^{-\lambda}$. This completes the proof of (8.1) in this special case.

Remark. Note that $u_{yy} = 0$ on $D_k, 0 \leq k \leq 4$. In fact, if (x_0, y_0) is in the interior of S and u is twice continuously differentiable on some open set containing (x_0, y_0), then either $u_{xx}(x_0, y_0) = 0$ or $u_{yy}(x_0, y_0) = 0$. This is the property that the least biconcave majorant must have; otherwise $u_{xx}(x_0, y_0)$ and $u_{yy}(x_0, y_0)$ are strictly negative so that u could be replaced by a biconcave majorant that is strictly smaller at the point (x_0, y_0). The extremal function must also satisfy (8.3). Moreover, any biconcave majorant must satisfy

$$u(\lambda - 1, \lambda - 1) \geq \tfrac{1}{2}[u(\lambda - 1, \lambda + 1) + u(\lambda - 1, \lambda - 3)] \geq \tfrac{1}{2}[1 + 0] = \tfrac{1}{2}.$$

These and similar considerations lead to the above choice of u. See Sections 4 and 6 of [Bur 84], where a related extremal problem is solved, for further insight and for the methods to use to establish that the function u given above is, indeed, the *least* biconcave majorant. The minimality of u is not used here. The minimality of its value at $(0, 0)$ is a consequence of an example that we now describe.

Sharpness. The sharpness of the inequality $P(g^* \geq \lambda) \leq \alpha e^{-\lambda}$ under the conditions of the theorem, follows from the sharpness of (8.1) in the above special case, which we now show. The idea is to find a martingale Z with values in S satisfying (8.2) and such that, for some positive integer n, the probability

$$P(|X_n + Y_n| \geq 2\lambda)$$

is near $\alpha e^{-\lambda}$. We shall also have to place a restriction on Z_0, for example, $Z_0 \equiv (1, 0)$. We can then define martingales f and g by

$$f_n = \frac{X_n - Y_n}{2},$$

$$g_n = \frac{X_n + Y_n}{2},$$

and these martingales will yield the sharpness of (8.1).

For example, what should be the behavior of Z in D_1? It is clear from the proof of the inequality that once in D_1, the martingale Z should then move vertically staying within D_1 since $u_{yy} = 0$ and $u_{xx} < 0$ there. The conditional form of Jensen's inequality will then yield equality. But this rule will have to be violated at least a little in order for Z to be able to move to the closure of D_4 with positive probability. Additional study of this kind leads to a candidate for Z: Let $Z_0 \equiv (1, 0)$. Let the next term Z_1 of the martingale have its values in the set $\{(1, -1), (1, 1 + \delta)\}$, where δ is a small positive number. On the set where $Z_1 = (1, -1)$, let the succeeding terms have the same value. On the set where $Z_1 = (1, 1 + \delta)$, let Z_2 have its values in the set $\{(-1 + \delta, 1 + \delta), (1 + 2\delta, 1 + \delta)\}$. A typical path crosses the line $y = x$ in small vertical or horizontal steps a number of times before hitting and then staying at the boundary of S. If the martingale hits either the line $y = \lambda - 1$ or the line $x = \lambda - 1$, as it can if $\lambda - 2$ is an integer multiple of δ, it should stay on this line to hit one of the two possible points on the boundary of S at the next step. This martingale suggests the following candidates for f and g.

It will be convenient to work on the Lebesgue unit interval $[0, 1)$. Let $\beta = (2 - \delta)/(2 + \delta)$ and $\gamma = 1/(2 + \delta)$, where $\delta = (\lambda - 2)/(n - 1)$ and $n \geq 3$ is a fixed positive integer exceeding $\lambda/2$. Using the same notation for a subinterval $[a, b)$ and its indicator function, set

$$e_0 = \tfrac{1}{2}[0,1),$$
$$e_1 = \tfrac{1}{2}(1+\delta)[0,\gamma) - \tfrac{1}{2}[\gamma,1),$$
$$e_k = \delta[0,\beta^{k-1}\gamma) - (1-\delta/2)[\beta^{k-1}\gamma, \beta^{k-2}\gamma), \qquad 2 \le k \le n-1,$$
$$e_n = (1+\delta/2)[0,\beta^{n-1}/4) - (1-\delta/2)[\beta^{n-1}/4, \beta^{n-2}\gamma),$$
$$e_k = 0, \qquad k > n.$$

Let $d_k = (-1)^k e_k$. Then (d_k) and (e_k) are the respective difference sequences of two martingales f and g satisfying the conditions assumed in the special case considered above. In particular, it is easy to check that $\|f\|_\infty \le 1$ and, with $\xi = (\lambda - 2)/2$, that

$$P(g_n \ge \lambda) = \beta^{n-1}/4$$
$$= \tfrac{1}{4}[1 - \xi/(n-1)]^{n-1}/[1 + \xi/(n-1)]^{n-1},$$

which converges to $\tfrac{1}{4}e^{-\xi}/e^{\xi} = \alpha e^{-\lambda}$ as $n \to \infty$. Therefore, the inequality (8.1) is sharp proving that the number $\alpha e^{-\lambda}$ in Theorem 8.1 cannot be replaced by a smaller bound.

Equality. Here we assume that $1 < \lambda \le 2$ and show that equality can hold in one of the inequalities considered earlier:

$$P(g^* \ge \lambda) \le 1/\lambda^2.$$

Let S be as above. Then the least biconcave function u on S majorizing the indicator function of $\{(x,y) \in S : |x+y| \ge 2\lambda\}$ satisfies

$$u(1,y) = (1+y)/\lambda^2$$

on the line segment joining the points $(1,-1)$ and $(1,\lambda-1)$. It is also linear on the line segment joining $(\lambda+1, \lambda-1)$ and $(-1, \lambda-1)$, on the line segment joining $(-1,1)$ and $(-1, -\lambda+1)$, and on the line segment joining $(-\lambda-1, -\lambda+1)$ and $(1, -\lambda+1)$.

Let Z be a martingale with $Z_0 \equiv (1,0)$ and such that Z_1 has its values in the set $\{(1,-1), (1, \lambda-1)\}$. Then, because of the linearity of u on the convex hull of this set, $Eu(Z_1) = Eu(Z_0)$. On the set where $Z_1 = (1,-1)$, let $Z_2 = \cdots = Z_\infty = (1,-1)$. On the set where $Z_1 = (1, \lambda-1)$, let Z_2 have its values in the set $\{(\lambda+1, \lambda-1), (-1, \lambda-1)\}$. Then $Eu(Z_2) = Eu(Z_1)$. Continue in this way. The martingale Z determines martingales f and g as above. It is easy to check that $\|f\|_\infty \le 1$, that g is the transform of f by the sequence $(\varepsilon_k)_{k \ge 0}$ with $\varepsilon_k = (-1)^k$, and that $P(Z_n = Z_\infty) \to 1$ as $n \to \infty$. Therefore,

$$P(g^* \ge \lambda) = P(Z_\infty \in \{(\lambda+1, \lambda-1), (-\lambda-1, -\lambda+1)\})$$
$$= Eu(Z_\infty) = \lim_{n \to \infty} Eu(Z_n)$$
$$= Eu(Z_0) = u(1,0) = 1/\lambda^2.$$

Note that in this proof of equality we do not need to know the value of u at every point of S. In fact, all we need is a function u that is defined on the union of the four line segments, that is linear on each of these segments, and is such that $u(1,0) = 1/\lambda^2$, $u(1,-1) = 0$, $u(\lambda+1, \lambda-1) = 1$, $u(-1,1) = 0$, and $u(-\lambda-1, -\lambda+1) = 1$. The solution is elementary.

A proof of (8.1). If the inequality holds for complex martingales, it will hold for real martingales. Therefore, we can assume in the proof that $\mathbf{H}$ has linear dimension over $\mathbf{R}$ of at least two. Alternatively, the one-dimensional case can be proved separately using an argument similar to, but not exactly the same as, the one used here.

Here it is convenient to let

$$S = \{(x,y) \in \mathbf{H} \times \mathbf{H} \colon |x| \leq 1\}.$$

Using the assumption that $\|f\|_\infty \leq 1$, we can assume that $(f_n(\omega), g_n(\omega)) \in S$ for all $\omega \in \Omega$ and $n \geq 0$.

We shall show that there is a function $u \colon S \to \mathbf{R}$ satisfying

$$(8.5) \qquad P(|g_n| \geq \lambda) \leq Eu(f_n, g_n),$$

$$(8.6) \qquad Eu(f_n, g_n) \leq Eu(f_{n-1}, g_{n-1}), \qquad n \geq 1,$$

$$(8.7) \qquad Eu(f_0, g_0) \leq \alpha e^{-\lambda}.$$

These three inequalities imply (8.1). Consider the function u defined by

$$(8.8) \qquad u(x,y) = u^0(|y| + |x|, |y| - |x|)$$

where u^0 denotes the function of the special case considered above. This function satisfies the above inequalites if $\mathbf{H} = \mathbf{R}$ but does not satisfy (8.6) for all f and g as in the theorem if the dimension is larger. For the higher dimensions, it does agree with the correct function on a large part of S but not on all of S.

If the linear dimension of $\mathbf{H}$ over $\mathbf{R}$ is at least two, as we assume, then we can define $u \colon S \to \mathbf{R}$ for a fixed $\lambda > 2$ as follows: $u(x,y) = 0$ if $|x| = 1$ and $|y| < \lambda$, $u(x,y) = 1$ if $|x| = 1$ and $|y| \geq \lambda$, the restriction of u to the interior of S is continuous, and

$$\begin{aligned}
u(x,y) &= \alpha(1 + |y|^2 - |x|^2)e^{-\lambda} & \text{if } (x,y) \in D_0, \\
&= 2\alpha(1 - |x|)e^{|x| + |y| - \lambda - 1} & \text{if } (x,y) \in D_1, \\
&= \frac{1 - |x|^2}{(\lambda - |y|)^2 + 1 - |x|^2} & \text{if } (x,y) \in D_2, \\
&= 1 - \frac{\lambda^2 - 1 - |y|^2 + |x|^2}{4(\lambda - 1)} & \text{if } (x,y) \in D_3, \\
&= 1 & \text{if } (x,y) \in D_4,
\end{aligned}$$

where, in this context,

$$\begin{aligned}
D_0 &= \{(x,y) \colon |x| + |y| < 1\}, \\
D_1 &= \{(x,y) \colon 1 < |x| + |y| < \lambda - 1 \text{ and } 0 < |x| < 1\}, \\
D_2 &= \{(x,y) \colon \lambda - 1 - |x| < |y| < \lambda - 1 + |x| \text{ and } |x| < 1\}, \\
D_3 &= \{(x,y) \colon \lambda - 1 + |x| < |y| < \sqrt{\lambda^2 - 1 + |x|^2}\}, \\
D_4 &= \{(x,y) \colon |y| > \sqrt{\lambda^2 - 1 + |x|^2} \text{ and } |x| < 1\}.
\end{aligned}$$

Although u is not continuous, it does satisfy

(8.9) $$\lim_{r \uparrow 1} u(rx, ry) = u(x,y), \qquad (x,y) \in S.$$

For example, let $|x| = 1$ and $|y| = \lambda$, the only case that is not immediately obvious. If $(\lambda - 1)/(\lambda + 1) < r < 1$, then $(rx, ry) \in D_2$ and

$$u(rx, ry) = \frac{1+r}{\lambda^2(1-r)+1+r} \uparrow 1 = u(x,y) \quad \text{as } r \uparrow 1.$$

Note that u majorizes the indicator function of the set of points $(x,y) \in S$ satisfying $|y| \geq \lambda$. This yields (8.5).

To prove (8.6), we shall use the H-valued functions φ and ψ defined on S as follows:

$$
\begin{aligned}
\varphi(x,y) &= -2\alpha e^{-\lambda} x && \text{if } (x,y) \in D_0, \\
&= -2\alpha e^{|x|+|y|-\lambda-1} x && \text{if } (x,y) \in D_1, \\
&= \frac{-2(\lambda - |y|)^2}{[(\lambda - |y|)^2 + 1 - |x|^2]^2} x && \text{if } (x,y) \in D_2, \\
&= -\frac{1}{2(\lambda - 1)} x && \text{if } (x,y) \in D_3, \\
&= 0 && \text{if } (x,y) \in D_4, \\
\psi(x,y) &= 2\alpha e^{-\lambda} y && \text{if } (x,y) \in D_0, \\
&= 2\alpha(1 - |x|)e^{|x|+|y|-\lambda-1} y' && \text{if } (x,y) \in D_1, \\
&= \frac{2(\lambda - |y|)(1 - |x|^2)}{[(\lambda - |y|)^2 + 1 - |x|^2]^2} y' && \text{if } (x,y) \in D_2, \\
&= \frac{1}{2(\lambda - 1)} y && \text{if } (x,y) \in D_3, \\
&= 0 && \text{if } (x,y) \in D_4,
\end{aligned}
$$

where $y' = y/|y|$. Now extend these two functions to S so their restrictions to $\bar{D}_3$, the closure of D_3, are continuous and so their restrictions to $S \setminus \bar{D}_3$ are also continuous. This is possible.

Suppose that $x, y, h, k \in H$ with $(x,y) \in S$, $(x+h, y+k) \in S$, and $|k| \leq |h|$. Then, as we shall show,

(8.10) $$u(x+h, y+k) \leq u(x,y) + \varphi(x,y) \cdot h + \psi(x,y) \cdot k$$

where $a \cdot b$ denotes $\text{Re}\langle a, b \rangle$ for $a, b \in H$. If $0 < r \leq 1$ and $n \geq 1$, then

$$u(rf_n, rg_n) \leq u(rf_{n-1}, rg_{n-1}) + r\varphi(rf_{n-1}, rg_{n-1}) \cdot d_n + r\psi(rf_{n-1}, rg_{n-1}) \cdot e_n.$$

This follows from (8.10) and the assumption that $|e_n(\omega)| \leq |d_n(\omega)|$. If $0 < r < 1$, then all four terms are bounded and integrable and, by the martingale condition on f and g, each of the last two terms has an expectation equal to zero so

$$Eu(rf_n, rg_n) \leq Eu(rf_{n-1}, rg_{n-1}).$$

Now let $r \uparrow 1$ and use (8.9) and the Lebesgue dominated convergence theorem to obtain (8.6).

Also use (8.10) to obtain

$$u(f_0, g_0) = u(d_0, e_0) \leq u(0,0) + \varphi(0,0) \cdot d_0 + \psi(0,0) \cdot e_0$$
$$= u(0,0) = \alpha e^{-\lambda}.$$

This implies (8.7) and, apart from verifying (8.10), completes the proof of (8.1).

A proof of (8.10). Let $(x, y) \in S$, $(x+h, y+k) \in S$, and $|k| \leq |h|$. If $h = 0$, then $k = 0$ and (8.10) holds. So assume that $h \neq 0$ and consider the function G defined by

$$G(t) = u(x + ht, y + kt)$$

on the interval $I = \{t \in \mathbf{R} : |x + ht| \leq 1\}$. Note that $[0,1] \subset I$ and that $|x + ht| < 1$ if t is in the interior of I.

It is easy to check (see the proof of (8.9)) that G is continuous on I. Also, if $(x, y) \in D_0 \cup \cdots \cup D_4$, then the derivative of G exists at 0 and

(8.11) $$G'(0) = \varphi(x, y) \cdot h + \psi(x, y) \cdot k.$$

In fact, by the continuity properties of φ, ψ, and G, this holds for all (x, y) in the interior of $S \setminus \partial D_3$. Furthermore,

(8.12) $$G''(0) \leq 0 \quad \text{if } (x, y) \in D_0 \cup \cdots \cup D_4.$$

For example, if $(x, y) \in D_0$, then

$$G''(0) = -2\alpha e^{-\lambda}(|h|^2 - |k|^2) \leq 0.$$

Elementary calculation also gives, for $(x, y) \in D_1$ and $x' = x/|x|$,

(8.13) $$G''(0) = 2\alpha e^{|x|+|y|-\lambda-1}\{-|x|[x' \cdot h + y' \cdot k]^2 - R\}$$

where $R = |h|^2 - |k|^2 + (|k|^2 - (y' \cdot k)^2)(|x| + |y| - 1)/|y|$ is nonnegative, and this yields the nonpositivity of $G''(0)$. If $(x, y) \in D_2$, then

$$G''(0) = -2[(\lambda - |y|)^2 + 1 - |x|^2]^{-3}(A + B + C) \leq 0$$

where

$$A = (|h|^2 - |k|^2)(\lambda - |y|)^2[(\lambda - |y|)^2 + 1 - |x|^2],$$
$$B = [\{(\lambda - |y|)^2 - (1 - |x|^2)\}y' \cdot k + 2(\lambda - |y|)x \cdot h]^2,$$
$$C = (\lambda - |y|)[(\lambda - |y|)^2 + 1 - |x|^2][|k|^2 - (y' \cdot k)^2][\lambda - |y| - (1 - |x|^2)/|y|],$$

and $A, B,$ and C are nonnegative: $|x|^2 \geq (\lambda - 1 - |y|)^2$ on D_2 and this implies that

$$\lambda - |y| - (1 - |x|^2)/|y| \geq (\lambda - 2)(\lambda - |y|)/|y| > 0.$$

If $(x, y) \in D_3$, then $G''(0) = -(|h|^2 - |k|^2)/2(\lambda - 1) \leq 0$. Finally, $G''(0) = 0$ for $(x, y) \in D_4$.

We shall now prove (8.10), but first in the special case that $|k| < |h|$. Under this temporary assumption, the set

$$K = \{t \in I : (x + ht, y + kt) \notin D_0 \cup \cdots \cup D_4\}$$

is finite: for example, the set of zeros of the mapping

$$t \mapsto |y + kt| - (\lambda - 1 + |x + ht|)$$

is a subset of the zero set of a fourth-degree polynomial with leading coefficient $|h|^2 - |k|^2$. By (8.11), (8.12), and a translation argument, the first two derivatives of G exist at t and satisfy

$$(8.14) \qquad G'(t) = \varphi(x + ht, y + kt) \cdot h + \psi(x + ht, y + kt) \cdot k$$

and

$$(8.15) \qquad G''(t) \leq 0$$

for all $t \in I \setminus K$. Furthermore, if t_0 is an interior point of I and $t_0 \in K$, then

$$(8.16) \qquad G'(t_0-) \geq G'(t_0+).$$

To show (8.16), we can assume that $t_0 = 0$, so $|x| < 1$, and that either (i) $|x| > 0$ and $|y| = \lambda - 1 + |x|$, or (ii) $|y|^2 = \lambda^2 - 1 + |x|^2$. For example, assume (ii). If $(x + ht, y + kt) \in D_4$ for all small positive t, then $G'(0+) = 0$ and the condition

$$|y + kt|^2 > \lambda^2 - 1 + |x + ht|^2$$

for such t implies that $2y \cdot k + t|k|^2 > 2x \cdot h + t|h|^2$, which in turn implies that $y \cdot k > x \cdot h$. By (8.14) and the definition of φ and ψ on D_3, the inequality $y \cdot k > x \cdot h$ implies that $G'(0-) > 0 = G'(0+)$. On the other hand, if $(x + ht, y + kt) \in D_4$ for all negative t near zero, an analogous argument gives $G'(0-) = 0 > G'(0+)$. Case (i) is proved in a similar way.

The function G is concave on I. This follows from (8.15), (8.16), and the continuity of G. Therefore,

$$G(1) \leq G(0) + G'(0+).$$

But $G'(0+) = \lim_{t \downarrow 0}[\varphi(x + ht, y + kt) \cdot h + \psi(x + ht, y + kt) \cdot k]$ and, by calculations similar to those used to prove (8.16), the last expression is not greater than $\varphi(x, y) \cdot h + \psi(x, y) \cdot k$. This gives (8.10) in the case $|k| < |h|$.

Therefore, returning to the general case with $|k| \leq |h|$, we have

$$u(x + h, y + kr) \leq u(x, y) + \varphi(x, y) \cdot h + \psi(x, y) \cdot kr, \qquad 0 < r < 1.$$

If $|x + h| < 1$ or $|y + k| \neq \lambda$, then (8.10) follows by taking $r \uparrow 1$.

Finally, consider (8.10) for the remaining possibility: $|x + h| = 1$ and $|y + k| = \lambda$. Then

$$|y| \geq |y + k| - |k| \geq \lambda - |h| \geq \lambda - |x + h| - |x| = \lambda - 1 - |x|.$$

So (a) $\lambda - 1 - |x| \leq |y| < \lambda - 1 + |x|$, (b) $\lambda - 1 + |x| \leq |y| \leq \sqrt{\lambda^2 - 1 + |x|^2}$, or (c) $|y| > \sqrt{\lambda^2 - 1 + |x|^2}$. If (c) holds, then both sides of (8.10) are equal to unity. If (b) holds, then

$$4(\lambda - 1)[u(x + h, y + k) - u(x, y) - \varphi(x, y) \cdot h - \psi(x, y) \cdot k]$$

is equal to $-(|h|^2 - |k|^2) \leq 0$: use $2x \cdot h = 1 - |x|^2 - |h|^2$ and $2y \cdot k = \lambda^2 - |y|^2 - |k|^2$. Also, use these and $|h| \geq |k| \geq \lambda - |y|$ to prove (8.10) in the case (a).

This completes the proof of (8.1).

Strictness. Let $\delta > 0$. We shall complete the proof of Theorem 8.1 by showing that, for $\lambda > 2 + \delta$,

$$P(g^* \geq \lambda) < \alpha e^{-\lambda}.$$

If $P(g^* > 1) = 0$, there is nothing to prove, so suppose that $P(g^* > 1) > 0$. Let $f_{-1} = g_{-1} \equiv 0$ and m be the least nonnegative integer n such that

$$P(|f_{n-1}| + |g_{n-1}| \leq 1) = 1 \quad \text{and} \quad P(|f_n| + |g_n| > 1) > 0.$$

Such an m exists. As we shall show,

$$(8.17) \qquad P(g^* > \lambda) \leq \alpha e^{-\lambda}[1 - 2EQ(f_{m-1}, g_{m-1}, d_m, e_m)]$$

where $EQ(f_{m-1}, g_{m-1}, d_m, e_m) > 0$ and the function Q, to be defined, does not depend on λ. As a consequence, we obtain the desired inequality:

$$P(g^* \geq \lambda) = \lim_{\eta \uparrow \lambda} P(g^* > \eta) < \alpha e^{-\lambda}.$$

To prove (8.17), consider the stopping time τ defined by

$$\tau(\omega) = \inf\{n \geq 0 : |g_n(\omega)| > \lambda\}$$

and the martingales $f^\tau = (f_{\tau \wedge n})$ and $g^\tau = (g_{\tau \wedge n})$. Note that $P(\tau \geq m) = 1$, $\|f^\tau\|_\infty \leq 1$, and g^τ is differentially subordinate to f^τ. Therefore, by the proof of (8.1), if $n > m$, then

$$P(|g_{\tau \wedge n}| > \lambda) \leq Eu(f_{\tau \wedge n}, g_{\tau \wedge n}) \leq Eu(f_{\tau \wedge m}, g_{\tau \wedge m}) = Eu(f_m, g_m).$$

It follows from the monotone convergence theorem that

$$(8.18) \qquad P(g^* > \lambda) \leq Eu(f_m, g_m).$$

Let G be any continuous function on $[0, 1]$ such that G' is continuous on $(0, 1)$ and $G''(t)$ exists and is nonpositive for all but a finite number of $t \in (0, 1)$. Then

$$G(1) = G(0) + G'(0+) + \int_0^1 (1 - t)G''(t)\, dt.$$

In particular, this holds if $G(t) = u(x + ht, y + kt)$ where

(8.19) $$|x| + |y| \leq 1, (x + h, y + k) \in S, |k| \leq |h|, h \neq 0.$$

This is because $N(t) < \lambda - 1 + M(t), 0 \leq t \leq 1$, where $M(t) = |x + ht|$ and $N(t) = |y + kt|$, and the function $M + N$ is strictly convex, as well as because of earlier calculations of G' and G''.

There is a number $\beta > 0$ such that $1 + 3\beta < 1 + \delta$ and

$$P(|f_m| + |g_m| > 1 + 3\beta) > 0.$$

So, in addition to (8.10), assume that

(8.20) $$|x + h| + |y + k| > 1 + 3\beta.$$

Let r and s satisfy $0 < r < s < 1$, $M(r) + N(r) = 1 + \beta$, and $M(s) + N(s) = 1 + 2\beta$. Define Q by

$$Q(x, y, h, k) = \int_r^s (1 - t)M(t)[M'(t) + N'(t)]^2 \, dt$$

if (8.19) and (8.20) hold, $Q(x, y, h, k) = 0$ otherwise. If (8.19) and (8.20) do hold, then $(x + ht, y + kt) \in D_2$, $r \leq t \leq s$, and, by (8.13),

$$\int_0^1 (1 - t)G''(t) \, dt \leq -2\alpha e^{-\lambda}Q(x, y, h, k) < 0.$$

The latter inequality follows from the strict positivity of the integrand on the interval $[r, s]$. Therefore, if (8.19) holds, then

$$u(x + h, y + k) \leq u(x, y) + \varphi(x, y) \cdot h + \psi(x, y) \cdot k - 2\alpha e^{-\lambda}Q(x, y, h, k).$$

This leads, as in the proof of (8.1), to

$$Eu(f_m, g_m) \leq Eu(f_{m-1}, g_{m-1}) - 2\alpha e^{-\lambda}EQ(f_{m-1}, g_{m-1}, d_m, e_m)$$
$$\leq u(0, 0) - 2\alpha e^{-\lambda}EQ(f_{m-1}, g_{m-1}, d_m, e_m)$$

which implies (8.17). This completes the proof of Theorem 8.1.

The inequality for stochastic integrals. Let M be a right-continuous martingale with limits from the left, and let U and V be predictable processes as in Chapter 3. Consider again (i) the case in which U and V are **H**-valued and M is scalar-valued, and (ii) the case in which U and V are scalar-valued and M is **H**-valued. Assume as before that **H** is separable and that $\int_{[0,\infty)} |U_t|^2 \, d[M, M]_t$ is finite a.s.

THEOREM 8.2. *Suppose that* $\|U \cdot M\|_\infty \leq 1$ *and that* $|V_t(\omega)| \leq |U_t(\omega)|$ *for all* $\omega \in \Omega$ *and* $t \geq 0$. *Then, for both* (i) *and* (ii),

$$P((V \cdot M)^* \geq \lambda) \leq 1 \qquad \text{if } 0 < \lambda \leq 1,$$

$$\leq 1/\lambda^2 \qquad \text{if } 1 < \lambda \leq 2,$$
$$\leq \alpha e^{-\lambda} \qquad \text{if } \lambda > 2,$$

where $\alpha = e^2/4$. *If λ is any positive number, then equality can hold.*

PROOF: The inequality follows from a slight modification of the proof of the corresponding L^p inequalities given in [Bur 88b]. Since $\|U \cdot M\|_\infty \leq 1$, Lemma 5.1 of that paper holds for $p = 2$, hence for $p = 1$. Therefore, by going to subsequences if necessary, we have elementary predictable processes U^n and V^n satisfying

$$(8.21) \qquad\qquad |V_t^n(\omega)| \leq |U_t^n(\omega)|,$$

$$(8.22) \qquad\qquad \lim_{n \to \infty} \|U^n \cdot M - U \cdot M\|_1 = 0,$$

$$(8.23) \qquad\qquad \lim_{n \to \infty} (V^n \cdot M - V \cdot M)^* = 0 \quad \text{a.s.}$$

By (8.21) and Theorem 8.1, the triplet (U^n, V^n, M) satisfies the desired inequality. So by (8.22) and (8.23), the triplet (U, V, M) does also.

We already know from the discrete-time case that equality can hold if $0 < \lambda \leq 2$. A slight modification of an example of Monroe (1987) for a different problem (see Chapter 12) will show that equality can hold also for $\lambda > 2$. Let $B = (B_t)_{t \geq 0}$ be a real Brownian motion. For $t > 1$, consider the local time

$$L_t = \lim_{\varepsilon \downarrow 0} \frac{1}{2\varepsilon} \int_{(1,t]} 1(|B_s - B_1| < \varepsilon)\, ds.$$

Let $S = \inf\{t > 1 : L_t = \lambda - 2\}$ and $T = \inf\{t > 1 : |B_t - B_1| = 1\}$, and define (U, V, M) as follows: if $t \geq 0$, then $U_t = 1(t \leq T)$,

$$V_t = 1(t = 0) - 1(0 < t \leq 1) - \operatorname{sgn}(B_t - B_1)1(1 < t \leq S) + 1(S < t \leq T),$$

and $M_t = \frac{1}{2}1(0 \leq t < 1) + (B_t - B_1)1(t \geq 1,\ B_1 \geq 0) + 1(t \geq 1,\ B_1 < 0)$. Then (U, V, M) satisfies the assumptions of Theorem 8.2. By Tanaka's formula,

$$\int_{(1,S]} \operatorname{sgn}(B_t - B_1)\, d(B_t - B_1) = L_S - |B_S - B_1| = L_S = \lambda - 2.$$

A little calculation shows that $N = V \cdot M$ satisfies

$$P(N_\infty = \lambda) = P(B_1 \geq 0,\ S < T,\ B_T - B_S = 1)$$
$$= P(B_1 \geq 0,\ |B_t - B_1| < 1 \text{ for } 1 < t \leq S,\ B_T - B_S = 1)$$
$$= (1/4)P(|B_t - B_1| < 1 \text{ for } 1 < t \leq S) = (1/4)e^{-(\lambda-2)}.$$

So here $P(N^* \geq \lambda) = \alpha e^{-\lambda}$. This completes the proof of Theorem 8.2.

9. A ONE-SIDED EXPONENTIAL INEQUALITY

Suppose that f is a real martingale satisfying $\|f\|_\infty \leq 1$. Let $\lambda > 0$ and $0 < \gamma < 1$. It is not always possible to find a nonnegative integer n and a real predictable sequence v uniformly bounded in absolute value by 1 such that the transform g of f by v satisfies

$$P(g_n \geq \lambda) \geq \gamma.$$

The following theorem, the analogue of Theorem 8.1 for the one-sided maximal function, can be used to find a necessary condition on λ and γ for this to be possible. The condition $-1 \leq f_n(\omega) \leq 1$ can be changed to $0 \leq f_n(\omega) \leq 1$ with an obvious change in the conclusion.

THEOREM 9.1. *Let f and g be real martingales with respect to the same filtration. If g is differentially subordinate to f and $\|f\|_\infty \leq 1$, then*

$$
\begin{aligned}
P(\sup_{n\geq 0} g_n \geq \lambda) &\leq 1 && \text{if } 0 < \lambda \leq 1, \\
&\leq (1 - \sqrt{\lambda - 1}/2)^2 && \text{if } 1 < \lambda \leq 2, \\
&< \alpha e^{-\lambda} && \text{if } \lambda > 2,
\end{aligned}
$$

where $\alpha = e^2/4$. For each λ, the bound on the right is sharp. If $0 < \lambda \leq 2$, then equality can hold.

The above inequality is already sharp for ± 1-transforms.

As does Theorem 8.1, this theorem implies an analogous inequality for stochastic integrals but one in which, for any positive number λ, equality can hold. See Theorem 9.2 below.

PROOF: For $0 < \lambda \leq 1$, the proof is as before. If $\lambda > 2$, then, by Theorem 8.1,

$$P(\sup_{n\geq 0} g_n \geq \lambda) \leq P(g^* \geq \lambda) < \alpha e^{-\lambda}.$$

The example in Chapter 8 showing that the second inequality is sharp also shows that the inequality

$$(9.1) \qquad\qquad P(\sup_{n\geq 0} g_n \geq \lambda) < \alpha e^{-\lambda}$$

is sharp. A proof of (9.1) that is independent of Theorem 8.1 and a proof of the remaining case $1 < \lambda \leq 2$ rest on methods similar to those used in the preceding chapter. Here let

$$S = \{(x, y) \in \mathbf{R}^2 : |x| \leq 1\}$$

and define $u: S \to \mathbf{R}$ so that the restriction of u to the interior of S is continuous,

$$u(x, y) = u(-x, y),$$

$u(x, y) = 0$ if $x = 1$ and $y < \lambda$, $u(x, y) = 1$ if $x = 1$ and $y \geq \lambda$, and

$$
\begin{aligned}
u(x, y) &= (1 - x)e^{x+y-\lambda+1}/2 && \text{if } (x, y) \in D_1, \\
&= \frac{(1 - x)(3 + x + y - \lambda)}{2(\lambda - x - y + 1)} && \text{if } (x, y) \in D_2, \\
&= 1 - (\lambda - y)/2 && \text{if } (x, y) \in D_3, \\
&= 1 && \text{if } (x, y) \in D_4,
\end{aligned}
$$

where

$$
\begin{aligned}
D_1 &= \{(x, y): y < \lambda - 1 - x \text{ and } 0 < x < 1\}, \\
D_2 &= \{(x, y): \lambda - 1 - x < y < \lambda - 1 + x \text{ and } x < 1\}, \\
D_3 &= \{(x, y): \lambda - 1 + |x| < y < \lambda\}, \\
D_4 &= \{(x, y): y > \lambda \text{ and } |x| < 1\}.
\end{aligned}
$$

Then, by the methods of Chapter 8,

(9.2) $$P(\sup_{n \geq 0} g_n \geq \lambda) \leq Eu(f_0, g_0).$$

But $Eu(f_0, g_0) \leq \sup_{|y| \leq |x| \leq 1} u(x, y)$ and if $1 < \lambda \leq 2$, for example, then

$$\sup_{|y| \leq |x| \leq 1} u(x, y) = u(x_0, x_0) = (1 - \sqrt{\lambda - 1}/2)^2$$

where $x_0 = (\lambda + 1)/2 - \sqrt{\lambda - 1}$. Using u as a guide, we can construct an example for which

(9.3) $$P(\sup_{n \geq 0} g_n \geq \lambda) = (1 - \sqrt{\lambda - 1}/2)^2.$$

First note that u is linear on the line segment joining the points $(1, \lambda - 2\sqrt{\lambda - 1})$ and $(1 - \sqrt{\lambda - 1}, \lambda - \sqrt{\lambda - 1})$. Therefore, if the martingale W defined by $W_n = (f_n, g_n)$ starts at the point (x_0, x_0), which is on the line segment, and moves on the first step to either end of the line segment with the appropriate probability, then $Eu(W_1) = Eu(W_0)$. Similarly, on the set $\{W_1 = (1 - \sqrt{\lambda - 1}, \lambda - \sqrt{\lambda - 1})\}$, the martingale W should move to either $(1, \lambda)$ or $(-1, \lambda - 2)$ so that $Eu(W_2) = Eu(W_1)$. These considerations lead to the following example on the Lebesgue unit interval where, as before, the same notation is used for a subinterval $[a, b]$ and its indicator function: Let $\beta = 1 - \sqrt{\lambda - 1}/2$ and

$$
\begin{aligned}
e_0 &= ((\lambda + 1)/2 - \sqrt{\lambda - 1})[0, 1), \\
e_1 &= ((\lambda - 1)/2)[0, \beta) + ((\lambda - 1)/2 - \sqrt{\lambda - 1})[\beta, 1), \\
e_2 &= \sqrt{\lambda - 1}[0, \beta^2) + (-2 + \sqrt{\lambda - 1})[\beta^2, \beta), \\
e_k &= 0, \quad k \geq 3.
\end{aligned}
$$

Then the martingales f and g determined by (d_k) and (e_k), respectively, where $d_k = (-1)^k e_k$, satisfy the conditions of the theorem and (9.3). This completes the proof of Theorem 9.1.

Remark. If $\lambda \in \mathbf{R}$, $(x_0, y_0) \in S$, $f_0 \equiv x_0$, and $g_0 \equiv y_0$, then, by (9.2),

$$(9.4) \qquad\qquad P(\sup_{n\geq 0} g_n \geq \lambda) \leq u(x_0, y_0).$$

Strict inequality holds if $y_0 < \lambda - 1 - |x_0|$. For example, if $f_0 = 0$, $g_0 = 0$, and $\lambda > 2$, then

$$P(\sup_{n\geq 0} g_n \geq \lambda) < \sqrt{\alpha}\, e^{-\lambda}.$$

Note that here $u(0,0)$ differs from the value at the origin of the function u that is extremal for the two-sided problem considered in the last chapter. However, both functions give the same value for $\sup_{|y|\leq|x|\leq 1} u(x, y)$.

An inequality for stochastic integrals. Let U, V, and M be real-valued but otherwise the same as in Chapter 3.

THEOREM 9.2. *Suppose* $\|U \cdot M\|_\infty \leq 1$ *and* $|V_t(\omega)| \leq |U_t(\omega)|$. *Then*

$$\begin{aligned} P(\sup_{t\geq 0} (V \cdot M)_t \geq \lambda) &\leq 1 && \text{if } 0 < \lambda \leq 1, \\ &\leq (1 - \sqrt{\lambda - 1}/2)^2 && \text{if } 1 < \lambda \leq 2, \\ &\leq \alpha e^{-\lambda} && \text{if } \lambda > 2, \end{aligned}$$

where $\alpha = e^2/4$. *If* λ *is any positive number, then equality can hold.*

See the proof of Theorem 8.2. The example contained therein is applicable here.

10. ON SUPERREFLEXIVITY

In the work of James, Enflo, Pisier, and others, there are many characterizations of superreflexive Banach spaces and we have seen a few of them in Chapter 4. Other examples can be found in the papers referred to there. Here is a new characterization.

THEOREM 10.1. *A Banach space* $\mathbf{B}$ *is superreflexive if and only if, for some* $q \in [2, \infty)$, *there is a function* $u : \mathbf{B} \times [0, \infty) \to \mathbf{R}$ *such that, for all* $x, a \in \mathbf{B}$ *and* $t \geq 0$,

$$(10.1) \qquad u(0,0) > 0,$$
$$(10.2) \qquad u(x, t) \leq |x| \quad \text{if } t \geq 1,$$
$$(10.3) \qquad u(x, t) \leq [u(x + a, t + |a|^q) + u(x - a, t + |a|^q)]/2.$$

The conditions on u are similar to the conditions on ζ in Chapter 1. Neither function can be too small, neither can be too large, and each must satisfy a convexity condition. Also, just as ζ has a simple martingale interpretation, so does u.

LEMMA 10.1. *If u satisfies (10.1), (10.2), and (10.3), and f is a B-valued conditionally symmetric martingale, then, for all $\lambda > 0$,*

$$(10.4) \qquad P(S(f,q) \geq \lambda) \leq \frac{\|f\|_1}{\lambda u(0,0)}.$$

If u is the greatest function satisfying (10.1), (10.2), and (10.3), then the inequality is sharp. In fact, it is sharp for the family of B-valued dyadic martingales on the Lebesgue unit interval.

Recall from Chapter 4 that $S(f,q) = (\sum_{k=0}^{\infty} |d_k|^q)^{1/q}$. The martingale f is *conditionally symmetric* if and only if

$$E\varphi(d_0, \ldots, d_{n-1}, -d_n) = E\varphi(d_0, \ldots, d_{n-1}, d_n)$$

for all bounded continuous functions $\varphi : \mathbf{B}^{n+1} \to \mathbf{R}$ and all $n \geq 1$. For example, dyadic martingales are conditionally symmetric.

Here is one way to prove the "if" part of Theorem 10.1. Assume that u exists but that B is not superreflexive. Let $\lambda > 1/u(0,0)$ and $0 < \varepsilon < 1$. Choose a positive integer n satisfying $n\varepsilon^q \geq \lambda^q$. Then there is a dyadic martingale f such that $\|f\|_1 \leq \|f\|_\infty \leq 1$ and $|d_k(\omega)| \geq \varepsilon$ for all $\omega \in \Omega$ and $k < n$. This is an immediate consequence of the work of James (1972a); a convenient reference is Pisier (1975b). So for this f the left-hand side of (10.4) is 1 but the right-hand side is strictly less than 1, a contradiction.

PROOF OF LEMMA 10.1: We can assume that u is the greatest function satisfying (10.2) and (10.3). Then

$$(10.5) \qquad u(x,t) \leq u(x,s) \quad \text{if} \quad 0 \leq s \leq t.$$

To see this, consider $v : \mathbf{B} \times [0,\infty) \to \mathbf{R}$ defined by $v(x,t) = \sup\{u(x,t') : t' \geq t\}$. Notice that $v \geq u$ and that v satisfies (10.2) and (10.3), so v also satisfies $v \leq u$ by the maximality of u. Therefore, $u = v$ and this implies (10.5).

The next step is to show that

$$(10.6) \qquad u(x,t) \leq u(0,t) + |x| \leq u(0,0) + |x|.$$

The inequality on the right follows from (10.5). To prove the one on the left, fix $y \in \mathbf{B}$ and note that the mapping $(x,t) \mapsto u(x+y,t) - |y|$ satisfies (10.2) and (10.3), so $u(x+y,t) - |y| \leq u(x,t)$. In particular, $u(y,t) \leq u(0,t) + |y|$, which is the desired inequality.

The function u is continuous. In fact, if (x,t) and (x',t') belong to $\mathbf{B} \times [0,\infty)$, then

$$(10.7) \qquad |u(x,t) - u(x',t')| \leq |x - x'| + |t - t'|^{1/q}.$$

For $t' = t$, this follows from the above discussion by setting $y = x' - x$. If $t' > t$ and $a \in \mathbf{B}$ is chosen to satisfy $|a| = (t' - t)^{1/q}$, then, by (10.5) and (10.3),

$$0 \leq u(x,t) - u(x,t') \leq [u(x+a,t') - u(x,t') + u(x-a,t') - u(x,t')]/2$$

which is less than or equal to $|a|$ by the special case of (10.7) with $t' = t$. So (10.7) holds also for the special case $x' = x$. These two special cases and the triangle inequality now yield (10.7).

There is one more property of u that we shall use and this also follows from its maximality:

$$(10.8) \qquad u(x, t) = u(-x, t).$$

To prove (10.4), we can assume that $\lambda = 1$. Let $S_n^q = S_n^q(f, q) = \sum_{k=0}^{n} |d_k|^q$. By (10.2), (10.6), and Chebyshev's inequality,

$$
\begin{aligned}
(10.9) \qquad P(S_n(f, q) \geq 1) &\leq P\left(|f_n| - u(f_n, S_n^q) + u(0, 0) \geq u(0, 0)\right) \\
&\leq E\left[|f_n| - u(f_n, S_n^q) + u(0, 0)\right] / u(0, 0) \\
&\leq E|f_n| / u(0, 0)
\end{aligned}
$$

where the last inequality follows from

$$(10.10) \qquad Eu(f_n, S_n^q) \geq Eu(f_{n-1}, S_{n-1}^q) \geq \cdots \geq Eu(f_0, S_0^q) \geq u(0, 0),$$

which we now prove. By the conditional symmetry of f, the expectation of $u(f_n, S_n^q)$ is given by

$$\left[Eu(f_{n-1} + d_n,\ S_{n-1}^q + |d_n|^q) + Eu(f_{n-1} - d_n,\ S_{n-1}^q + |d_n|^q)\right] / 2$$

which, by (10.3), is greater than or equal to $Eu(f_{n-1}, S_{n-1}^q)$. Since $S_0 = |f_0|$, the inequality on the right in (10.10) follows in a similar way from (10.8) and (10.3). So (10.9) holds. Now use homogeneity and take limits to obtain

$$P(S(f, q) > \lambda) \leq \frac{\|f\|_1}{\lambda u(0, 0)}.$$

To finish the proof of (10.4), replace λ by λ' in this inequality and let $\lambda' \uparrow \lambda$.

To complete the proof of Lemma 10.1, we need to show that (10.4) is sharp for the greatest function u satisfying (10.2) and (10.3). If $(x, t) \in \mathbf{B} \times [0, \infty)$, let $\mathbf{S}(x, t)$ be the set of all simple (see Chapter 2) dyadic martingales f on the Lebesgue unit interval such that $f_0 \equiv x$ and

$$P(t - |x|^q + S^q(f, q) \geq 1) = 1.$$

The set $\mathbf{S}(x, t)$ is nonempty, the function u is given by

$$(10.11) \qquad u(x, t) = \inf\{\|f\|_1 : f \in \mathbf{S}(x, t)\},$$

and (10.11) implies that (10.4) is sharp: take $\lambda = 1$ without loss of generality. To prove (10.11), let $v(x, t)$ denote its right-hand side. Then v satisfies (10.2): if $t \geq 1$ and $f_n \equiv x$ for all $n \geq 0$, then $f \in \mathbf{S}(x, t)$ and $\|f\|_1 = |x|$, which implies that $v(x, t) \leq |x|$. To check that v has the property (10.3), let $g \in \mathbf{S}(x + a,\ t + |a|^q)$ and $h \in \mathbf{S}(x - a,\ t + |a|^q)$. The splice (see Chapter 2) of g and h with weight $1/2$ is a martingale f in $\mathbf{S}(x, t)$ satisfying

$$v(x, t) \leq \|f\|_1 = (\|g\|_1 + \|h\|_1)/2$$

so v satisfies (10.3) also and the maximality of u implies that $v \leq u$. To see that $v \geq u$, let $f \in \mathbf{S}(x, t)$ and choose n so that $P(t - |x|^q + S_n^q \geq 1) = 1$. Then, by (10.2) and the same reasoning that led to (10.10),

$$\|f\|_1 \geq E|f_n| \geq Eu(f_n,\ t - |x|^q + S_n^q) \geq u(f_0,\ t - |x|^q + S_0^q) = u(x, t),$$

which implies that $v \geq u$. Therefore, $u = v$ and the proof of Lemma 10.1 is complete.

We shall need the following observation (cf. [Hit 90]).

LEMMA 10.2. *If B is a Banach space and f is a B-valued conditionally symmetric martingale, then f is also a martingale relative to the filtration $\mathcal{G}$ defined by*

$$\mathcal{G}_n = \sigma\{d_0,\ldots,d_n,|d_{n+1}|\}, \quad n \geq 0.$$

PROOF: Certainly, f_n is $\mathcal{G}_n$-measurable (as is $|d_{n+1}|$, which is the point of the lemma). Suppose that $n \geq 1$ and $\varphi : B^n \times R \to R$ is bounded and continuous. The conditional symmetry of f implies that

$$E\varphi(d_0,\ldots,d_{n-1},|d_n|)d_n = E\varphi(d_0,\ldots,d_{n-1},|d_n|)(-d_n).$$

Therefore, $E\varphi(d_0,\ldots,d_{n-1},|d_n|)d_n = 0$ and the lemma is proved.

Remark. By going to a new probability space, one can find a martingale F with the same distribution as f where F is the transform by a real predictable sequence V of a martingale with all of its differences E_k satisfying $|E_k| \equiv 1$. To see this, let F and R be independent where F has the same distribution as f and R has the same distribution as the Rademacher sequence. For a fixed $a \in B$ with $|a| = 1$, let

$$\begin{aligned} E_n &= D_n/|D_n| && \text{on } \{|D_n| > 0\}, \\ &= aR_n && \text{on } \{|D_n| = 0\}, \end{aligned}$$

and $V_n = |D_n|$ where D is the difference sequence of F. Then, $D_n = V_n E_n$ and F is a transform of the desired type relative to the filtration given by

$$\sigma\{D_0,\ldots,D_n,|D_{n+1}|,R_0,\ldots,R_n\}, n \geq 0.$$

See Remark 8.2 of [Bur-Gun 70] for a special case.

LEMMA 10.3. *Suppose that $u : B \times [0,\infty) \to R$ satisfies (10.1), (10.2), and (10.3). If f is a B-valued conditionally symmetric martingale, then, for all $\delta > 0, \beta > 2\delta + 1$, and $\lambda > 0$,*

(10.12) $$P(S(f,q) > \beta\lambda, f^* \leq \delta\lambda) \leq \frac{4\delta}{(\beta - 2\delta - 1)u(0,0)}P(S(f,q) > \lambda).$$

PROOF: Let g be the martingale relative to the filtration $\mathcal{G}$ of Lemma 10.2 defined by

$$g_n = \sum_{k=0}^{n} 1(\mu < k \leq \nu \wedge \sigma)d_k$$

where μ, ν, and σ are the stopping times relative to $\mathcal{G}$ defined by

$$\begin{aligned} \mu &= \inf\{n \geq 0 : S_n(f,q) > \lambda\}, \\ \nu &= \inf\{n \geq 0 : S_n(f,q) > \beta\lambda\}, \\ \sigma &= \inf\{n \geq 0 : |f_n| > \delta\lambda \text{ or } |d_{n+1}| > 2\delta\lambda\}. \end{aligned}$$

Then $S_\mu(f,q) = |f_0| \le \delta\lambda$ on the set $\{\mu = 0, \sigma = \infty\}$ and, if n is a positive integer, then $S_\mu(f,q) \le S_{n-1}(f,q) + |d_n| \le \lambda + 2\delta\lambda$ on $\{\mu = n, \sigma = \infty\}$. So, by Minkowski's inequality, on the set $\{\nu < \infty, \sigma = \infty\}$,

$$S(g,q) \ge S_\nu(f,q) - S_\mu(f,q) \ge \beta\lambda - 2\delta\lambda - \lambda.$$

The inequality $d^* \le 2f^*$ implies that $\{f^* \le \delta\lambda\} = \{\sigma = \infty\}$. Thus, by Lemma 10.1,

$$
\begin{aligned}
P(S(f,q) > \beta\lambda, \, f^* \le \delta\lambda) &= P(\nu < \infty, \sigma = \infty) \\
&\le P(S(g,q) > \lambda(\beta - 2\delta - 1)) \\
&\le \|g\|_1 / \lambda(\beta - 2\delta - 1) u(0,0).
\end{aligned}
$$
(10.13)

To obtain an upper bound on $\|g\|_1$, note that if n is a nonnegative integer, then $g_n = 0$ on the set $\{\mu \ge \nu \wedge \sigma\}$ and

$$|g_n| \le |f_{\nu \wedge \sigma \wedge n}| + |f_{\mu \wedge \sigma \wedge n}| \le (\delta\lambda + 2\delta\lambda) + \delta\lambda = 4\delta\lambda$$

on the set $\{\mu < \nu \wedge \sigma\}$. Therefore,

$$
\begin{aligned}
\|g\|_1 &\le 4\delta\lambda P(\mu < \nu \wedge \sigma) \\
&\le 4\delta\lambda P(\mu < \infty) \\
&= 4\delta\lambda P(S(f,q) > \lambda).
\end{aligned}
$$

To complete the proof of Lemma 10.3, substitute this bound for $\|g\|_1$ in (10.13) to obtain (10.12).

Now recall the definition of $\gamma_{p,q}(\mathbf{B})$ in Chapter 4: $\gamma_{p,q}(\mathbf{B})$ is the least $\gamma \le \infty$ such that

(10.14)
$$\|S(f,q)\|_p \le \gamma \|f\|_p$$

for all $\mathbf{B}$-valued martingales. Let $\gamma_{p,q}^0(\mathbf{B})$ be the least $\gamma \le \infty$ such that (10.14) holds for all conditionally symmetric $\mathbf{B}$-valued martingales.

PROOF OF THEOREM 10.1: Let u satisfy (10.1), (10.2), and (10.3). By Lemma 10.3, the inequality (10.12) must hold. By extrapolation (see [Bur-Gun 70] and, in particular, Lemma 7.1 of [Bur 73]), the constant $\gamma_{p,q}^0(\mathbf{B})$ is finite for all real $p > 1$. As Pisier (1975b) has shown, the finiteness of this constant for the dyadic case with $p = q$ already implies that the Banach space $\mathbf{B}$ is superreflexive.

To prove the other half of Theorem 10.1, suppose that $\mathbf{B}$ is superreflexive. Then the constant $\gamma_{q,q}(\mathbf{B})$ is finite [Pis 75b] and, again by extrapolation, so is $\gamma_{p,q}(\mathbf{B})$ for all real $p > 1$. Adapting the proof of (20) in [Bur 89b] to this simpler case, we have

$$P(S(f,q) \ge \lambda) \le \frac{2\gamma_{2,q}(\mathbf{B})}{\lambda} \|f\|_1$$

for all $\mathbf{B}$-valued martingales f. Let u be the greatest function satisfying (10.2) and (10.3). Then Lemma 10.1 yields the inequality $1/u(0,0) \le 2\gamma_{2,q}(\mathbf{B}) < \infty$ or, equivalently, that

(10.15)
$$u(0,0) \ge 1/2\gamma_{2,q}(\mathbf{B}) > 0.$$

So here the greatest function u satisfying (10.2) and (10.3) must also satisfy (10.1). This completes the proof of Theorem 10.1.

The following lemma gives further information about the constant $\gamma_{p,q}^0(\mathbf{B})$.

LEMMA 10.4. *Suppose that* **B** *is a Banach space*, $p \in (1, \infty)$, $q \in [2, \infty)$, *and* $\gamma \in [1, \infty)$. *Then*

$$(10.16) \qquad \gamma_{p,q}^{0}(\mathbf{B}) \leq \gamma$$

if and only if there is a function $u : \mathbf{B} \times [0, \infty) \to \mathbf{R}$ *such that, for all* $x, a \in \mathbf{B}$ *and* $t \geq 0$,

$$(10.17) \qquad u(x, t) \geq F(x, t),$$
$$(10.18) \qquad u(x, t) \geq [u(x + a, t + |a|^q) + u(x - a, t + |a|^q)] / 2$$

where $F(x, t) = t^{p/q} - \gamma^p |x|^p$.

PROOF: The proof is similar to that of Theorem 2.1. Suppose there is a function $u : \mathbf{B} \times [0, \infty) \to \mathbf{R}$ satisfying (10.17) and (10.18). Then we can assume that u also satisfies

$$u(\alpha x, |\alpha|^q t) = |\alpha|^p u(x, t), \qquad \alpha \in \mathbf{R}.$$

In particular, $u(0, 0) = 0, u(x, t) = u(-x, t)$, and, by (10.18), $u(x, |x|^q) \leq 0$. To show that (10.16) holds for the γ in the definition of F, we must prove that (10.14) holds for all conditionally symmetric **B**-valued martingales. To do this, we can assume that f is simple. Then (see the proof of Lemma 10.1)

$$EF(f_n, S_n^q) \leq Eu(f_n, S_n^q) \leq \cdots \leq Eu(f_0, S_0^q).$$

But $u(f_0, S_0^q) = u(f_0, |f_0|^q) \leq 0$ so $\|S_n(f, q)\|^p - \gamma^p \|f_n\|_p^p = EF(f_n, S_n^q) \leq 0$ and (10.14) follows.

Now assume that (10.16) holds. If $x \in \mathbf{B}$, let $S(x)$ be the set of all **B**-valued simple dyadic martingales f on the Lebesgue unit interval satisfying $f_0 \equiv x$. The filtration may vary with f. Define $U : \mathbf{B} \times [0, \infty) \to (-\infty, \infty]$ by

$$U(x, t) = \sup \{ EF(f_\infty, t - |x|^q + S^q(f, q)) : f \in S(x) \}$$

where f_∞ denotes the pointwise limit of the simple martingale f. The function U satisfies (10.17): consider the $f \in S(x)$ given by $f_n \equiv x, n \geq 0$. To show that U satisfies (10.18), let $x_i = x + (-1)^i a, m_i \in \mathbf{R}$, and

$$U(x_i, t + |a|^q) > m_i, \qquad i = 0, 1.$$

Then $U(x, t) > (m_0 + m_1)/2$. To see this, choose $f^i \in S(x_i)$ so that

$$EF(f_\infty^i, t + |a|^q - |x_i|^q + S^q(f^i, q)) > m_i, \qquad i = 0, 1.$$

Let $f \in S(x)$ be the splice (see Chapter 2) of f_0 and f_1 with weight $1/2$. Then

$$U(x, t) \geq EF(f_\infty, t - |x|^q + S^q(f, q))$$

$$= \frac{1}{2} \sum_{i=0,1} EF(f^i_\infty, t + |a|^q - |x_i|^q + S^q(f^i, q)).$$

Therefore, $U(x, t) > (m_0 + m_1)/2$ and U satisfies (10.18).

The function U never takes the value ∞. With γ as in (10.16), inequality (10.14) holds for all $f \in \mathbf{S}(x)$ and this implies that $EF(f_\infty, S^q(f, q)) \leq 0$. Therefore, $U(0, 0) \leq 0$. If $t \leq t'$, then $F(x, t) \leq F(x, t')$ and $U(x, t) \leq U(x, t')$. Using the definition of U and (10.18) applied to U, we see that $U(x, |x|^p) = U(-x, |x|^p) \leq U(0, 0) \leq 0$. Moreover, $U(0, t) < \infty$: let x satisfy $t \leq 2|x|^q$ and note that

$$U(0, t) + F(2x, 2|x|^q) \leq U(0, 2|x|^q) + U(2x, 2|x|^q) \leq 2U(x, |x|^q) \leq 0.$$

Thus, $U(x, t) + F(-x, t) \leq U(x, t + |x|^q) + U(-x, t + |x|^q) \leq 2U(0, t) < \infty$. Therefore, U is a real-valued function satisfying (10.17) and (10.18). This completes the proof of Lemma 10.4.

Remark. The above proof shows that the least $\gamma \leq \infty$ such that (10.14) holds for all dyadic **B**-valued martingales cannot be strictly less than $\gamma^0_{p,q}(\mathbf{B})$. Therefore, $\gamma^0_{p,q}(\mathbf{B})$, the best constant in the conditionally symmetric case, is also the best constant in the dyadic case.

A question. If f is a real martingale, then

(10.19) $$P(S(f) \geq \lambda) \leq \frac{\sqrt{e}}{\lambda} \|f\|_1.$$

The proof of the original inequality [Bur 66], which implies that the square function of a real L^1-bounded martingale is in weak-L^1, gave no information about the best constant. Theorem 3.6 implies that it is no greater than 2 even for **H**-valued martingales. Cox (1982) proved that $\sqrt{e}$ is the best constant for real martingales. (Littlewood's letter reproduced on page 18 of [Bol 86] suggests that Bollobas and Littlewood knew in 1975 that $\sqrt{e}$ was a good possibility.) Recently, Cox and Kertz (1985) have shown that if $\|f\|_1 > 0$, then there is strict inequality in (10.19).

The best constant in the analogous weak-type inequality for real conditionally symmetric martingales (the same as the best constant for the dyadic case, which, according to [Bol 80], must belong to the interval $[1.44, 1.463]$) is not yet known as far as we are aware. The question is: What is the greatest function $u : \mathbf{R} \times [0, \infty) \to \mathbf{R}$ such that, for all $x, a \in \mathbf{R}$ and $t \geq 0$, the inequalities (10.2) and (10.3) hold with $q = 2$? By Lemma 10.4, the best constant would then be $1/u(0, 0)$.

11. MORE SHARP INEQUALITIES FOR THE SQUARE FUNCTION

Let $(\Omega, \mathcal{F}_\infty, P)$ be a complete probability space and $\mathcal{F} = (\mathcal{F}_t)_{t \geq 0}$ a right-continuous filtration such that $\mathcal{F}_0$ contains all $A \in \mathcal{F}_\infty$ with $P(A) = 0$. Suppose that M is a right-continuous **H**-valued martingale. Then [Bur 88b], for $1 < p < \infty$,

(11.1) $$(p^* - 1)^{-1} \|S(M)\|_p \leq \|M\|_p \leq (p^* - 1) \|S(M)\|_p$$

where $S(M) = [M, M]_\infty^{1/2}$ and $[M, M]$ is the square bracket of M. This inequality is sharp on the left-hand side for $1 < p \leq 2$ and on the right-hand side for $2 \leq p < \infty$ as is the inequality (3.6) from which it follows. The best constants in the remaining cases are as yet unknown.

If M is continuous, then $\|S(M)\|_p$ can also be bounded above and below for smaller values of p. Specifically, for $0 < p < \infty$,

$$(11.2) \qquad c_p\|S(M)\|_p \leq \|M^*\|_p \leq C_p\|S(M)\|_p$$

where $M^*(\omega) = \sup_{t \geq 0} |M_t(\omega)|$ and the constants do not depend on M. The key to this inequality is the special case in which M is stopped Brownian motion [Bur-Gun 70] and this special case yields (11.2) for real M. The proof of (11.2) in the Hilbert-space case requires only a slight modification of the proof of (6.3) and (6.4) in [Bur 73].

Now suppose that M is real and continuous. Then, by (11.1) and (11.2),

$$(11.3) \qquad a_p\|S(M)\|_p \leq \|M\|_p, \qquad 1 < p < \infty,$$
$$(11.4) \qquad \|M\|_p \leq A_p\|S(M)\|_p, \qquad 0 < p < \infty,$$

and Davis (1976) has found the best constants a_p and A_p in these two inequalities. Let M_p be the solution y of the differential equation

$$(11.5) \qquad y''(x) - xy'(x) + py(x) = 0$$

satisfying the conditions $y(0) = 1$ and $y'(0) = 0$. This is a confluent hypergeometric function. Let D_p be the parabolic cylinder function of order p. (See [Abr-Ste 64] for example, and for some applications of these functions to stopping time problems with square-root boundaries, see [She 67] and [Nov 71].) Let μ_p be the largest positive zero of D_p and ν_p the smallest positive zero of M_p. Then $a_p = \mu_p$ if $1 < p \leq 2$ and $a_p = \nu_p$ if $2 \leq p < \infty$. Also, $A_p = \nu_p$ if $0 < p \leq 2$ and $A_p = \mu_p$ if $2 \leq p < \infty$. Davis also shows that if f is a real conditionally symmetric martingale (see Chapter 10), then

$$(11.6) \qquad \nu_p\|S(f)\|_p \leq \|f\|_p, \qquad 2 \leq p < \infty,$$

and ν_p is the best constant. His proof of (11.6) rests on his sharp inequality for the continuous case and on Skorohod embedding. He proves in a similar way that

$$(11.7) \qquad \|f\|_p \leq \nu_p\|S(f)\|_p, \qquad 0 < p \leq 2.$$

Again, f is conditionally symmetric and the constant is best possible.

The embedding method does not seem to carry over to the cases not covered by (11.6) and (11.7). Neither does it carry over to H-valued martingales. Wang (1989) has studied these cases and has obtained a number of results including the following: Both (11.6) and (11.7) do hold for all H-valued conditionally symmetric martingales f as does the inequality

$$(11.8) \qquad \|f\|_p \leq \mu_p\|S(f)\|_p, \qquad 3 \leq p < \infty,$$

and the constant in the last inequality is best possible (as are, of course, the constants in the other two inequalities).

How does Wang prove these inequalities, for example, (11.6) in the Hilbert space setting? If (10.17) and (10.18) hold for some function u where $\gamma = \nu_p^{-1}$ and $q = 2$, then, by (10.16), inequality (11.6) must hold. There is a natural candidate for u that is implicit in the real continuous case. Wang shows that, indeed, the natural candidate does satisfy (10.17) and (10.18). Inequalities (11.7) and (11.8) can be proved in a similar way but there is a difference. The natural choice for u is the right choice to prove (11.8) in the case $3 \leq p < \infty$ but is not the right choice, as Wang has shown, for the case $2 \leq p < 3$. See [Wan 89] and his forthcoming paper in the Transactions.

If $q = 2$ and $\mathbf{B} = \mathbf{R}$, then (10.18) is the discrete form of the heat inequality

$$(11.9) \qquad \frac{\partial u(x,t)}{\partial t} + \frac{1}{2}\frac{\partial^2 u(x,t)}{\partial x^2} \leq 0.$$

This differential inequality can be a guide, but is not always a reliable guide, to (10.18): a solution of (11.9) is not necessarily a solution of (10.18). Even if it is, to check that it does satisfy (10.18) for all (x,t,a) can require a good bit of enlightened calculation.

12. ON THE OPTIMAL CONTROL OF MARTINGALES

In this chapter, M is a real right-continuous martingale with left limits and V is a real predictable process. The probability space and filtration are as in Chapter 1. Assume that V is bounded or, more generally, that $\int_{[0,t]} |V_s|^2 \, d[M,M]_s$ is finite almost surely for all $t > 0$. Controlling M by V gives a right-continuous martingale N with left limits such that

$$N_t = V_0 M_0 + \int_{(0,t]} V_s \, dM_s \quad \text{a.s.}$$

Let $\beta \in \mathbf{R}$ and suppose that the goal is to find a V in some given class of predictable processes such that

$$(12.1) \qquad P(N_t \geq \beta \text{ for some } t \geq 0) = 1.$$

The example $M \equiv 0$ shows that this is not always possible. The following theorem gives a necessary condition for the existence of such a V.

THEOREM 12.1. Let $a, b, \alpha, \beta \in \mathbf{R}$ with $a \leq 0$ and $b \geq 1$. Suppose that $M_0 \equiv \alpha$ and V is a predictable process satisfying $a \leq V_t(\omega) \leq b$ with $V_0 \equiv 1$. If (12.1) is also satisfied, then

$$(12.2) \qquad \|M\|_1 \geq |\alpha| \vee \left[\frac{(a+b-2)\alpha + 2\beta}{b-a}\right].$$

This inequality is sharp and equality can hold.

For example, if $a = -1$ and $b = 1$, the bound on the right becomes $|\alpha| \vee (\beta - \alpha)$. To prove this theorem, reduce it to the discrete-time case and adapt the proof of Theorem 7.3 of [Bur 84] to this setting.

Let $0 \leq \gamma \leq 1$ and suppose that (12.1) is replaced by the less stringent requirement

$$(12.3) \qquad P(N_t \geq \beta \text{ for some } t \geq 0) \geq \gamma.$$

What is the analogous necessary condition for the existence of V in this control problem? Choi (1988) has discovered the following sharp lower bound on $\|M\|_1$:

THEOREM 12.2. *Let $\alpha, \beta \in \mathbf{R}$ and $M_0 \equiv \alpha$. If V is a predictable process satisfying (12.3) and $-1 \leq V_t(\omega) \leq 1$, then*

$$(12.4) \qquad \|M\|_1 \geq |\alpha| \vee \left\{ \beta - \alpha - [\beta^+ (\beta - 2\alpha)^+ (1 - \gamma)]^{1/2} \right\}.$$

Equality can hold.

Here $\beta^+ = \beta \vee 0$. For the proof in the discrete-time case to which it reduces and related results, see Choi (1988). The main lemma requires finding the greatest function L among the functions $u : \mathbf{R} \times \mathbf{R} \times [0, 1] \to \mathbf{R}$ such that the mappings

$$(x, t) \mapsto u(x, y, t) \quad \text{and} \quad (y, t) \mapsto u(x, y, t)$$

are convex on $\mathbf{R} \times [0, 1]$ and more.

Suppose that the one-sided condition (12.1) is replaced by the two-sided condition

$$(12.5) \qquad P(|N_t| \geq 1 \text{ for some } t \geq 0) = 1$$

or by

$$(12.6) \qquad P(|N_t| \leq 1 \text{ for all } t \geq 0) = 1.$$

Then the inequalities (16.2) and (16.3) for stochastic integrals in [Bur 84] yield the following necessary conditions for optimal control. If $|V_t(\omega)| \leq 1$ and (12.5) is satisfied, then

$$(12.7) \qquad \|M\|_p^p \geq \Gamma(p + 1)/2, \qquad 1 \leq p \leq 2,$$

and the inequality is sharp. If $|V_t(\omega)| \geq 1$ and (12.6) is satisfied, then

$$(12.8) \qquad \|M\|_p^p \leq \Gamma(p + 1)/2, \qquad 2 \leq p \leq \infty,$$

and this inequality is also sharp. Strict inequality holds in the special case of discrete time if p is in the open interval $(1, 2)$ or in the open interval $(2, \infty)$; see [Bur 84]. For the continuous-time case, as Monroe (1987) has shown using local times, equality can hold for all p. A slight modification of his example is used at the end of Chapter 8.

APPENDIX

A decomposition of martingale transforms. Suppose that B is, as usual, a real or complex Banach space with norm $|\cdot|$.

LEMMA A.1. *Let g be the transform of a B-valued martingale f by a real-valued predictable sequence v uniformly bounded in absolute value by 1. Then there exist B-valued martingales $F^j = (F_n^j)_{n \geq 0}$ and Borel measurable functions $\varphi_j \colon [-1,1] \to \{1,-1\}$ such that, for $j \geq 1$ and $n \geq 0$,*

$$f_n = F_{2n+1}^j,$$

$$g_n = \sum_{j=1}^{\infty} 2^{-j} \varphi_j(v_0) G_{2n+1}^j,$$

where G^j is the transform of F^j by $\varepsilon = (\varepsilon_k)_{k \geq 0}$ with $\varepsilon_k = (-1)^k$.

This strengthens Lemma 2.1 of [Bur 84].

PROOF: For the special case in which each term v_n of the predictable sequence v has its values in $\{1, -1\}$, let

$$D_{2n} = \frac{1 + v_0 v_n}{2} d_n,$$

$$D_{2n+1} = \frac{1 - v_0 v_n}{2} d_n.$$

Then $D = (D_n)_{n \geq 0}$ is a martingale difference sequence relative to the filtration it generates. To see this, note that $D_0 = d_0$, $D_1 = 0$, and if $n \geq 1$, then

(A.1) $$E\varphi(D_0, \ldots, D_{2n-1})D_{2n} = 0$$

and

(A.2) $$E\psi(D_0, \ldots, D_{2n})D_{2n+1} = 0$$

for all bounded and continuous functions $\varphi \colon \mathbf{B}^{2n} \to \mathbf{R}$ and $\psi \colon \mathbf{B}^{2n+1} \to \mathbf{R}$. It is clear that (A.1) holds and (A.2) follows from

(A.3) $$\psi(D_0, \ldots, D_{2n})D_{2n+1} = \psi(D_0, \ldots, D_{2n-1}, 0)D_{2n+1}.$$

To see that (A.3) holds, note that both sides vanish on the set where $v_0 v_n = 1$ and D_{2n} vanishes on its complement.

Let F be the martingale corresponding to D and G its transform by ε. We see from the definition of D that $d_n = D_{2n} + D_{2n+1}$ and $v_0 v_n d_n = D_{2n} - D_{2n+1}$ so $f_n = F_{2n+1}$ and $g_n = v_0 G_{2n+1}$, where we have used $v_0^2 = 1$.

Now consider the general case in which the v_k have their values in the interval $[-1, 1]$. There exist Borel measurable functions $\varphi_j \colon [-1,1] \to \{1,-1\}$ such that

$$t = \sum_{j=1}^{\infty} 2^{-j} \varphi_j(t), \qquad t \in [-1, 1].$$

Let $v^j = (\varphi_j(v_n))_{n \geq 0}$. This is a predictable sequence relative to the original filtration. The special case discussed above now yields, for each positive integer j, a martingale F^j and its transform G^j by ε such that

$$f_n = F^j_{2n+1},$$

$$\sum_{k=0}^{n} \varphi_j(v_k)d_k = \varphi_j(v_0)G^j_{2n+1}.$$

Multiplying both sides by 2^{-j} and summing, we obtain the desired decomposition. This completes the proof of Lemma A.1.

Consider a typical application. Let g be the transform of a $\mathbf{B}$-valued martingale f by a real predictable sequence v uniformly bounded in absolute value by 1. Let $1 < p < \infty$ and suppose that $\beta_p(\mathbf{B})$ is as in Chapter 1. Then, by Lemma A.1,

$$\|g\|_p \leq \sum_{j=1}^{\infty} 2^{-j}\|G^j\|_p$$

$$\leq \beta_p(\mathbf{B}) \sum_{j=1}^{\infty} 2^{-j}\|F^j\|_p$$

$$= \beta_p(\mathbf{B}) \sum_{j=1}^{\infty} 2^{-j}\|f\|_p$$

$$= \beta_p(\mathbf{B})\|f\|_p.$$

It is easy to see from this argument that if strict inequality holds in the case of ± 1-sequences, then strict inequality holds also for the sequences v.

REFERENCES

M. Abramowitz and I. A. Stegun, editors, "Handbook of mathematical functions," Dover, New York, 1970.

D. J. Aldous, *Unconditional bases and martingales in $L_p(F)$*, Math. Proc. Cambridge Phil. Soc. **85** (1979), 117–123.

T. Ando, *Contractive projections in L_p spaces*, Pacific J. Math. **17** (1966), 391–405.

A. Baernstein, *Some sharp inequalities for conjugate functions*, Indiana Univ. Math. J. **27** (1978), 833–852.

R. Bañuelos, *A sharp good-λ inequality with an application to Riesz transforms*, Mich. Math. J. **35** (1988), 117-125.

R. Bass, *A probabilistic approach to the boundedness of singular integral operators*, Séminaire de Probabilités XXIV 1988/89, Lecture Notes in Mathematics **1426** (1990), 15–40.

A. Benedek, A. P. Calderón, and R. Panzone, *Convolution operators on Banach space valued functions*, Proc. Nat. Acad. Sci. **48** (1962), 356–365.

E. Berkson, T. A. Gillespie, and P. S. Muhly, *Théorie spectrale dans les espaces UMD*, C. R. Acad. Sci. Paris **302** (1986), 155–158.

_____, *Generalized analyticity in UMD spaces*, Arkiv för Math. **27** (1989), 1–14.

K. Bichteler, *Stochastic integration and L^p-theory of semimartingales*, Ann. Prob. **9** (1981), 49–89.

O. Blasco, *Hardy spaces of vector-valued functions: Duality*, Trans. Amer. Math. Soc. **308** (1988a), 495–507.

—————, *Boundary values of functions in vector-valued Hardy spaces and geometry on Banach spaces*, J. Funct. Anal. **78** (1988b), 346–364.

G. Blower, *A multiplier characterization of analytic UMD spaces*, Studia Math. **96** (1990), 117–124.

B. Bollobás, *Martingale inequalities*, Math. Proc. Cambridge Phil. Soc. **87** (1980), 377–382.

—————, editor, "Littlewood's Miscellany," Cambridge University Press, Cambridge, 1986.

J. Bourgain, *Some remarks on Banach spaces in which martingale difference sequences are unconditional*, Ark. Mat. **21** (1983), 163–168.

—————, *Extension of a result of Benedek, Calderón, and Panzone*, Ark. Mat. **22** (1984), 91–95.

—————, *Vector valued singular integrals and the H^1-BMO duality*, in "Probability Theory and Harmonic Analysis," edited by J. A. Chao and W. A. Woyczynski, Marcel Dekker, New York, 1986, pp. 1–19.

A. V. Bukhvalov, *Hardy spaces of vector-valued functions*, J. Sov. Math. **16** (1981), 1051–1059.

—————, *Continuity of operators in spaces of vector functions, with applications to the theory of bases*, J. Sov. Math. **44** (1989), 749–762.

D. L. Burkholder, *Martingale transforms*, Ann. Math. Statist. **37** (1966), 1494–1504.

—————, *Distribution function inequalities for martingales*, Ann. Prob. **1** (1973), 19–42.

—————, *Exit times of Brownian motion, harmonic majorization, and Hardy spaces*, Advances in Math. **26** (1977), 182–205.

—————, *A sharp inequality for martingale transforms*, Ann. Prob. **7** (1979), 858–863.

—————, *A geometrical characterization of Banach spaces in which martingale difference sequences are unconditional*, Ann. Prob. **9** (1981a), 997–1011.

—————, *Martingale transforms and the geometry of Banach spaces*, Proceedings of the Third International Conference on Probability in Banach Spaces, Tufts University, 1980, Lecture Notes in Mathematics **860** (1981b), 35–50.

—————, *A nonlinear partial differential equation and the unconditional constant of the Haar system in L^p*, Bull. Amer. Math. Soc. **7** (1982), 591–595.

—————, *A geometric condition that imples the existence of certain singular integrals of Banach-space-valued functions*, in "Conference on Harmonic Analysis in Honor of Antoni Zygmund (Chicago, 1981)," edited by William Beckner, Alberto P. Calderón, Robert Fefferman, and Peter W. Jones. Wadsworth, Belmont, California, 1983, pp. 270–286.

—————, *Boundary value problems and sharp inequalities for martingale transforms*, Ann. Prob. **12** (1984), 647–702.

—————, *An elementary proof of an inequality of R. E. A. C. Paley*, Bull. London Math. Soc. **17** (1985), 474–478.

—————, *An extension of a classical martingale inequality*, in "Probability Theory and Harmonic Analysis," edited by J. A. Chao and W. A. Woyczynski. Marcel Dekker, New York, 1986a, pp. 21–30.

—————, *Martingales and Fourier analysis in Banach spaces*, C.I.M.E. Lectures, Varenna (Como), Italy, 1985, Lecture Notes in Mathematics **1206** (1986b), 61–108.

—————, *A sharp and strict L^p-inequality for stochastic integrals*, Ann. Prob. **15** (1987), 268–273.

—————, *A proof of Pełczyński's conjecture for the Haar system*, Studia Math. **91** (1988a), 79–83.

—————, *Sharp inequalities for martingales and stochastic integrals*, Colloque Paul Lévy, Palaiseau, 1987, Astérisque **157-158** (1988b), 75–94.

—————, *Differential subordination of harmonic functions and martingales*, Harmonic Analysis and Partial Differential Equations (El Escorial, 1987), Lecture Notes in Mathematics **1384** (1989a), 1–23.

—————, *On the number of escapes of a martingale and its geometrical significance*, in "Almost Everywhere Convergence," edited by Gerald A. Edgar and Louis Sucheston. Academic Press, New York, 1989b, pp. 159–178.

D. L. Burkholder and R. F. Gundy, *Extrapolation and interpolation of quasi-linear operators on martingales*, Acta. Math. **124** (1970), 249–304.

A. P. Calderón and A. Zygmund, *On the existence of certain singular integrals*, Acta Math. **88** (1952), 85–139.

—————, *On singular integrals*, Amer. J. Math. **78** (1956), 289–309.

S. D. Chatterji, *Les martingales et leurs applications analytiques*, Lecture Notes in Mathematics **307** (1973), 27–164.

K. P. Choi, *Some sharp inequalities for martingale transforms*, Trans. Amer. Math. Soc. **307** (1988), 279–300.

F. Cobos, *Some spaces in which martingale difference sequences are unconditional*, Bull. Polish Acad. of Sci. Math. **34** (1986), 695–703.

—————, *Duality, UMD-property and Lorentz-Marcinkiewicz operator spaces*, in "16 Colóquio Brasileiro de Matemática," Rio de Janeiro, 1988, pp. 97–106.

F. Cobos and D. L. Fernandez, *Hardy-Sobolev spaces and Besov spaces with a function parameter*, Lecture Notes in Mathematics **1302** (1988), 158–170.

T. Coulhon and D. Lamberton, *Régularité L^p pour les équations d'évolution*, in "Séminaire d'Analyse Fonctionnelle, 1984/1985," Publ. Math. Univ. Paris VII 26 (1986), 155–165.

D. C. Cox, *The best constant in Burkholder's weak-L^1 inequality for the martingale square function*, Proc. Amer. Math. Soc. **85** (1982), 427–433.

D. C. Cox and R. P. Kertz, *Common strict character of some sharp infinite-sequence martingale inequalities*, Stochastic Process. Appl. **20** (1985), 169–179.

B. Davis, *A comparison test for martingale inequalities*, Ann. Math. Statist. **40** (1969), 505–508.

—————, *On the weak type (1,1) inequality for conjugate functions*, Proc. Amer. Math. Soc. **44** (1974), 307–311.

—————, *On the L^p norms of stochastic integrals and other martingales*, Duke Math. J. **43** (1976), 697–704.

M. Defant, *On the vector-valued Hilbert transform*, Math. Nachr. **141** (1989), 251–265.

C. Dellacherie and P. A. Meyer, "Probabilités et potentiel: théorie des martingales," Hermann, Paris, 1980.

J. Diestel and J. J. Uhl, "Vector Measures," Math. Surveys 15, American Mathematical Society, Providence, Rhode Island, 1977.

C. Doléans, *Variation quadratique des martingales continues à droite*, Ann. Math. Statist. **40** (1969), 284–289.

J. L. Doob, "Stochastic Processes," Wiley, New York, 1953.

—————, *Remarks on the boundary limits of harmonic functions*, J. SIAM Numer. Anal. **3** (1966), 229–235.

—————, "Classical Potential Theory and Its Probabilistic Counterpart," Springer, New York, 1984.

L. E. Dor and E. Odell, *Monotone bases in L_p*, Pacific J. Math. **60** (1975), 51–61.

G. Dore and A. Venni, *On the closedness of the sum of two closed operators*, Math. Z. **196** (1987), 189–201.

I. Doust, *Contractive projections on Banach spaces*, Proc. Centre for Math. Anal., Australian National University **20** (1988), 50–58.

—————, *Well-bounded and scalar-type spectral operators on L^p spaces*, J. London Math. Soc. **39** (1989), 525–534.

L. E. Dubins, *Rises and upcrossings of nonnegative martingales*, Illinois J. Math. **6** (1962), 226–241.

P. Enflo, *Banach spaces which can be given an equivalent uniformly convex norm*, Israel J. Math. **13** (1972), 281–288.

M. Essén, *A superharmonic proof of the M. Riesz conjugate function theorem*, Ark. Math. **22** (1984), 241–249.

D. L. Fernandez, *Vector-valued singular integral operators on L^p-spaces with mixed norms and applications*, Pacific J. Math. **129** (1987), 257–275.

—————, *On Fourier multipliers of Banach-lattice valued functions*, Rev. Roumaine Math. Pures Appl. **34** (1989), 635–642.

D. L. Fernandez and J. B. Garcia, *Interpolation of Orlicz-valued function spaces and U.M.D. property*, 26° Semi'ario Brasileiro de Análise (Rio de Janeiro, 1987), Trabalhos Apresentados, 269–281.

T. Figiel, *On equivalence of some bases to the Haar system in spaces of vector-valued functions*, Bull. Polon. Acad. Sci. **36** (1988), 119–131.

————, *Singular integral operators: a martingale approach*, to appear in the Proceedings of the Conference on the Geometry of Banach Spaces (Strobl, Austria, 1989).

T. W. Gamelin, "Uniform Algebras and Jensen Measures," Cambridge University Press, London, 1978.

D. J. H. Garling, *Brownian motion and UMD-spaces*, Conference on Probability and Banach Spaces, Zaragoza, 1985, Lecture Notes in Mathematics **1221** (1986), 36–49.

Y. Giga and H. Sohr, *Abstract L^p estimates for the Cauchy problem with applications to the Navier-Stokes equations in exterior domains*, preprint.

D. Gilat, *The best bound in the $L \log L$ inequality of Hardy and Littlewood and its martingale counterpart*, Proc. Amer. Math. Soc. **97** (1986), 429–436.

S. Guerre, *On the closedness of the sum of closed operators on a UMD space*, in "Banach Space Theory," American Mathematical Society, Providence, Rhode Island, 1989, pp. 239–251.

————, *Complex powers of operators and UMD spaces*, manuscript.

R. F. Gundy, "Some Topics in Probability and Analysis," Regional Conference Series in Mathematics **70**, American Mathematical Society, Providence, Rhode Island, 1989.

U. Haagerup, *The best constants in the Khintchine inequality*, Studia Math. **70** (1982), 231–283.

U. Haagerup and G. Pisier, *Factorization of analytic functions with values in non-commutative L_1-spaces and applications*, Can. J. Math. **41** (1989), 882–906.

G. H. Hardy, J. E. Littlewood, and G. Pólya, "Inequalities," Cambridge University Press, Cambridge, 1934.

W. Hensgen, *On complementation of vector-valued Hardy spaces*, Proc. Amer. Math. Soc. **104** (1988), 1153–1162.

————, *On the dual space of $H^p(X)$, $1 < p < \infty$*, J. Funct. Anal. **92** (1990), 348–371.

P. Hitczenko, *Comparison of moments for tangent sequences of random variables*, Probab. Th. Rel. Fields **78** (1988), 223–230.

————, *On tangent sequences of UMD-space valued random vectors*, manuscript.

————, *Upper bounds for the L_p-norms of martingales*, Probab. Th. Rel. Fields **86** (1990), 225–238.

————, *Best constants in martingale version of Rosenthal's inequality*, Ann. Probab. **18** (1990), 1656–1668.

R. C. James, *Some self dual properties of normed linear spaces*, Ann. Math. Studies **69** (1972a), 159–175.

————, *Super-reflexive spaces with bases*, Pacific J. Math. **41** (1972b), 409–419.

————, *Super-reflexive Banach spaces*, Can. J. Math. **24** (1972c), 896–904.

W. B. Johnson and G. Schechtman, *Martingale inequalities in rearrangement invariant function spaces*, Israel J. Math. **64** (1988), 267–275.

N. J. Kalton, *Differentials of complex interpolation processes for Köthe function spaces*, a paper delivered at the Conference on Function Spaces (Auburn University, 1989).

G. Klincsek, *A square function inequality*, Ann. Prob. **5** (1977), 823–825.

A. N. Kolmogorov, *Sur les fonctions harmoniques conjuguées et les séries de Fourier*, Fund. Math. **7** (1925), 24–29.

H. König, *Vector-valued multiplier theorems*, in "Séminaire d'analyse fonctionnelle, 1985-1987," Publications mathématique de l'université Paris VII, 1988, pp. 131–140.

H. Kunita, *Stochastic integrals based on martingales taking values in Hilbert space*, Nagoya Math. J. **38** (1970), 41–52.

S. Kwapień, *Isomorphic characterizations of inner product spaces by orthogonal series with vector valued coefficients*, Studia Math. **44** (1972), 583–595.

S. Kwapień and W. A. Woyczynski, *Tangent sequences of random variables: Basic inequalities and their applications*, in "Almost Everywhere Convergence," edited by Gerald A. Edgar and Louis Sucheston. Academic Press, New York, 1989, pp. 237–265.

J. Lindenstrauss and A. Pełczyński, *Contributions to the theory of the classical Banach spaces*, J. Funct. Anal. **8** (1971), 225–249.

J. Lindenstrauss and L. Tzafriri, "Classical Banach Spaces I: Sequence Spaces," Springer, New York, 1977.

_____, "Classical Banach Spaces II: Function Spaces," Springer, New York, 1979.

A. Mandelbaum, L. A. Shepp, and R. Vanderbei, *Optimal switching between a pair of Brownian motions*, Ann. Prob. **18** (1990), 1010–1033.

J. Marcinkiewicz, *Quelques théorèmes sur les séries orthogonales*, Ann. Soc. Polon. Math. **16** (1937), 84–96.

B. Maurey, *Système de Haar*, in "Séminaire Maurey-Schwartz, 1974-1975," École Polytechnique, Paris, 1975.

T. R. McConnell, *On Fourier multiplier transformations of Banach-valued functions*, Trans. Amer. Math. Soc. **285** (1984), 739–757.

_____, *A Skorohod-like representation in infinite dimensions*, Probability in Banach Spaces V, Lecture Notes in Mathematics **1153** (1985), 359-368.

_____, *Decoupling and stochastic integration in UMD Banach spaces*, Prob. Math. Stat. **10** (1989), 283–295.

H. P. McKean, *Geometry of differential space*, Ann. Prob. **1** (1973), 197–206.

I. Monroe, *Martingale operator norms and local times*, manuscript.

A. A. Novikov, *On stopping times for the Wiener process*, (Russian, English summary), Teor. Verojatnost. i Primenen **16** (1971), 458–465.

A. M. Olevskiĭ, *Fourier series and Lebesgue functions*, (Russian), Uspehi Mat. Nauk **22** (1967), 237–239.

_____, "Fourier Series with Respect to General Orthogonal Systems," Springer, New York, 1975.

R. E. A. C. Paley, *A remarkable series of orthogonal functions I.*, Proc. London Math. Soc. **34** (1932), 241–264.

A. Pełczyński, *Structural theory of Banach spaces and its interplay with analysis and probability*, in "Proceedings of the International Congress of Mathematicians (Warsaw, 1983)," PWN, Warsaw, 1984, pp. 237–269.

———, *Norms of classical operators in function spaces*, Colloque Laurent Schwartz, Astérisque **131** (1985), 137–162.

A. Pełczyński and H. Rosenthal, *Localization techniques in L^p spaces*, Studia Math. **52** (1975), 263–289.

S. K. Pichorides, *On the best values of the constants in the theorems of M. Riesz, Zygmund and Kolmogorov*, Studia Math. **44** (1972), 165–179.

G. Pisier, *Un exemple concernant la super-réflexivité*, in "Séminaire Maurey-Schwartz, 1974-75," École Polytechnique, Paris, 1975a.

_____, *Martingales with values in uniformly convex spaces*, Israel J. Math. **20** (1975b), 326–350.

A. O. Pittenger, *Note on a square function inequality*, Ann. Prob. **7** (1979), 907–908.

M. Riesz, *Sur les fonctions conjuguées*, Math. Z. **27** (1927), 218–244.

J. L. Rubio de Francia, *Martingale and integral transforms of Banach space valued functions*, Conference on Probability and Banach Spaces, Zaragoza, 1985, Lecture Notes in Mathematics **1221** (1986), 195–222.

J. L. Rubio de Francia and J. L. Torrea, *Some Banach techniques in vector valued Fourier analysis*, Colloq. Math. **54** (1987), 271–284.

J. L. Rubio de Francia, F. J. Ruiz, and J. L. Torrea, *Calderón-Zygmund theory for operator-valued kernels*, Advances in Math. **62** (1986), 7–48.

J. Schwartz, *A remark on inequalities of Calderón-Zygmund type for vector-valued functions*, Comm. Pure Appl. Math. **14** (1961), 785–799.

L. A. Shepp, *A first passage problem for the Wiener process*, Ann. Math. Statist. **38** (1967), 1912-1914.

E. M. Stein, "Singular Integrals and Differentiability Properties of Functions," Princeton University Press, Princeton, 1970.

E. M. Stein and G. Weiss, *On the theory of harmonic functions of several variables: I. The theory of H^p-spaces*, Acta Math. **103** (1960), 25–62.

S. J. Szarek, *On the best constants in the Khinchin inequality*, Studia Math. **58** (1976), 197–208.

B. Tomaszewski, *Sharp weak-type inequalities for analytic functions on the unit disc*, Bull. London Math. Soc. **18** (1986), 355-358.

L. Tzafriri, *Remarks on contractive projections in L_p-spaces*, Israel J. Math. **7** (1969), 9–15.

G. Wang, "Some Sharp Inequalities for Conditionally Symmetric Martingales," doctoral thesis, University of Illinois, Urbana, Illinois, 1989.

_____, *Sharp square-function inequalities for conditionally symmetric martingales*, Trans. Amer. Math. Soc. (to appear).

_____, *Sharp maximal inequalities for conditionally symmetric martingales and Brownian motion*, Proc. Amer. Math. Soc. (to appear).

_____, *Sharp inequalities for the conditional square function of a martingale*, Ann. Prob. (to appear).

T. M. Wolniewicz, *The Hilbert transform in weighted spaces of integrable vector-valued functions*, Colloq. Math. **53** (1987), 103–108.

M. Yor, *Sur les intégrales stochastique à valeurs dans un Banach*, C. R. Acad. Sci. Paris **277** (1973), 467–469.

F. Zimmermann, *On vector-valued Fourier multiplier theorems*, Studia Math. **93** (1989), 201–222.

J. Zinn, *Comparison of martingale differences*, Lecture Notes in Mathematics **1153** (1985), 453–457.

A. Zygmund, "Trigonometric Series I, II," Cambridge University Press, Cambridge, 1959.

FILTRAGE NON LINEAIRE

ET EQUATIONS AUX DERIVEES PARTIELLES

STOCHASTIQUES ASSOCIEES

Etienne PARDOUX

TABLE DES MATIERES

E. PARDOUX : "FILTRAGE NON LINEAIRE ET EQUATIONS AUX DERIVEES PARTIELLES STOCHASTIQUES ASSOCIEES"

4. Continuité du filtre par rapport à l'observation

5. Deux applications du calcul de Malliavin au filtrage non linéaire

6. Filtres de dimension finie et filtres de dimension finie approchés

Introduction

Le filtrage non linéaire est une partie de la théorie des processus stochastiques qui est fortement motivée par les applications, et qui se situe au carrefour de nombreuses théories mathématiques. Il a motivé aussi bien l'étude des changements de probabilité et de filtration en théorie générale des processus, que de nombreux travaux sur les équations aux dérivées partielles stochastiques. Il a posé le célèbre problème de l'innovation (cf. section 2.2 ci-dessous) qui n'est toujours pas complètement résolu. Il a été un des domaines privilégiés d'application du calcul de Malliavin. Il a produit des résultats qui sont essentiels pour le contrôle stochastique des systèmes partiellement observés, et l'analogie avec les problèmes de contrôlabilité des systèmes déterministes a conduit à des conditions de non existence de filtres de dimension finie.

Pendant que le filtrage non linéaire suscitait des travaux théoriques riches et variés, la conception d'algorithmes efficaces utilisables en pratique butait sur d'énormes difficultés. D'un côté le filtrage de Kalman étendu des ingénieurs ne reposait jusque très récemment sur aucune mathématique sérieuse et son efficacité est très aléatoire. Par ailleurs, la résolution numérique des équations du filtrage non linéaire soulève de grosses difficultés en dehors des cas d'école en dimension un ou deux. Cependant, quelques progrès ont été enregistrés dans ce domaine ces dernières années.

Le but de ce cours est de présenter la théorie du filtrage non linéaire, ainsi que des éléments de théorie des équations aux dérivées partielles stochastiques et du calcul de Malliavin, avec leurs applications au filtrage. Enfin, outre le filtre de Kalman-Bucy et ses généralisations, on présente des algorithmes de calcul approché du filtre dans deux cas particuliers.

Le premier chapitre présente trois exemples, précise la classe des problèmes de filtrage qui sera considérée dans les sections suivantes, et rappelle quelques liens entre équations différentielles stochastiques et équations aux dérivées partielles du second ordre.

Le second chapitre établit les équations générales du filtrage non linéaire, et accessoirement de la prédiction et du lissage. Il se termine par une application en statistique des processus.

Le troisième chapitre présente des résultats sur les équations aux dérivées partielles stochastiques et leur application au filtrage, à savoir des théorèmes d'unicité et de régularité de la solution de l'équation de Zakai.

Le quatrième chapitre donne des résultats de continuité du filtre par rapport à l'observation.

Le cinquième chapitre présente les idées essentielles du calcul de Malliavin, et deux applications (très différentes l'une de l'autre) en filtrage : l'absolue continuité de la loi conditionnelle, et la non existence d'un "filtre de dimension finie". Ce chapitre se termine par un résultat de non existence d'un filtre de dimension finie démontré sans le calcul de Malliavin.

Enfin le dernier chapitre présente une partie des filtres de dimension finie connus (le filtre de Kalman-Bucy, et sa généralisation au cas conditionnellement gaussien) et deux filtres de dimension finie approchés : l'un dans le cas d'un grand rapport signal sur bruit, l'autre dans une situation "sans bruit de dynamique".

La lecture de ce texte nécessite une bonne connaissance du calcul stochastique d'Itô (par rapport au processus de Wiener) et des équations différentielles stochastiques, ainsi que des connaissances en analyse fonctionnelle.

Je remercie Paul-Louis Hennequin de m'avoir invité à donner ce cours à St Flour et l'auditoire pour l'intérêt qu'il a manifesté. La frappe du texte a été effectuée par Ephie Deriche et Noëlle Tabaracci. Qu'elles en soient remerciées, ainsi que Fabien Campillo qui m'a beaucoup aidé à corriger et à fignoler le texte.

Chapitre 1

Le problème du filtrage stochastique

1.1   Des exemples

Exemple 1.1.1 *Estimation de la position d'un satellite au cours de son orbite de trans-fert*   L'orbite de transfert est une orbite elliptique, qui est une transition entre le lancement du satellite et l'orbite géostationnaire. Le mouvement du satellite est décrit en première approximation par l'action du champ de gravitation de la terre. Cela donne une équation de la mécanique du type "$F = m\gamma$", qui peut s'écrire sous la forme :

$$\frac{dX_t}{dt} = f(X_t)$$

avec $X_t \in \mathbb{R}^6$ (trois paramètres de position, trois paramètres de vitesse). Cependant, le satellite ne suit pas exactement le mouvement correspondant à la solution de cette équation, car en écrivant l'équation on a négligé :

- la non sphéricité de la terre,
- l'influence d'autres corps (lune, soleil),
- le frottement atmosphérique,
- la pression de radiation,...

Signalons que certains de ces phénomènes (en particulier le 1er et le 3^{è}) sont plus sensibles au voisinage du périgée que dans les autres phases du mouvement. On est donc amené, pour prendre en compte à la fois les perturbations aléatoires et l'imperfection de la modélisation, à rajouter des termes stochastiques dans l'équation du mouvement :

$$\frac{dX_t}{dt} = f(X_t) + g(X_t)\frac{dW_t}{dt}$$

que l'on interprète sous la forme d'une EDS au sens de Stratonovich :

$$dX_t = f(X_t)\,dt + g(X_t) \circ dW_t \,.$$

Pour suivre un satellite, on dispose de n stations radar (dans le cas des vols d'Ariane, trois stations radar situées à Kourou, Toulouse et Pretoria) qui mesurent suivant les cas soit seulement la distance station–satellite, soit en outre des angles de site et de gisement. La i–ième station radar reçoit le signal :

$$y_{i,t} = h_i(t, X_t) + \eta_{i,t} \, , \; 1 \leq i \leq n$$

où $\eta_{i,t}$ est un bruit de mesure. Notons qu'en pratique chaque station ne reçoit des signaux que lorsque le satellite est dans une portion restreinte de la trajectoire. Le reste du temps, on peut considérer que la fonction h_i correspondante est nulle (la station ne reçoit que du bruit). Signalons qu'en pratique on reçoit des mesures en temps discret, i.e. à des instants $t_1 < t_2 < \cdots$. Nous n'étudierons que des modèles en temps continu, mais bien entendu tous les algorithmes que l'on utilise sont en temps discret.

Le problème de filtrage, ou d'"estimation" de la position du satellite se résume de la façon suivante : à chaque instant t, on cherche à "estimer" X_t au vu des observations jusqu'à l'instant t, i.e. connaissant $\mathcal{Y}_t = \sigma\{y_{i,s} \, ; \, 1 \leq i \leq n, 0 \leq s \leq t\}$. En fait on va calculer la loi conditionnelle de X_t sachant $\mathcal{Y}_t$ (dans certains cas, on se contente de chercher à déterminer l'espérance conditionnelle). Dans ce problème particulier, le but de ce filtrage est de commander au bon moment la manœuvre de passage de l'orbite de transfert à l'orbite géostationnaire.

Exemple 1.1.2 *Trajectographie passive* Dans ce problème, le "porteur" "écoute" un "bruiteur" qu'il cherche à localiser. La situation envisagée n'étant pas nécessairement pacifique, le "porteur" écoute de façon purement passive, sans envoyer de signal, afin de ne pas se faire repérer. Le résultat est que les seules quantités mesurées sont des angles. Dans le cas où le bruiteur est un navire, il est raisonnable de supposer qu'il suit un mouvement rectiligne et uniforme, i.e.

$$\frac{dX_t}{dt} = V \, , \quad X_0 = P$$

et on observe

$$y_t = h(t, X_t) + \eta_t \, .$$

Si l'on considère (P, V) comme un paramètre déterministe inconnu, on tombe sur un problème de statistique classique. On peut proposer des estimateurs pour (P, V), par exemple l'estimateur du maximum de vraisemblance. Mais ces estimateurs ne sont pas récursifs : une fois que l'on a estimé (P, V) au vu des observations $(y_s ; 0 \leq s \leq t)$, si l'on veut "rafraîchir l'estimation" en utilisant les observations $(y_s ; t \leq s \leq t + h)$, il faut recommencer les calculs depuis le début. Si l'on choisit une approche bayésienne, c'est à dire que l'on choisit une loi a priori pour (P, V), alors cette loi apparaît comme la loi initiale (i.e. à l'instant $t = 0$) du couple $\{(X_t, V_t)\}$ solution de

$$\begin{cases} \dfrac{dX_t}{dt} = V_t \, , \\[2mm] \dfrac{dV_t}{dt} = 0 \, , \\[2mm] (X_0, V_0) \text{ de loi donnée} \, , \end{cases}$$

qui est le processus non observé, l'observation étant de la forme :

$$y_t = h(t, X_t) + \eta_t .$$

La solution du problème sera alors "récursive", comme on le constatera au chapitre 2, au sens où, connaissant la loi conditionnelle de X_t sachant $\mathcal{Y}_t = \sigma\{y_s; 0 \leq s \leq t\}$, on n'a plus besoin de réutiliser les observations faites aux instants antérieurs à t pour calculer la loi conditionnelle de X_{t+h} sachant $\mathcal{Y}_{t+h} = \sigma\{y_s; 0 \leq s \leq t + h\}$.

Le problème de filtrage non linéaire que nous venons d'énoncer peut paraître "trivial". Il est vrai que du point de vue de la théorie qui va suivre, il est assez pauvre. Mais du point de vue algorithmique, il possède essentiellement les difficultés des problèmes de filtrage non linéaires plus généraux que nous considèrerons dans la suite.

Exemple 1.1.3 *Un problème d'estimation en radio–astronomie*   Afin d'estimer certaines caractéristiques d'une étoile, on effectue une expérience d'interférométrie à l'issue de laquelle on recueille un signal qui admet la représentation suivante :

$$y_t = a \exp\left[i\,(b + X_t)\right] + \eta_t , \quad t \geq 0$$

où $i = \sqrt{-1}$, $\{\eta_t\}$ est un bruit de mesure complexe, et $\{X_t, t \geq 0\}$ est une perturbation aléatoire de moyenne nulle, qui provient de la turbulence atmosphérique. Le problème est d'estimer au mieux les paramètres a et b caractéristiques de l'étoile visée au vu des observations. L'approche la plus simple consiste à négliger la perturbation $\{X_t\}$. Mais elle peut conduire à de mauvais résultats lorsque cette perturbation est importante. Le Gland [53] a proposé de modéliser le processus X_t comme un processus d'Ornstein–Uhlenbeck stationnaire du type :

$$dX_t = -\beta\,X_t\,dt + \sigma\,\sqrt{2\beta}\,dW_t$$

où $\{W_t\}$ est un processus de Wiener standard réel. Notons que la mesure invariante de $\{X_t\}$ est la loi $N(0, \sigma^2)$, et que β est une constante de temps. Les deux paramètres β et σ ont donc une interprétation "physique" simple. En outre, on peut les estimer, par exemple en visant au préalable une étoile dont les paramètres caratéristiques (a, b) sont connus.

Le problème de filtrage associé au problème que nous venons d'énoncer consisterait à calculer à chaque instant t la loi conditionnelle de X_t sachant $\mathcal{Y}_t = \sigma\{y_s; 0 \leq s \leq t\}$. En tant que tel, ce problème ne nous intéresse pas. Mais le problème de l'estimation des paramètres a et b, sur la base de l'observation *partielle* de y_t (X_t n'est pas observé) est très lié au problème de filtrage. En fait, pour calculer la vraisemblance du couple (a, b), il faut résoudre les équations du filtrage (voir ci–dessous la section 2.6).   □

Deux conclusions peuvent être tirées de ces quelques exemples. La première est qu'il existe des problèmes appliqués qui se formulent comme problèmes de filtrage. La seconde est que le filtrage est utile comme étape dans des problèmes de statistique de processus partiellement observés. Pour d'autres applications du filtrage et du lissage en statistique, voir Campillo, Le Gland [15]. C'est aussi une étape essentielle dans le contrôle des processus partiellement observés, voir Fleming, Pardoux [27], El Karoui, Hu Nguyen, Jeanblanc–Picqué [25], Bensoussan [8]

1.2 La classe de problèmes considérés

Il existe beaucoup de "familles" de problèmes de filtrage, suivant que le problème est en temps discret ou continu, et que les processus considérés sont à valeurs dans un ensemble dénombrable, un espace euclidien, ou un espace de dimension infinie, suivant aussi le type de processus que l'on considère.

Nous nous limiterons dans ce cours à considérer le filtrage de processus de diffusion (à valeurs dans un espace euclidien), en temps continu. Plus précisément, reprenons l'exemple 1.1.1 ci–dessus. Le processus non observé $\{X_t\}$ est un processus M–dimensionel, solution d'une EDS (que nous écrirons désormais au sens d'Itô) :

$$(1.1) \qquad X_t = X_0 + \int_0^t f(X_s)\,ds + \int_0^t g(X_s)\,dB_s$$

et on observe le processus N–dimensionel :

$$y_t = h(X_t) + \eta_t \ .$$

Une hypothèse essentielle dans toute la théorie du filtrage est que le processus bruit de mesure $\{\eta_t\}$ est un *"bruit blanc"*, i.e. la dérivée (au sens des distributions) d'un processus de Wiener, *de covariance non dégénérée*. Comme il est équivalent d'observer $\{y_s; 0 \leq s \leq t\}$ ou $\{\int_0^s y_r\,dr; 0 \leq s \leq t\}$, on appellera dorénavant observation le processus $\{Y_t\}$ donné par

$$(1.2) \qquad Y_t = \int_0^t h(X_s)\,ds + W_t$$

où $\{W_t\}$ est un processus de Wiener. La transformation qui vient d'être faite a pour but d'éviter de faire appel à des processus généralisés. Il y a cependant certains avantages (mais aussi des inconvénients !) à travailler directement avec le processus $\{y_t\}$. C'est ce qu'ont proposé récemment Kallianpur, Karandikar [42].

Reprenons le modèle (1.1)–(1.2), et réécrivons–le de façon plus générale, en tenant compte du fait que les Wiener $\{B_t\}$ et $\{W_t\}$ ne sont pas nécessairement indépendants, et que les coefficients peuvent dépendre du processus $\{Y_t\}$.

$$\begin{cases} X_t &= X_0 + \displaystyle\int_0^t b(s, Y, X_s)\,ds + \int_0^t f(s, Y, X_s)\,dV_s + \int_0^t g(s, Y, X_s)\,dW_s \ , \\ Y_t &= \displaystyle\int_0^t h(s, Y, X_s)\,ds + \int_0^t k(s, Y)\,dW_s \end{cases}$$

où $\{V_t\}$ et $\{W_t\}$ sont des processus de Wiener standard indépendants à valeurs dans $\mathbb{R}^M$ et $\mathbb{R}^N$ respectivement, globalement indépendants de X_0. Les coefficients peuvent dépendre à chaque instant s de toute la portion de trajectoire $\{Y_r; 0 \leq r \leq s\}$. Cette hypothèse est fondamentale pour les applications en contrôle stochastique, où les coefficients dépendent d'un contrôle qui lui même est une fonction arbitraire du passé des observations. Par contre, on ne fait dépendre les coefficients que du présent de X, ce qui fait que le processus $\{X_t\}$ est "conditionnellement markovien". Cette propriété est fondamentale pour que l'on puisse obtenir une équation d'évolution pour la loi conditionnelle de X_t sachant $\mathcal{Y}_t$.

Remarquons enfin que le coefficient devant le bruit d'observation ne dépend pas de X. S'il dépendait de X, alors on aurait une observation non bruitée de X, à savoir la variation quadratique de $\{Y_t\}$. Or on ne sait pas écrire les équations du filtrage dans une telle situation.

1.3 Liens entre EDS et EDP. Quelques rappels

Nous chercherons au chapitre 2 une équation qui régit l'évolution de la loi conditionnelle de X_t, sachant $\mathcal{Y}_t$. Il est utile de rappeler les résulats que l'on a dans le cas beaucoup plus simple où l'on n'a pas d'observation, et où on s'intéresse à l'évolution de la loi "a priori" de X_t.

Supposons que $\{X_t\}$ est un processus M–dimensionnel solution de l'EDS :

$$(1.3) \qquad X_t = X_0 + \int_0^t f(s, X_s)\, ds + \int_0^t g(s, X_s)\, dW_s$$

où $\{W_t\}$ est un Wiener standard M–dimensionnel, $f : \mathbb{R}_+ \times \mathbb{R}^M \to \mathbb{R}^M$, $g : \mathbb{R}_+ \times \mathbb{R}^M \to \mathbb{R}^{M^2}$ sont mesurables et localement bornées. On supposera que l'EDS (1.3) possède une unique solution (soit au sens "fort", soit au sens "faible"), ce qui fait que $\{X_t\}$ est un processus de Markov. Son générateur infinitésimal est l'opérateur aux dérivées partielles :

$$L_t = \frac{1}{2} a^{ij}(t, x) \frac{\partial^2}{\partial x^i \partial x^j} + f^i(t, x) \frac{\partial}{\partial x^i}$$

où $a(t, x) = gg^*(t, x)$ et nous avons utilisé, comme nous le ferons toujours dans la suite, la convention de sommation sur indices répétés. Remarquons qu'au moins si $g \in C^{0,1}(\mathbb{R}_+ \times \mathbb{R}^M)$, on peut écrire (1.3) au sens de Stratonovich :

$$X_t = X_0 + \int_0^t \bar{f}(s, X_s)\, ds + \int_0^t g_i(s, X_s) \circ dW_s^i$$

avec $\bar{f}(t, x) = f(t, x) - \frac{1}{2} \frac{\partial g_i}{\partial x}(s, x) g_i(s, x)$, et g_i est le i–ème vecteur colonne de la matrice g, $\frac{\partial g_i}{\partial x}$ désigne la matrice $\left(\frac{\partial g_i^j}{\partial x^k} \right)_{j,k}$. Considérons les opérateurs aux dérivées partielles de 1er ordre :

$$U_{0,t} = \bar{f}^j(t, x) \frac{\partial}{\partial x^j}, \; U_{1,t} = g_1^j(t, x) \frac{\partial}{\partial x^j}, \ldots, U_{M,t} = g_M^j(t, x) \frac{\partial}{\partial x^j}, \; t \geq 0 \; .$$

Il est utile de noter que l'opérateur L_t peut se réécrire sous la forme :

$$L_t = \frac{1}{2} \sum_{i=1}^M U_{i,t}^2 + U_{0,t} \; .$$

Soit maintenant $\varphi \in C_c^\infty(\mathbb{R}^M)$ (l'espace des fonctions C^∞ à support compact de $\mathbb{R}^M$ dans $\mathbb{R}$). Il résulte de la formule d'Itô :

$$\varphi(X_t) = \varphi(X_0) + \int_0^t L_s\varphi(X_s)\, ds + M_t^\varphi$$

où $\{M_t^\varphi\}$ est une martingale. Pour $t \geq 0$, notons μ_t la loi de probabilité de X_t. En prenant l'espérance dans l'égalité ci–dessus, on obtient l'équation de Fokker–Planck :

$$(1.4) \qquad \mu_t(\varphi) = \mu_0(\varphi) + \int_0^t \mu_s(L_s\varphi)\, ds \; .$$

Cette équation peut se réécrire, au sens des distributions :

$$\frac{\partial \mu_t}{\partial t} = L_t^* \mu_t , \quad t \geq 0 .$$

Dans le cas où pour tout $t \geq 0$ μ_t possède une densité $p(t, x)$, cette équation devient une EDP "usuelle" :

$$\frac{\partial p}{\partial t}(t,x) = \frac{1}{2} \frac{\partial^2}{\partial x^i \partial x^j}(a^{ij} p)(t,x) - \frac{\partial}{\partial x^i}(f^i p)(t,x) .$$

Nous allons maintenant énoncer une formule de Feynman–Kac. Considérons l'EDP parabolique rétrograde :

$$(1.5) \quad \begin{cases} \dfrac{\partial v}{\partial s}(s,x) + L_s v(s,x) + \rho\, v(s,x) = 0 , \quad 0 \leq s \leq t , \\[2mm] v(t,x) = \varphi(x) \end{cases}$$

où $\rho \in C_b([0,t] \times {\rm I\!R}^M)$, $\varphi \in C_c({\rm I\!R}^M)$. Sous des hypothèses ad hoc sur les coefficients de L_t, cette équation admet une unique solution dans un espace convenable. Supposons en outre que cette solution soit la limite des solutions obtenues en régularisant les coefficients de L, ρ et φ. On a alors la formule suivante :

$$(1.6) \quad v(s,x) = E \left(\varphi\left(X_t^{sx}\right) \exp\left[\int_s^t \rho(r, X_r^{sx})\, dr \right] \right)$$

où

$$(1.7) \quad X_t^{sx} = x + \int_s^t f(r, X_r^{sx})\, dr + \int_s^t g(r, X_r^{sx})\, dW_r , \quad t \geq s .$$

Il suffit d'établir la formule (1.6) dans le cas où tous les coefficients sont réguliers; on passe ensuite à la limite à la fois dans l'EDP (1.5) et dans l'EDS (1.7). Dans le cas des coefficients réguliers, $v \in C_b^{1,2}([0,t] \times {\rm I\!R}^M)$, et on peut appliquer la formule d'Itô au processus $v(r, X_r^{sx}) \exp[\int_s^r \rho(u, X_u^{sx})\, du]$, $s \leq r \leq t$:

$$v(s,x) + \int_s^t \left(\frac{\partial v}{\partial r} + L_r v + \rho\, v \right)(r, X_r^{sx})\, e^{\int_s^r \rho(u, X_u^{sx})\, du}\, dr + M_t^{v,\rho} =$$

$$= v(t, X_t^{sx})\, e^{\int_s^t \rho(r, X_r^{sx})\, dr}$$

où $\{M_r^{v,\rho}, s \leq r \leq t\}$ est une martingale. Il reste à utiliser le fait que v satisfait (1.5) et à prendre l'espérance pour obtenir (1.6).

L'approche que nous venons de décrire permet de montrer que "la" solution de l'EDP rétrograde (1.5) satisfait (1.6). On pourrait aussi définir $v(s,x)$ par (1.6), et montrer que cette quantité satisfait l'EDP (1.5). Cette dernière démarche est peut-être plus classique. Elle sera exposée dans un cadre plus complexe au chapitre 2.

On vient de voir certaines connexions entre les processus de diffusion et les EDP paraboliques du deuxième ordre. Dans la suite du cours, on verra le lien entre "diffusions conditionnelles" et EDP paraboliques stochastiques du deuxième ordre.

Chapitre 2

Les équations du filtrage non linéaire, de la prédiction et du lissage

2.1 Formulation du problème

Soit $\{(X_t, Y_t); \ t \geq 0\}$ un processus à valeurs dans $\mathbb{R}^M \times \mathbb{R}^N$, solution du système différentiel stochastique :

$$(2.1) \quad \begin{cases} X_t = X_0 + \int_0^t b(s, Y, X_s) \, ds + \int_0^t f(s, Y, X_s) \, dV_s + \int_0^t g(s, Y, X_s) \, dW_s \\ Y_t = \int_0^t h(s, Y, X_s) \, ds + \int_0^t k(s, Y) \, dW_s \end{cases}$$

où X_0 est un v.a. de dimension M indépendant du processus de Wiener standard $\{(V_t, W_t)\}$ à valeurs dans $\mathbb{R}^M \times \mathbb{R}^N$, tous définis sur un espace de probabilité filtré $(\Omega, \mathcal{F}, \mathcal{F}_t, P)$. On peut remarquer que le coefficient k du bruit d'observation ne dépend pas de X (cf. chapitre 1).

On supposera pour fixer les idées que $(\Omega, \mathcal{F}, \mathcal{F}_t, P)$ est l'espace canonique du processus $\{(X_t, Y_t)\}$, c'est à dire que :

$$\begin{aligned} \Omega &= \Omega_1 \times \Omega_2 \,, \\ \Omega_1 &= C(\mathbb{R}_+; \mathbb{R}^M), \ \Omega_2 = C(\mathbb{R}_+; \mathbb{R}^N) \,, \\ X_t(\omega) &= \omega_1(t), \ Y_t(\omega) = \omega_2(t) \,, \\ \mathcal{F} &= \text{la tribu borélienne de } \Omega \vee \mathcal{N} \,, \\ \mathcal{F}_t &= \sigma\{(X_s, Y_s); 0 \leq s \leq t\} \vee \mathcal{N} \end{aligned}$$

où $\mathcal{N}$ est la classe des ensembles de P-mesure nulle.

P est donc la loi de probabilité du processus (X, Y). On notera P^X et P^Y les lois marginales.

b, f, g et h sont des applications de $\mathbb{R}_+ \times C(\mathbb{R}_+; \mathbb{R}^N) \times \mathbb{R}^M$ à valeurs respectivement dans $\mathbb{R}^M$, $\mathbb{R}^{M \times M}$, $\mathbb{R}^{M \times N}$ et $\mathbb{R}^N$. On suppose qu'elles sont mesurables, l'espace de départ étant muni de la tribu $\mathcal{P}_2 \otimes B_M$, et l'espace d'arrivée de la tribu borélienne correspondante, et que $k : \mathbb{R}_+ \times C(\mathbb{R}_+; \mathbb{R}^N) \to \mathbb{R}^{N \times N}$ est $\mathcal{P}_2 / \mathcal{B}_{N \times N}$ mesurable. $\mathcal{P}_2$ désigne la tribu des parties progressivement mesurables de $\mathbb{R}_+ \times \Omega_2$, et $\mathcal{B}_M$ désigne la tribu borélienne de $\mathbb{R}^M$. Rappelons que la tribu $\mathcal{P}_2$ est la plus petite tribu qui rend mesurable toutes les applications $\varphi : \mathbb{R}_+ \times \Omega_2 \to \mathbb{R}$ qui sont telles que leur restriction à $]0, t[\times \Omega_2$ est $\mathcal{B}([0, t]) \otimes \mathcal{F}_t$ mesurable, pour tout $t \geq 0$.

Remarque 2.1.1 Rappelons que notre motivation pour permettre une dépendance arbitraire des coefficients par rapport au passé de $\{Y_t\}$ vient du contrôle stochastique. $\quad \square$

On pourra dans la suite supposer que le problème de martingales associé à (2.1) est bien posé (i.e. que le système différentiel stochastique (2.1) admet une solution unique en loi). On peut trouver dans la littérature plusieurs jeux d'hypothèses sur les coefficients qui entraînent cette propriété. Pour l'instant, nous supposerons que $\{(X_t, Y_t); t \geq 0\}$ est un processus continu et $\mathcal{F}_t$ adapté satisfaisant (2.1).

Nous allons maintenant préciser les hypothèses sur les coefficients.

On suppose

$$(H.1) \quad k(t, y) = k^*(t, y) > 0.$$

et on pose :

$$a(t, y, x) = f f^*(t, y, x) + g g^*(t, y, x) ,$$
$$e(t, y) = k k(t, y), ,$$
$$t \in \mathbb{R}_+, y \in C(\mathbb{R}_+; \mathbb{R}^N), x \in \mathbb{R}^M .$$

Désignons par Λ (resp. Σ) la collection des fonctions de $\mathbb{R}_+ \times \Omega_2 \times \mathbb{R}^M$ (resp. $\mathbb{R}_+ \times \Omega_2$) dans $\mathbb{R}$ qui sont des coordonnées de l'un des vecteurs $b, a, k^{-1}h$ (resp. de la matrice e). On suppose :

$$(H.2) \quad \lambda \text{ (resp. } \sigma\text{) est localement bornée sur } \mathbb{R}_+ \times \Omega_2 \times \mathbb{R}^M$$
$$\text{(resp. sur } \mathbb{R}_+ \times \Omega_2\text{) } \forall \lambda \in \Lambda \text{ (resp. } \forall \sigma \in \Sigma\text{)}.$$

On pose enfin, pour $t \geq 0$:

$$Z_t = \exp \left(\int_0^t \left(e^{-1}(s, Y)h(s, Y, X_s), dY_s \right) - \frac{1}{2} \int_0^t | k^{-1}(s, Y)h(s, Y, X_s) |^2 \, ds \right)$$

et on pose les hypothèses suivantes (qui ne seront pas toujours supposées être satisfaites) :

$(H.3) \quad$ pour tous $t > 0$, $n \in \mathbb{N}$, pour toute fonction mesurable
$\rho : \Omega_2 \to [0, 1]$ tels que $\rho(y) = 0$ si $\sup_{0 \leq s \leq t} |y(s)| > n$,

$$E \left[\rho(Y) \int_0^t | k^{-1}(s, Y)h(s, Y, X_s) |^2 \, ds \right] < \infty ,$$

$(H.4) \quad E(Z_t^{-1}) = 1 , \quad \forall \, t \geq 0.$

Lorsque l'hypothèse $(H.4)$ est satisfaite, on définit une nouvelle probabilité $\overset{\circ}{P}$, appelée "probabilité de référence", sur $(\Omega, \mathcal{F})$, caractérisée par :

$$\frac{d\overset{\circ}{P}}{dP}\bigg|_{\mathcal{F}_t} = Z_t^{-1}, \quad t \geq 0 \,.$$

Il resulte alors du théorème de Girsanov que, sous $\overset{\circ}{P}$, $\{(V_t, \overline{Y}_t);\ t \geq 0\}$ est un $\mathcal{F}_t$-processus de Wiener standard à valeurs dans $\mathbb{R}^M \times \mathbb{R}^N$, où :

$$\overline{Y}_t = \int_0^t k^{-1}(s, Y)\, dY_s \,.$$

Afin d'assurer l'indépendance sous $\overset{\circ}{P}$ de $\{Y_t\}$ et de $(X_0, \{V_t\})$, on va supposer que, si $\mathcal{Y}_t = \sigma(Y_s;\ 0 \leq s \leq t) \vee \mathcal{N}$ et $\overline{\mathcal{Y}}_t = \sigma(\overline{Y}_s;\ 0 \leq s \leq t) \vee \mathcal{N}$:

$$\mathcal{Y}_t = \overline{\mathcal{Y}}_t, \quad t \geq 0 \,.$$

Remarquons que l'on a toujours $\overline{\mathcal{Y}}_t \subset \mathcal{Y}_t$, et que l'inclusion inverse est vraie si l'EDS

$$\xi_t = \int_0^t k(s, \xi)\, d\overline{Y}_s$$

admet une unique solution forte, donc par exemple dès que :

(H.5)   l'application $y \to k(t, y)$ est localement lipchitzienne, uniformément par rapport à t dans un compact.

Notre premier but est d'établir les équations du filtrage. Afin que la technique n'obscurcisse pas les idées générales, nous allons tout d'abord considérer un cas particulièrement simple.

2.2 Les équations du filtrage dans le cas où $k = I$ et tous les coefficients sont bornés.

Dans cette section, on considèrera le modèle :

$$(2.2) \quad \begin{cases} X_t = X_0 + \displaystyle\int_0^t b(s, Y, X_s)\,ds + \int_0^t f(s, Y, X_s)\,dV_s + \int_0^t g(s, Y, X_s)\,dW_s \\[2mm] Y_t = \displaystyle\int_0^t h(s, Y, X_s)\,ds + W_t \end{cases}$$

et on suppose que les coefficients $b, f, g,$ et h sont *bornés* par une constante uniforme c.

Dans ce cas, $(H.4)$ est évidemment satisfaite, et en outre pour tout $t \geq 0$ les restrictions de P et $\overset{\circ}{P}$ à $\mathcal{F}_t$ sont équivalentes, et :

$$\frac{dP}{d\overset{\circ}{P}}\bigg|_{\mathcal{F}_t} = Z_t, \quad t \geq 0 \,.$$

Proposition 2.2.1 *Pour tout $t \geq 0$ et $\xi \in L^1(\Omega, \mathcal{F}_t, P)$, $\xi Z_t \in L^1(\Omega, \mathcal{F}_t, \overset{\circ}{P})$ et*

$$E(\xi \,/\, \mathcal{Y}_t) = \frac{\overset{\circ}{E}(\xi Z_t \,/\, \mathcal{Y}_t)}{\overset{\circ}{E}(Z_t \,/\, \mathcal{Y}_t)} \,.$$

Preuve La première affirmation est évidente. Comme $Z_t > 0$ P p.s., donc aussi $\overset{\circ}{P}$ p.s., $\overset{\circ}{E}(Z_t/\mathcal{Y}_t) > 0$ $\overset{\circ}{P}$ p.s. et P p.s., donc le membre de droite de l'égalité est bien défini p.s. Il suffit d'établir le résultat pour $\xi \geq 0$. Soit η une v.a.r. ≥ 0 et $\mathcal{Y}_t$ mesurable,

$$
\begin{aligned}
E(\xi\eta) &= \overset{\circ}{E}(\xi\eta Z_t) \\
&= \overset{\circ}{E}(\eta\,\overset{\circ}{E}(\xi Z_t/\mathcal{Y}_t)) \\
&= \overset{\circ}{E}\left(\eta\,\frac{Z_t}{\overset{\circ}{E}(Z_t/\mathcal{Y}_t)}\,\overset{\circ}{E}(\xi Z_t/\mathcal{Y}_t)\right) \\
&= E\left(\eta\,\frac{\overset{\circ}{E}(\xi Z_t/\mathcal{Y}_t)}{\overset{\circ}{E}(Z_t/\mathcal{Y}_t)}\right)
\end{aligned}
$$

$\square$

L'identité ci–dessus est souvent appelée en filtrage la "formule de Kallianpur–Striebel". Elle permet de ramener le calcul d'espérances conditionnelles sous P à des calculs d'espérance conditionnelle sous $\overset{\circ}{P}$. Quel est l'intérêt de $\overset{\circ}{E}(\cdot/\mathcal{Y}_t)$ par rapport à $E(\cdot/\mathcal{Y}_t)$?

Remarquons qu'une v.a. $\mathcal{F}_t$ mesurable est (en gros) une fonction de $(X_0;\ V_s,\ 0 \leq s \leq t)$ et de $(Y_s,\ 0 \leq s \leq t)$, et ces deux "objets" sont indépendants sous $\overset{\circ}{P}$. Donc $\overset{\circ}{E}(\cdot/\mathcal{Y}_t)$ est en fait une intégrale par rapport à la loi de $(X_0;\ V_s,\ 0 \leq s \leq t)$.

Remarquons en outre que, avec ξ comme dans l'énoncé de la Proposition 2.2.1, pour tout $s \geq 0$,

$$
\overset{\circ}{E}(\xi Z_t/\mathcal{Y}_t) = \overset{\circ}{E}(\xi Z_t/\mathcal{Y}_{t+s}).
$$

En effet, $\mathcal{Y}_{t+s} = \mathcal{Y}_t \vee \sigma(Y_{t+u} - Y_t;\ 0 \leq u \leq s) = \mathcal{Y}_t \vee \mathcal{Y}_{t+s}^t$, et sous $\overset{\circ}{P}$ $\mathcal{Y}_{t+s}^t$ et $\mathcal{F}_t$ sont indépendantes. On a donc avec

$$
\mathcal{Y}_\infty \triangleq \bigvee_{t>0} \mathcal{Y}_t,
$$

(2.3)
$$
\overset{\circ}{E}(\xi Z_t/\mathcal{Y}_t) = \overset{\circ}{E}(\xi Z_t/\mathcal{Y}_\infty).
$$

Dans la suite, on écrira $\overset{\circ}{E}(\xi Z_t/\mathcal{Y})$ pour $\overset{\circ}{E}(\xi Z_t/\mathcal{Y}_\infty)$. Notons que (2.3) résulte de ce que l'on conditionne par rapport à la filtration d'un $\overset{\circ}{P}$–processus de Wiener.

Avant d'établir l'équation de Zakai, indroduisons quelques familles d'opérateurs aux dérivées partielles, indexées par $(t,y) \in \mathbb{R}_+ \times \Omega_2$. Pour $\varphi \in C_b^2(\mathbb{R}^M)$, on note (avec la convention de sommation sur indices répétés) :

$$
\begin{aligned}
L_{ty}\varphi(x) &= \frac{1}{2}a^{ij}(t,y,x)\frac{\partial^2\varphi}{\partial x^i\,\partial x^j}(x) + b^i(t,y,x)\frac{\partial\varphi}{\partial x^i}(x), \\
A_{ty}^j\varphi(x) &= f^{lj}(t,y,x)\frac{\partial\varphi}{\partial x^l}(x), \quad j = 1,\dots,M, \\
B_{ty}^i\varphi(x) &= g^{li}(t,y,x)\frac{\partial\varphi}{\partial x^l}(x), \quad i = 1,\dots,N, \\
L_{ty}^i\varphi(x) &= h^i(t,y,x)\varphi(x) + B_{ty}^i\varphi(x), \quad i = 1,\dots,N.
\end{aligned}
$$

(Attention : $y \in C(\mathbb{R}_+;\ \mathbb{R}^N)$, $x \in \mathbb{R}^M$!).

On définit maintenant deux processus à valeurs dans l'espace $\mathcal{M}_+(\mathbb{R}^M)$ des mesures finies sur $\mathbb{R}^M$.

Définition 2.2.2 *Soit* $\{\sigma_t,\ t \geq 0\}$ *et* $\{\Pi_t,\ t \geq 0\}$ *les processus à valeurs dans* $\mathcal{M}_+(\mathbb{R}^M)$ *définis par :*

$$\begin{aligned}
\sigma_t(\varphi) &= \overset{\circ}{E}\left(\varphi(X_t)Z_t\,/\,\mathcal{Y}_t\right), \\
\Pi_t(\varphi) &= E(\varphi(X_t)\,/\,\mathcal{Y}_t),
\end{aligned}$$

$t \geq 0,\ \varphi \in C_b(\mathbb{R}^M)$.

Le fait que les formules ci-dessus définissent bien des mesures aléatoires σ_t et Π_t résulte de ce que l'application $\varphi \to (\sigma_t(\varphi),\ \Pi_t(\varphi))$ est p.s. continue, si l'on munit $C_b(\mathbb{R}^M)$ de la topologie de la convergence uniforme sur tout compact.

Notons que $\sigma_0 = \Pi_0 = $ loi de X_0. On peut maintenant établir l'équation de Zakai (aussi appelée équation de Duncan–Mortensen–Zakai).

Théorème 2.2.3 *Si tous les coefficients de (2.2) sont bornés, alors pour tout* $\varphi \in C_b^2(\mathbb{R}^M)$,

$$(Z) \qquad \sigma_t(\varphi) = \sigma_0(\varphi) + \int_0^t \sigma_s(L_{sY}\,\varphi)ds + \int_0^t \sigma_s(L_{sY}^i\,\varphi)\,dY_s^i.$$

Preuve Remarquons tout d'abord que

$$\begin{aligned}
X_t &= X_0 + \int_0^t [\,b(s,Y,X_s) - g\,h(s,Y,X_s)\,]ds + \\
&\quad + \int_0^t f(s,Y,X_s)\,dV_s + \int_0^t g(s,Y,X_s)\,dY_s.
\end{aligned}$$

Utilisons la formule d'Itô :

$$\begin{aligned}
\varphi(X_t) &= \varphi(X_0) + \int_0^t L_{sY}\,\varphi(X_s)ds - \int_0^t h^i(s,Y,X_s)\,B_{sY}^i\,\varphi(X_s)ds + \\
&\quad + \int_0^t A_{sY}^l\,\varphi(X_s)\,dV_s^l + \int_0^t B_{sY}^i\,\varphi(X_s)\,dY_s^i, \\
Z_t &= 1 + \int_0^t Z_s\,h^i(s,Y,X_s)\,dY_s^i, \\
Z_t\varphi(X_t) &= \varphi(X_0) + \int_0^t Z_s\,L_{sY}\,\varphi(X_s)ds + \int_0^t Z_s\,A_{sY}^l\,\varphi(X_s)\,dV_s^l + \\
&\quad + \int_0^t Z_s\,L_{sY}^i\,\varphi(X_s)\,dY_s^i.
\end{aligned}$$

Il reste à prendre $\overset{\circ}{E}(\,\cdot\,/\,\mathcal{Y})$ des deux membres de cette égalité, à commuter l'espérance conditionnelle et l'intégrale de Lebesgue et à utiliser le Lemme 2.2.4. $\qquad\square$

Lemme 2.2.4 *Soit* $\{U_t,\ t \geq 0\}$ *un processus* $\mathcal{F}_t$*–progressif t.q.*

$$E\int_0^T U_t^2\,dt < \infty,\ \forall\,T \geq 0,$$

alors

$$\begin{aligned}
\overset{\circ}{E}\left(\int_0^t U_s\,dV_s^j\,/\,\mathcal{Y}\right) &= 0, \quad t \geq 0, \quad j = 1,\ldots,M, \\
\overset{\circ}{E}\left(\int_0^t U_s\,dY_s^i\,/\,\mathcal{Y}\right) &= \int_0^t \overset{\circ}{E}(U_s\,/\,\mathcal{Y})\,dY_s^i, \quad t \geq 0, \quad i = 1,\ldots,N.
\end{aligned}$$

Preuve Notons que si $\xi \in L^2(\Omega, \mathcal{F}_t, \overset{\circ}{P})$, pour calculer $\overset{\circ}{E}(\xi / \mathcal{Y})$, il suffit de calculer $\overset{\circ}{E}(\xi\eta)$, pour tout $\eta \in \mathbf{S}_t$, où $\mathbf{S}_t \subset L^2(\Omega, \mathcal{Y}_t, \overset{\circ}{P})$ et $\mathbf{S}_t$ est total dans $L^2(\Omega, \mathcal{Y}_t, \overset{\circ}{P})$. On choisit

$$
\mathbf{S}_t = \left\{ \eta = \exp\left(\int_0^t \rho_s^i \, dY_s^i - \frac{1}{2} \int_0^t \mid \rho_s \mid^2 ds \right), \ \rho \in L^2(0, t; \ \mathbb{R}^N) \right\},
$$

$$
\eta = 1 + \int_0^t \eta_s \, \rho_s^i \, dY_s^i,
$$

$$
\overset{\circ}{E}\left(\eta \int_0^t U_s \, dV_s^j \right) = 0,
$$

$$
\overset{\circ}{E}\left(\eta \int_0^t U_s \, dY_s^i \right) = \overset{\circ}{E} \int_0^t \eta_s \, \rho_s^i \, U_s \, ds
$$

$$
= \overset{\circ}{E} \int_0^t \eta_s \, \rho_s^i \, \overset{\circ}{E}(U_s / \mathcal{Y}) ds
$$

$$
= \overset{\circ}{E}\left(\eta \int_0^t \overset{\circ}{E}(U_s / \mathcal{Y}) \, dY_s^i \right)
$$

$\square$

L'équation de Zakai que nous venons d'établir a l'avantage d'être une équation linéaire. Remarquons que l'on a :

$$
\Pi_t = \sigma_t(1)^{-1} \sigma_t
$$

où 1 désigne la fonction constante égale à 1.

Nous allons maintenant établir l'équation de Kushner-Stratonovich satisfaite par Π_t. Pour cela, il nous faut d'abord donner une expression pour $\sigma_t(1)$.

Proposition 2.2.5 $\sigma_t(1) = \overset{\circ}{E}(Z_t / \mathcal{Y})$ *est donnée par :*

$$
\overset{\circ}{E}(Z_t / \mathcal{Y}) = \exp\left[\int_0^t \Pi_s(h^i(s, Y, \cdot)) \, dY_s^i - \frac{1}{2} \int_0^t \mid \Pi_s(h(s, Y, \cdot)) \mid^2 ds \right].
$$

Preuve

$$
Z_t = 1 + \int_0^t Z_s \, h^i(s, Y, X_s) \, dY_s^i.
$$

D'après le Lemme 2.2.4 et la Proposition 2.2.1,

$$
\overset{\circ}{E}(Z_t / \mathcal{Y}) = 1 + \int_0^t \overset{\circ}{E}(Z_s \, h^i(s, Y, X_s) / \mathcal{Y}) \, dY_s^i
$$

$$
= 1 + \int_0^t \overset{\circ}{E}(Z_s / \mathcal{Y}) \Pi_s(h^i(s, Y, \cdot)) \, dY_s^i.
$$

Théorème 2.2.6 *Si tous les coefficients de (2.2) sont bornés, alors pour tout* $\varphi \in C_b^2(\mathbb{R}^M)$,

$$
\Pi_t(\varphi) = \Pi_0(\varphi) + \int_0^t \Pi_s(L_{sY}\,\varphi) ds +
$$

(KS)

$$
+ \int_0^t [\Pi_s(L_{sY}^i\,\varphi) - \Pi_s(h^i(s, Y, \cdot))\Pi_s(\varphi)] \, [\, dY_s^i - \Pi_s(h^i(s, Y, \cdot) ds].
$$

Preuve Il résulte de la Proposition 2.2.5 et de la formule d'Itô :

$$\sigma_t(1)^{-1} \; = \; 1 - \int_0^t \sigma_s^{-1}(1)\,\Pi_s(h^i(s,Y,\cdot))\,dY_s^i$$
$$+ \int_0^t \sigma_s^{-1}(1)\,|\,\Pi_s(h(s,Y,\cdot))\,|^2\,ds\,.$$

On utilise maintenant (Z) et à nouveau Itô :

$$\sigma_t(1)^{-1}\sigma_t(\varphi) \; = \; \sigma_0(\varphi) + \int_0^t \sigma_s^{-1}(1)\,\sigma_s(L_{sY}\,\varphi)ds$$
$$+ \int_0^t \sigma_s^{-1}(1)\,\sigma_s(L_{sY}^i\,\varphi)\,dY_s^i$$
$$- \int_0^t \sigma_s^{-1}(1)\,\sigma_s(\varphi)\,\Pi_s(h^i(s,Y,\cdot))\,dY_s^i$$
$$+ \int_0^t \sigma_s^{-1}(1)\,\sigma_s(\varphi)\,|\,\Pi_s(h(s,Y,\cdot))\,|^2\,ds$$
$$- \int_0^t \sigma_s^{-1}(1)\,\sigma_s(L_{sy}^i\,\varphi)\Pi_s(h^i(s,Y,\cdot))ds\,.$$

Il reste à se souvenir que $\Pi_t = \sigma_t(1)^{-1}\sigma_t$

Remarquons que si l'on pose :

$$I_t = Y_t - \int_0^t \Pi_s(h(s,Y,\cdot))ds$$

l'équation (KS) se réécrit :

$$\Pi_t(\varphi) \; = \; \Pi_0(\varphi) + \int_0^t \Pi_s(L_{sY}\,\varphi)\,ds +$$
$$+ \int_0^t [\Pi_s(L_{sY}^i\,\varphi) - \Pi_s(h^i(s,Y,\cdot))\Pi_s(\varphi)]\,dI_s^i\,.$$

On remarque que l'équation de $\Pi_t(\varphi)$ contient le terme que l'on retrouve dans l'équation pour la loi a priori de $\{X_t\}$, $\Pi_s(L_{sY}\,\varphi)$, plus un terme "dirigé" par l'innovation $\{I_t\}$.

Remarquons que l'on a :

$$Y_t = \int_0^t h(s,Y,X_s)\,ds + W_t = \int_0^t \Pi_s(h(s,Y,\cdot))\,ds + I_t\,.$$

On va voir que la seconde écriture est la décomposition de $\{Y_t\}$ comme $\mathcal{Y}_t$–semi-martingale. Essayons tout d'abord de donner une interprétation intuitive de la terminologie "processus d'innovation".

$$I_{t+dt} - I_t \simeq Y_{t+dt} - Y_t - \Pi_t(h(t,Y,\cdot))\,dt\,,$$

$I_{t+dt} - I_t$ est la partie "innovante" de la nouvelle observation obtenue entre t et $t+dt$, puisque c'est la différence entre la nouvelle observation et ce que l'on s'attendait à observer au vu des observations précédentes. Cette interprétation serait plus claire en temps discret.

Proposition 2.2.7 $\{I_t\,,\,t \geq 0\}$ *est un* $P - \mathcal{Y}_t$ *processus de Wiener standard.*

Preuve Il est clair que I_t est une semi–martingale $\mathcal{Y}_t$–adaptée, et que $< I >_t = t$. Il reste à montrer que c'est une $\mathcal{Y}_t$–martingale. Soit $0 \leq s < t$.

$$
\begin{aligned}
E(I_t - I_s \,/\, \mathcal{Y}_s) &= E(E(W_t - W_s \,/\, \mathcal{F}_s) \,/\, \mathcal{Y}_s) + \\
&\quad + E\left[\int_s^t (h(r, Y, X_r) - E(h(r, Y, X_r) \,/\, \mathcal{Y}_r)\, dr \,/\, \mathcal{Y}_s \right] \\
&= 0
\end{aligned}
$$

$\square$

Remarquons que l'égalité :

$$
Y_t = \int_0^t \Pi_s(h(s, Y, \cdot))\, ds + I_t
$$

est en fait une équation différentielle stochastique du type :

$$
(2.4) \qquad\qquad Y_t = \int_0^t \Lambda_s(Y)\, ds + I_t
$$

avec $\Lambda : \mathbb{R}_+ \times \Omega_2 \to \mathbb{R}^N$ progressivement mesurable. D'où la conjecture naturelle que $\mathcal{Y}_t \subset \mathcal{F}_t^I$, soit $\mathcal{Y}_t = \mathcal{F}_t^I$, appelée "conjecture de l'innovation".

Mais Tsirel'son a montré par un contre–exemple que le fait que Y était solution d'une équation du type (2.4) n'impliquait pas que $\mathcal{Y}_t \subset \mathcal{F}_t^I$. Plusieurs démonstrations fausses de la conjecture de l'innovation ont été publiées, dont nous tairons les références. Par contre, il semble bien que le résultat d'Alinger et Mitter [2] soit correct. Les hypothèses d'Alinger et Mitter sont les suivantes : les coefficients b, f et h ne dépendent pas de y (mais ne sont pas nécessairement bornés) et $g \equiv 0$. En outre,

$$
E \int_0^t |\, h(s, X_s)\,|^2 \, ds < \infty, \quad \forall t > 0 \,.
$$

Remarquons que – indépendamment de la réponse à la conjecture de l'innovation – on sait que toute $\mathcal{Y}_t - P$ martingale de carré intégrable est une intégrale stochastique par rapport à $\{I_t\}$. Ce résultat est la clé d'une dérivation directe de l'équation (KS) – voir Fujisaki–Kallianpur–Kunita [31] et Pontier–Stricker–Szpirglas [78]

2.3 Les équations du filtrage dans le cas général

On revient au modèle (2.1), et on supposera dans toute la suite que les hypothèses $(H.1)$, $(H.2)$ et $(H.5)$ sont satisfaites sans éprouver le besoin de le rappeler. On a tout d'abord la :

Proposition 2.3.1 *Sous l'hypothèse $(H.4)$, $\{Z_t, \; t \geq 0\}$ et une $\overset{\circ}{P} - \mathcal{F}_t$ martingale et $\{\overset{\circ}{E}(Z_t \,/\, \mathcal{Y}); \; t \geq 0\}$ est une $\overset{\circ}{P} - \mathcal{Y}_t$ martingale.*

Preuve Remarquons que $\forall\, t \geq 0$,

$$\overset{\circ}{E}\,(Z_t) = E(Z_t\, Z_t^{-1}) = 1 \;.$$

La première affirmation résulte alors de ce que Z_t est une surmartingale d'espérance constante. Il est clair que $\overset{\circ}{E}\,(Z_t\,/\,\mathcal{Y})$ est $\mathcal{Y}_t$–mesurable, et $\overset{\circ}{P}$–intégrable puisque Z_t l'est. Si ξ est $\mathcal{Y}_s$ mesurable et bornée,

$$
\begin{aligned}
\overset{\circ}{E}\,[\overset{\circ}{E}\,(Z_t\,/\,\mathcal{Y})\xi] \;&=\; \overset{\circ}{E}\,[Z_t\,\xi]\\
&=\; \overset{\circ}{E}\,[Z_s\,\xi]\\
&=\; \overset{\circ}{E}\,[\overset{\circ}{E}\,(Z_s\,/\,\mathcal{Y})\xi]
\end{aligned}
$$

On a utilisé le fait que $\{Z_t\}$ est une $\overset{\circ}{P} - \mathcal{F}_t$ martingale. $\qquad\qquad\square$

Il résulte alors de résultats "bien connus" sur les martingales par rapport à la filtration d'un processus de Wiener (bien connus dans le cas des martingales de carré intégrable, mais le résultat s'étend à toutes les martingales) :

Corollaire 2.3.2 *Le processus* $\{\overset{\circ}{E}\,(Z_t\,/\,\mathcal{Y});\;\; t \geq 0\}$ *possède une version à trajectoires continues.*

Nous pourrons donc supposer dorénavant, sous l'hypothèse $(H.4)$, que les trajectoires du processus $\{\overset{\circ}{E}\,(Z_t\,/\,\mathcal{Y});\;\; t \geq 0\}$ sont bornées sur tout intervalle compact. Toujours en supposant $(H.4)$ vérifiée, on définit Π_t et σ_t comme à la section précédente. On reprend les autres notations de cette section, à ceci près que les opérateurs L^i_{ty}, $i = 1 \ldots, N$, sont maintenant donnés par :

$$L^i_{ty}\,\varphi(x) = (e^{-1}h)^i\,(t,y,x)\,\varphi\,(x) + (k^{-1})^{ji}\,B^j_{iy}\,\varphi(x)$$

et on pose en outre $\overline{L}^i_{ty} = k^{ij}(t,y)L^j_{ty}$.

On a à nouveau le :

Théorème 2.3.3 *Sous l'hypothèse* $(H.4)$, *pour tout* $\varphi \in C^2_c(\mathbb{R}^M)$,

$$(Z)\qquad \sigma_t(\varphi) = \sigma_0(\varphi) + \int_0^t \sigma_s(L_{sY}\,\varphi)\,ds + \int_0^t \sigma_s(L^i_{sY}\,\varphi)\,dY^i_s\;.$$

Preuve Le début de la preuve suit celle du Théorème 2.2.3. On obtient :

$$
\begin{aligned}
Z_t\,\varphi(X_t) \;=\; &\varphi(X_0) + \int_0^t Z_s\,L_{sY}\,\varphi(X_s)\,ds +\\
&+ \int_0^t Z_s\,L^i_{sY}\,\varphi(X_s)\,dY^i_s + \int_0^t Z_s\,A^j_{sY}\,\varphi(X_s)\,dV^j_s\;,
\end{aligned}
$$

on ne peut pas appliquer directement le Lemme 2.2.4 pour prendre $\overset{\circ}{E}\,(\cdot\,/\,\mathcal{Y})$ dans l'égalité ci–dessus. On pose :

$$S_n = \inf\,\{t;\; |\,X_t\,| \vee |\,Y_t\,| \geq n\},\quad \chi_n(t) = 1_{[0,S_n]}(t)\;,$$

$$Z_{t \wedge S_n} \varphi(X_{t \wedge S_n}) = \varphi(X_0) + \int_0^t \chi_n(s) Z_s \, L_{sY} \, \varphi(X_s) \, ds$$
$$+ \int_0^t \chi_n(s) \, L_{sY}^i \, \varphi(X_s) \, dY_s^i$$
$$+ \int_0^t \chi_n(s) Z_s \, A_{sY}^j \, \varphi(X_s) \, dY_s^j \, ,$$

on peut maintenant prendre $\overset{\circ}{E}\,(\,\cdot\,/\,\mathcal{Y})$ comme dans la preuve du Théorème 2.2.3, d'où :

$$\overset{\circ}{E}\,(Z_{t \wedge S_n} \varphi(X_{t \wedge S_n})\,/\,\mathcal{Y}) = \overset{\circ}{E}\,\varphi(X_0)$$
$$+ \int_0^t \overset{\circ}{E}\,(\chi_n(s) Z_s \, L_{sY} \, \varphi(X_s)\,/\,\mathcal{Y}) \, ds$$
$$+ \int_0^t \overset{\circ}{E}\,(\chi_n(s) Z_s \, L_{sY}^i \, \varphi(X_s)\,/\,\mathcal{Y}) \, dY_s^i \, .$$

On déduit de la convergence dans $L^1(\overset{\circ}{P})$:

$$\overset{\circ}{E}\,(Z_{t \wedge S_n} \varphi(X_{t \wedge S_n})\,/\,\mathcal{Y}) \to \overset{\circ}{E}\,(Z_t \varphi(X_t)\,/\,\mathcal{Y})$$

en $\overset{\circ}{P}$ probabilité. En outre,

$$|\overset{\circ}{E}\,(\chi_n(s) Z_s \, L_{sY} \, \varphi(X_s)\,/\,\mathcal{Y})\,| \leq \overset{\circ}{E}\,(Z_s \,|\, L_{sY} \, \varphi(X_s)\,|\,/\,\mathcal{Y})$$
$$\leq \sup_{s \leq t} \overset{\circ}{E}\,(Z_s\,/\,\mathcal{Y})\,E\,(|\,L_{sY} \, \varphi(X_s)\,|\,/\,\mathcal{Y})$$

et cette dernière quantité est ds–intégrable sur $[0, t]$ p.s. On peut donc passer à la limite p.s. dans l'intégrale de Lebesgue. Enfin

$$[\overset{\circ}{E}\,(\chi_n(s) Z_s \, L_{sY}^i \, \varphi(X_s)\,/\,\mathcal{Y})]^2 \leq [\overset{\circ}{E}\,(Z_s \,|\, L_{sY}^i \, \varphi(X_s)\,|\,/\,\mathcal{Y})]^2$$
$$\leq \sup_{s \leq t} \left([\overset{\circ}{E}\,(Z_s\,/\,\mathcal{Y})\,]^2 \,[E\,(|\,L_{sY}^i \, \varphi(X_s)\,|\,/\,\mathcal{Y})]^2 \right)$$

et cette dernière quantité est ds–intégrable sur $[0, t]$ p.s. On peut donc finalement passer à la limite en $\overset{\circ}{P}$ probabilité dans l'intégrale stochastique, grâce au Lemme 2.3.4. $\qquad \Box$

Lemme 2.3.4 *Soit* $(\Omega, \mathcal{F}, \mathcal{F}_t, P, W_t)$ *un processus de Wiener réel standard,*

$$\{\varphi_n(t), \ t \geq 0, \ n \in \mathbb{N}\} \ et \ \{\varphi(t), \ t \geq 0\}$$

des processus stochastiques progressivement mesurables à valeurs dans $\mathbb{R}$ *tels que*

$$\int_0^t \varphi(s)^2 \, ds < \infty \ p.s. \, ,$$
$$\int_0^t |\,\varphi_n(s) - \varphi(s)\,|^2 \, ds \to 0 \ en \ probabilité \ quand \ n \to \infty \, .$$

Alors

$$\sup_{s \leq t} \left| \int_0^s \varphi(r) \, dW_r - \int_0^s \varphi_n(r) \, dW_r \right| \to 0$$

en probabilité, quand $n \to \infty$.

Etablissons tout d'abord le :

Lemme 2.3.5 *Soit* $(\Omega, \mathcal{F}, \mathcal{F}_t, P, W_t)$ *un processus de Wiener standard et* $\{\varphi(t); \; t \geq 0\}$ *un processus progressivement mesurable, t.q*

$$\int_0^t \varphi^2(s)ds < \infty \quad p.s.$$

Alors pour tout $t, \varepsilon, N > 0$,

$$P\left(\sup_{0 \leq s \leq t} \left|\int_0^s \varphi(r)\, dW_r\right| > \varepsilon\right) \leq P\left(\int_0^t \varphi^2(s)ds > N\right) + \frac{N}{\varepsilon^2}\,.$$

Preuve Posons $\tau_N = \inf\{t; \; \int_0^t \varphi^2(s)ds \geq N\}$,

$$\varphi^N(s) = \varphi(s)\, 1_{[0,\tau_N]}(s)\,.$$

$$
\begin{aligned}
P\left(\sup_{0 \leq s \leq t} \left|\int_0^s \varphi(r)\, dW_r\right| > \varepsilon\right) &\leq P(\tau_N < T) + P\left(\sup_{0 \leq s \leq t} \left|\int_0^s \varphi^N(r)\, dW_r\right| > \varepsilon\right) \\
&\leq P\left(\int_0^T \varphi^2(s)ds > N\right) + \frac{N}{\varepsilon^2}
\end{aligned}
$$

$\square$

Preuve du Lemme 2.3.4 En utilisant le Lemme 2.3.5, il suffit de remarquer que $\forall \eta > 0$,

$$\limsup_{n \to \infty} P\left(\int_0^t |\varphi_n(s) - \varphi(s)|^2\, ds > \eta\right) = 0$$

$\square$

Nous allons maintenant établir l'équation (KS). Pour cela, il nous faut tout d'abord généraliser la Proposition 2.2.5 :

Proposition 2.3.6 *Sous les hypothèses* (H.3) *et* (H.4), *pour tout* $t > 0$,

$$
\begin{aligned}
\overset{\circ}{E}(Z_t/\mathcal{Y}) = \exp\Bigg[&\int_0^t (e^{-1}(s,Y)\Pi_s(h(s,Y,\cdot)),\; dY_s) - \\
&-\frac{1}{2}\int_0^t |k^{-1}(s,Y)\Pi_s(h(s,Y,\cdot))|^2\; ds\Bigg]\,.
\end{aligned}
$$

Preuve S_n et χ_n étant définis comme dans la démonstration précédente,

$$Z_{t \wedge S_n} = 1 + \int_0^t \chi_n(s)Z_s(e^{-1}(s,Y)h(s,Y,X_s),\; dY_s)\,,$$

$\chi_n\, b^{-1}(s,Y)h(s,Y,X_s)$ et $E\int_0^t \chi_n Z_s^2\, ds$ sont bornés par des constantes dépendantes de n. On obtient donc comme à la Proposition 2.2.5 :

$$\overset{\circ}{E}(Z_{t \wedge S_n}/\mathcal{Y}) = 1 + \int_0^t \overset{\circ}{E}(Z_s/\mathcal{Y})(e^{-1}(s,Y)E[h(s,Y,X_s)\chi_n(s)/\mathcal{Y}],\; dY_s)\,.$$

A nouveau, $Z_{t \wedge S_n} \to Z_t$ dans $L^1(\overset{\circ}{P})$ quand $n \to \infty$, ce qui permet de passer à la limite dans le membre de gauche de l'égalité. En outre

$$\int_0^t \mid e^{-1}(s,Y)E[h(s,Y,X_s)\chi_n(s) / \mathcal{Y}] \mid^2 ds$$

$$\leq E\left(\int_0^t \mid e^{-1}(s,Y)h(s,Y,X_s)\chi_n(s) \mid^2 ds / \mathcal{Y}\right)$$

$$\leq E\left(\int_0^t \mid e^{-1}(s,Y)h(s,Y,X_s) \mid^2 ds / \mathcal{Y}\right)$$

$$< \infty$$

grâce à $(H.3)$. On conclut comme au Théorème 2.3.3. $\square$

Théorème 2.3.7 *Sous l'hypothèse* $(H.3)$, *pour tout* $\varphi \in C_C^2(\mathbb{R}^M)$,

$$\Pi_t(\varphi) = \Pi_0(\varphi) + \int_0^t \Pi_s(L_{sY}\varphi)\,ds +$$

$$(KS) \qquad \int_0^t \left[\Pi_s(\overline{L}_{sY}\varphi) - \Pi_s((k^{-1}\,h)^i(s,Y,\cdot))\,\Pi_s(\varphi)\right] \times$$

$$\times \left[(k^{-1}(s,Y)dY_s)^i - \Pi_s((k^{-1}h)^i(s,Y,\cdot))ds\right] \ .$$

Preuve On va approcher notre problème de filtrage par un problème qui satisfasse l'hypothèse $(H.4)$. Remarquons que $\chi_n(t) = \chi_n(t,Y,X)$ est progressivement mesurable, de $\mathbb{R}_+ \times \Omega$ à valeurs dans $[0,1]$. On pose

$$\begin{aligned}
b_n(t,Y,X) &= \chi_n(t,Y,X)\,b(t,Y,X_t)\,, \\
f_n(t,Y,X) &= \chi_n(t,Y,X)\,f(t,Y,X_t)\,, \\
g_n(t,Y,X) &= \chi_n(t,Y,X)\,g(t,Y,X_t)\,, \\
h_n(t,Y,X) &= \chi_n(t,Y,X)\,h(t,Y,X_t)\,, \\
k_n(t,y) &= (k(t,Y) - I)\,\rho_n(t,Y) + I\,, \\
e_n(t,y) &= k_n(t,Y)\,k_n(t,Y)
\end{aligned}$$

où $\rho_n(t,Y) = \alpha_n(\sup_{s \leq t} |Y_s|)$, avec $\alpha_n \in C^\infty(\mathbb{R}_+; [0,1])$, $\alpha_n(z) = 1$ pour $0 \leq z \leq n$, et $\alpha_n(z) = 0$ pour $z \geq 2n$.

Considérons le processus $\{(X_t^n, Y_t^n), t \geq 0\}$ défini par $X_t^n = X_{t \wedge S_n}$, et $\{Y_t^n\}$ est l'unique solution (grâce à $(H.5)$) de l'EDS :

$$Y_t^n = Y_{t \wedge S_n} + \int_{t \wedge S_n}^t k_n(s,Y^n)\,dW_s \ .$$

Alors

$$\begin{cases}
X_t^n &= X_0 + \displaystyle\int_0^t b_n(s,Y^n,X^n)\,ds + \int_0^t f_n(s,Y^n,X^n)\,dV_s \\
&\qquad + \displaystyle\int_0^t g_n(s,Y^n,X^n)\,dW_s\,, \\[2mm]
Y_t^n &= \displaystyle\int_0^t h_n(s,Y^n,X^n)\,ds + \int_0^t k_n(s,Y^n)\,dW_s \ .
\end{cases}$$

On pose

$$Z_t^n = Z_{t \wedge S_n} = \exp\left[\int_0^t (e_n^{-1}(s, Y^n) h_n(s, Y^n, X^n), dY_s^n)\right.$$
$$\left. - \frac{1}{2} \int_0^t \left|k_n^{-1}(s, Y^n) h_n(s, Y^n, X^n)\right|^2 ds\right] .$$

Il est facile de montrer que $E[(Z_t^n)^{-1}] = 1$, et il existe une unique probabilité $\overset{\circ}{P}_n$ sur $(\Omega, \mathcal{F})$ telle que

$$\left. \frac{d\overset{\circ}{P}_n}{dP} \right|_{\mathcal{F}_t} = (Z_t^n)^{-1} , \quad t \geq 0 .$$

On pose $\mathcal{Y}_t^n = \sigma(Y_s^n; 0 \leq s \leq t) \vee \mathcal{N}$, $\mathcal{Y}^n = \bigvee_{t \geq 0} \mathcal{Y}_t^n$, et pour $\psi : \mathbb{R}_+ \times \Omega \to \mathbb{R}$ progressivement mesurable et tel que $E|\psi(t, X, Y)| < \infty$, on pose

$$\sigma_t^n(\psi) = \overset{\circ}{E}_n\left(\psi(t, X^n, Y^n) Z_t^n / \mathcal{Y}^n\right) ,$$
$$\Pi_t^n(\psi) = E\left(\psi(t, X^n, Y^n) / \mathcal{Y}_t^n\right) .$$

On montre alors comme à la section précédente que pour $\varphi \in C_C^2(\mathbb{R}^M)$,

$$\Pi_t^n(\varphi) = \Pi_0(\varphi) + \int_0^t \Pi_s^n(L_n \varphi) ds +$$
$$+ \int_0^t \left[\Pi_s^n(\overline{L}_n^i \varphi) - \Pi_s^n(\varphi) \Pi_s^n((k_n^{-1} h_n(s, \cdot, Y))^i)\right] \times$$
$$\times \left[k_n^{-1}(s, Y) dY_s^n\right]^i - \Pi_s\left((k_n^{-1} h_n)^i(s, Y^n, \cdot) ds\right] .$$

Il reste à prendre la limite quand $n \to \infty$ dans cette égalité. Le passage à la limite repose sur les lemmes suivants :

Lemme 2.3.8

(i) Si $\theta \in L^2(\Omega, \mathcal{F}, P)$ et $t \geq 0$, $\Pi_t^n(\theta) \to \Pi_t(\theta)$ dans $L^2(\Omega, \mathcal{F}, P)$ quand $n \to \infty$.

(ii) Soit $\{\theta_t, \theta_t^n ; n \in \mathbb{N}\}$ des processus progressivement mesurables tels que

   (ii.a) $\theta_t^n = \theta_t$, $0 \leq t \leq S_n$, p.s.

   (ii.b) $\{\theta_t^n; n \in \mathbb{N}\}$ est uniformément de carré intégrable par rapport à $P \times \lambda$ sur $\Omega \times [0, t]$.

Alors

$$E \int_0^t |\Pi_s^n(\theta_s^n) - \Pi_s(\theta_s)|^2 ds \to 0 .$$

Lemme 2.3.9 Soit ψ et $\{\psi_n, n \in \mathbb{N}\}$ des processus progressivement mesurables t.q. pour tout $t > 0$, $p \in \mathbb{N}$, l'ensemble $\bar{B}_{tp} = B^{tp} \cup (\cup_n B_n^{tp})$ soit borné, avec

$$B^{tp} = \left\{\psi(s, x, y); s \in [0, t], \sup_{0 \leq s \leq t} |y(s)| \leq p\right\} ,$$
$$B_n^{tp} = \left\{\psi_n(s, x, y); s \in [0, t], \sup_{0 \leq s \leq t} |y(s)| \leq p\right\} ,$$

et tels que $\psi_n(s, x, y) = \psi(s, x, y)$ sur $[0, S_n]$. Alors pour tout $q \geq 1$, $t > 0$,

$$\int_0^t |\Pi_s^n(\psi_n) - \Pi_s(\psi)|^q \, ds \to 0$$

en probabilité, quand $n \to \infty$.

Lemme 2.3.10 *Soit $\{\alpha_t, \alpha_t^n \,;\, t \geq 0,\, n \in \mathbb{N}\}$ des processus progressivement mesurables à valeurs dans $\mathbb{R}^N$, tels que pour tout $t > 0$,*

(i) $\displaystyle\int_0^t \left(|\alpha_s|^2 + |\alpha_s^n|^2\right) ds < \infty$ *p.s., $n \in \mathbb{N}$,*

(ii) $\displaystyle\int_0^t (\alpha_s - \alpha_s^n)^2 \, ds \to 0$ *en probabilité quand $n \to \infty$.*

Alors pour tout $t > 0$,

$$\int_0^t \alpha_s^n \, \rho_n^{-1}(s, Y^n) \, dY_s^n \to \int_0^t \alpha_s \, \rho^{-1}(s, Y) \, dY_s$$

en probabilité, quand $n \to \infty$.

Les détails des démonstrations du Théorème et des Lemmes, qui sont dûs à D. Michel et l'auteur, seront publiés ultérieurement.

2.4   Le problème de la prédiction

Supposons que l'on veuille calculer la loi conditionnelle de X_t, sachant $\mathcal{Y}_s$ (avec $0 < s < t$), ou, ce qui revient au même, $E(\varphi(X_t)/\mathcal{Y}_s)$. Nous allons voir que ce problème, dit de prédiction, se ramène aisément à un problème de filtrage, à condition que les coefficients du système ne dépendent pas de l'observation Y. Considérons donc le processus $\{(X_t, Y_t); t \geq 0\}$ solution du système différentiel stochastique :

$$(2.5) \quad \begin{cases} X_t = X_0 + \displaystyle\int_0^t b(s, X_s) \, ds + \int_0^t f(s, X_s) \, dV_s + \int_0^t g(s, X_s) \, dW_s \,, \\[2mm] Y_t = \displaystyle\int_0^t h(s, X_s) \, ds + W_t \,, \end{cases}$$

les coefficients étant tous supposés mesurables, et pour simplifier bornés. Soit maintenant $\{U_t, t \geq 0\}$ un processus de Wiener standard à valeurs dans $\mathbb{R}^N$ indépendant de tous les processus ci-dessus, et soit le processus :

$$\bar{Y}_t = \int_0^{t \wedge s} h(r, X_r) \, dr + W_{t \wedge s} + U_t - U_{t \wedge s} \,.$$

Alors, pour $s < t$, il est clair que (avec des notations évidentes) :

$$E\left(\varphi(X_t) \,/\, \mathcal{Y}_s\right) = E\left(\varphi(X_t) \,/\, \bar{\mathcal{Y}}_t\right) .$$

Mais le membre de droite de cette égalité est une quantité à laquelle on peut appliquer les résultats des sections précédentes. Si l'on définit les mesures $\bar{\sigma}_t$ et $\bar{\Pi}_t$ par analogie avec les notations ci–dessus (mais avec $\mathcal{Y}.$ remplacé par $\bar{\mathcal{Y}}.$), on obtient l'équation de Zakai de la prédiction :

$$(ZP) \qquad \bar{\sigma}_t(\varphi) = \sigma_0(\varphi) + \int_0^t \bar{\sigma}_r(L_r\varphi)\,dr + \int_0^{s\wedge t} \bar{\sigma}_r(L_r^i\varphi)\,dY_r^i$$

et l'équation de Kushner–Stratonovich de la prédiction :

$$\bar{\Pi}_t(\varphi) = \bar{\Pi}_0(\varphi) + \int_0^t \bar{\Pi}_r(L_r\varphi)\,dr +$$

$$(KSP) \qquad + \int_0^{s\wedge t} \left[\bar{\Pi}_r(L_r^i\varphi) - \bar{\Pi}_r(\varphi)\bar{\Pi}_r((h(r,\cdot,Y))^i) \right] \times$$

$$\times \left[dY_r^i - \bar{\Pi}_r^i(h^i)\,dr \right] .$$

Notons que $\bar{\Pi}_t(\varphi) = \bar{\sigma}_t^{-1}(1)\,\bar{\sigma}_t(\varphi) = \bar{\sigma}_s^{-1}(1)\,\bar{\sigma}_t(\varphi)$, $0 \leq s \leq t$, et que au delà de l'instant s, ces deux équations se ramènent à l'équation de Fokker–Planck. Remarquons que l'on peut aussi obtenir (ZP) en prenant $\overset{\circ}{E}(\cdot / \mathcal{Y}_s)$ dans (Z), et (KSP) en prenant $E(\cdot / \mathcal{Y}_s)$ dans (KS). On peut traiter le cas où les coefficients dépendent à chaque instant t de l'observation courante Y_t, en considérant $\{Y_r\}$ comme un processus observé jusqu'à l'instant s, et non observé au delà de l'instant s (on écrira alors entre s et t l'evolution de la loi conditionnelle du couple (X_r, Y_r), sachant $\mathcal{Y}_s$).

2.5   Le problème du lissage

Nous allons maintenant considérer le problème du calcul de $E[\varphi(X_s) / \mathcal{Y}_t]$, avec $0 \leq s < t$. Pour alléger les notations, on supposera que $t = 1$, et on cherchera à calculer $E[\varphi(X_t) / \mathcal{Y}_1]$, avec $0 \leq t < 1$. Les résultats de cette section proviennent pour l'essentiel de Pardoux [69]. Pour simplifier, on va d'abord considérer le cas où les coefficients ne dépendent pas de l'observation, i.e. on considère le système :

$$(2.6) \quad \begin{cases} X_t = X_0 + \int_0^t b(s, X_s)\,ds + \int_0^t f(s, X_s)\,dV_s + \int_0^t g(s, X_s)\,dW_s \,, \\[2mm] Y_t = \int_0^t h(s, X_s)\,ds + W_t \end{cases}$$

et on suppose pour simplifier que tous les coefficients sont mesurables et *bornés*.

On pose comme ci–dessus :

$$L_t\varphi(x) = \frac{1}{2}\sum_{i,j=1}^M a^{ij}(t,x)\frac{\partial^2\varphi}{\partial x^i \partial x^j}(x) + \sum_{i=1}^M b^i(t,x)\frac{\partial\varphi}{\partial x^i}(x) \,,$$

et on suppose en outre que pour toute condition initiale $(s, (x, 0))$, le problème de martingales associé au système différentiel stochastique :

$$\begin{cases} dX_t = [b(t,X_t) - gh(t,X_t)]\,dt + f(t,X_t)\,dV_t + g(t,X_t)\,d\overset{\circ}{W}_t \,, \\[2mm] dY_t = d\overset{\circ}{W}_t \end{cases}$$

possède une unique solution $\overset{\circ}{P}_{sx}$, qui est une mesure de probabilité sur $C([s,1];\mathbb{R}^{M+N})$.

Avec les notations du paragraphe 2.2, on voit que le problème de lissage se ramène à la détermination d'une loi conditionnelle "non normalisée" μ_t donnée par :

$$\mu_t(\varphi) = \overset{\circ}{E}\left(\varphi(X_t)\, Z_1 \,/\, \mathcal{Y}\right).$$

Rappelons que $\sigma_t(\varphi) = \overset{\circ}{E}\left(\varphi(X_t)\, Z_t \,/\, \mathcal{Y}\right)$. Contrairement à ce qui se passait pour le problème de filtrage, cette définition de σ_t (où l'on conditionne par $\mathcal{Y}$ plutôt que par $\mathcal{Y}_t$) est cruciale. On pose :

$$v(t,x) = \overset{\circ}{E}_{tx}\left(Z_1^t \,/\, \mathcal{Y}\right)$$

où $Z_1^t = (Z_t)^{-1}\, Z_1$, et $\overset{\circ}{E}_{tx}$ est l'espérance sous $\overset{\circ}{P}_{tx}$. On notera ci-dessous $\mathcal{Y}_1^t = \sigma\{Y_r - Y_t\,;\, t \le r \le 1\} \vee \mathcal{N}$.

On a le :

Théorème 2.5.1

$$\mu_t(\varphi) = \sigma_t\left(\varphi\, v(t,\cdot)\right).$$

Corollaire 2.5.2

$$E[\varphi(X_t)\,/\,\mathcal{Y}_1] = \mu_t(1)^{-1}\,\mu_t(\varphi).$$

Preuve Le corollaire résulte immédiatement du théorème et de la proposition 2.2.1. $\qquad \square$

Le théorème est une conséquence immédiate des deux lemmes :

Lemme 2.5.3

$$\overset{\circ}{E}\left[\varphi(X_t)\, Z_1 \,/\, \mathcal{Y}\right] = \overset{\circ}{E}^{\mathcal{Y}}\left[\varphi(X_t)\, Z_t\, \overset{\circ}{E}^{\mathcal{Y}}_{t,X_t}\left(Z_1^t\right)\right].$$

Preuve Il suffit de démontrer que $\forall \theta$ v.a.r. $\mathcal{Y}_t$-mesurable et bornée, et $\forall \lambda$ v.a.r. $\mathcal{Y}_1^t$ mesurable et bornée, les produits scalaires dans $L^2(\Omega, \overset{\circ}{P})$ avec $\theta\lambda$ des deux membres de l'égalité ci-dessus coïncident. On va utiliser ci-dessous la propriété de Markov du processus $\{(X_r, Y_{r\vee t} - Y_t)\,;\, r \ge 0\}$, dont l'état à l'instant t ne dépend que de X_t.

$$\begin{aligned}
\overset{\circ}{E}\left[\varphi(X_t)\, Z_1\, \theta\, \lambda\right] &= \overset{\circ}{E}\left[\varphi(X_t)\, Z_t\, \theta\, \overset{\circ}{E}^{\mathcal{F}_t}\left(Z_1^t\, \lambda\right)\right] \\
&= \overset{\circ}{E}\left[\varphi(X_t)\, Z_t\, \theta\, \overset{\circ}{E}_{t,X_t}\left(Z_1^t\, \lambda\right)\right] \\
&= \overset{\circ}{E}\left[\theta\, \lambda\, \overset{\circ}{E}^{\mathcal{Y}}\left[\varphi(X_t)\, Z_t\, \overset{\circ}{E}^{\mathcal{Y}}_{t,X_t}\left(Z_1^t\right)\right]\right]
\end{aligned}$$

$\qquad \square$

Lemme 2.5.4 *Soit* $(x,\omega) \to G(x,\omega)$ *une application* $\mathcal{B}_N \otimes \mathcal{Y}_1^t$ *mesurable, de* $\mathbb{R}^N \times \Omega_2$ *à valeurs dans* $\mathbb{R}_+$. *Alors :*

$$\overset{\circ}{E}\left[G(X_t)\, Z_t \,/\, \mathcal{Y}\right] = \sigma_t(G).$$

Preuve Par le théorème des classes monotones, il suffit d'établir l'égalité pour une fonction G de la forme :

$$G(x, \omega) = \mathbf{1}_B(x)\, \mathbf{1}_A(\omega)$$

avec $B \in \mathcal{B}_N$, $A \in \mathcal{Y}_1^t$. On utilise les notations du lemme précédent.

$$
\begin{aligned}
\mathring{E}\left[\mathbf{1}_B(X_t)\, Z_t\, \mathbf{1}_A\, \theta\, \lambda\right] &= \mathring{E}\left[\mathbf{1}_B(X_t)\, \mathring{Z}_t\, \theta\right]\, \mathring{E}\left[\mathbf{1}_A\, \lambda\right] \\
&= \mathring{E}\left[\sigma_t(\mathbf{1}_B)\, \theta\right]\, \mathring{E}\left[\mathbf{1}_A\, \lambda\right] \\
&= \mathring{E}\left[\sigma_t(\mathbf{1}_B)\, \mathbf{1}_A\, \theta\, \lambda\right]
\end{aligned}
$$

et $\sigma_t(\mathbf{1}_B)\, \mathbf{1}_A = \sigma_t(\mathbf{1}_B\, \mathbf{1}_A)$. $\qquad\square$

Le théorème se réécrit :

$$\frac{d\mu_t}{d\sigma_t}(x) = v(t, x)\,.$$

Notre prochaine étape consiste à établir l'équation satisfaite par $v(t, x)$. Notons tout de suite que $v(1, x) = 1$, $\forall x$, et que $v(t, x)$ est $\mathcal{Y}_1^t$ mesurable. Nous allons d'abord donner une dérivation "formelle" de l'équation satisfaite par v (dérivation qui peut se justifier sous des hypothèses ad hoc !).

Supposons que $\forall t \in [0, 1]$, $v(t, \cdot) \in \mathbf{C}^2(\mathbb{R}^M)$ p.s. Il résulte alors de la formule d'Itô que pour $0 \le s < r \le 1$,

$$
\begin{aligned}
Z_r^s\, v(r, X_r) = {}& v(r, X_s) + \int_s^r Z_\theta^s\, L_\theta v(r, X_\theta)\, d\theta + \\
&+ \int_s^r Z_\theta^s\, \nabla v(r, X_\theta)\, f(\theta, X_\theta)\, dV_\theta + \int_s^r Z_\theta^s\, L_\theta^i v(r, X_\theta)\, dY_\theta^i\,.
\end{aligned}
$$

Notons que le fait que $v(r, \cdot)$ est aléatoire ne pose pas de problème, puisqu'il est $\mathcal{Y}_1^r$ adapté, et que $\mathcal{Y}_1^r$ et $\mathcal{Y}_r$ sont indépendants.

Discrétisons l'intervalle $[t, 1]$ par $t = t_0^n < t_1^n < \cdots < t_n^n = 1$, avec $t_k^n = t + \frac{1-t}{n}k$. Dans la suite on écrira t_k pour t_k^n. D'après la formule ci-dessus,

$$
\begin{aligned}
Z_{t_{k+1}}^{t_k}\, v(t_{k+1}, X_{t_{k+1}}) = {}& v(t_{k+1}, X_{t_k}) + \int_{t_k}^{t_{k+1}} Z_\theta^{t_k}\, L_\theta v(t_{k+1}, X_\theta)\, d\theta \\
&+ \int_{t_k}^{t_{k+1}} Z_\theta^{t_k}\, \nabla v(t_{k+1}, X_\theta)\, f(\theta, X_\theta)\, dV_\theta \\
&+ \int_{t_k}^{t_{k+1}} Z_\theta^{t_k}\, L_\theta^i v(t_{k+1}, X_\theta)\, dY_\theta^i\,.
\end{aligned}
$$

Prenons $\mathring{E}_{t_k, x}^y$ dans cette égalité :

$$
\begin{aligned}
v(t_k, x) = {}& v(t_{k+1}, x) + \int_{t_k}^{t_{k+1}} \mathring{E}_{t_k, x}^y \left(Z_\theta^{t_k}\, L_\theta v(t_{k+1}, X_\theta)\right) d\theta \\
&+ \int_{t_k}^{t_{k+1}} \mathring{E}_{t_k, x}^y \left(Z_\theta^{t_k}\, L_\theta^i v(t_{k+1}, X_\theta)\right) dY_\theta^i\,.
\end{aligned}
$$

Sommant de $k = 0$ à $n - 1$, et en prenant la limite quand $n \to \infty$ dans l'égalité ainsi obtenue (cette étape demande bien sûr une justification pour devenir rigoureuse), on obtient

$$v(t, x) = 1 + \int_t^1 L_s v(s, x)\, ds + \int_t^1 L_s^i v(s, x)\, dY_s^i$$

ou encore

$$(2.7) \quad \begin{cases} d_t v(t,x) + L_t v(t,x)\, dt + L_t^i v(t,x)\, dY_t^i = 0 \ , \\[2mm] v(1,x) = 1 \ . \end{cases}$$

Remarquons que v étant adapté à $\mathcal{Y}_1^t$, l'intégrale stochastique

$$\int_t^1 L_s^i v(s,x)\, dY_s^i$$

est une "intégrale de la formule d'Itô rétrograde", i.e. une intégrale par rapport au "processus de Wiener rétrograde" $\{Y_s^i - Y_1^i; t \le s \le 1\}$ (faire le changement de variable $u = 1-s$ pour retrouver une intégrale "progressive"). Une telle intégrale est — comme une intégrale d'Itô usuelle — un cas particulier de l'intégrale de Skorohod, voir ci-dessous Définition 5.2.1. Il n'y a donc pas lieu d'utiliser une notation différente de celle de l'intégrale d'Itô.

On va maintenant réécrire l'équation (2.7). On considère maintenant $\{v(t,x); 0 \le t \le 1, x \in \mathbb{R}^d\}$ comme un processus $\{v(t); 0 \le t \le 1\}$ à valeurs dans un espace de fonctions de x (qui est bien sûr un espace de dimension infinie, et qui en pratique sera un espace de Hilbert — ce pourrait être plus généralement un espace de Banach).

$$(2.8) \quad \begin{cases} dv(t) + L_t v(t)\, dt + L_t^i v(t)\, dY_t^i = 0 \ , \quad 0 \le t \le 1 \ , \\[2mm] v(1) = 1 \ . \end{cases}$$

Posons :

$$\begin{aligned} \rho(x) &= (1+|x|^2)^{-M} \ , \\ L_\rho^2 &= L^2(\mathbb{R}^M; \rho(x)\, dx) \ , \\ H_\rho^1 &= \{u \in L_\rho^2 \, ; \, \partial u/\partial x^i \in L_\rho^2, \, 1 \le i \le M\} \ . \end{aligned}$$

En outre $M_r^2(0,1; V)$ désignera l'espace $L^2((0,1) \times \Omega, \mathcal{P}_r, dt \times dP; V)$, ou $\mathcal{P}_r$ est la tribu qui rend mesurable les processus continus à gauche et $\mathcal{Y}_1^t$ adaptés.

Il résulte des résultats du chapitre 3 ci-dessous le :

Théorème 2.5.5 *Supposons satisfaite l'une des deux conditions suivantes :*

(i) $ff^*(t,x) \ge \alpha I > 0$, $\forall x$; et f est de classe C^1 en x, $\frac{\partial f}{\partial x^1}, \ldots, \frac{\partial f}{\partial x^M}$ étant bornées.

(ii) f est de classe C^2 en x, b,g,h sont de classe C^1 en x, toutes les dérivées étant bornées.

Alors l'équation (2.8) possède une unique solution :

$$v \in M_r^2(0,1; H_\rho^1) \cap L^2(\Omega; C([0,1]; L_\rho^2)) \ .$$

On a alors la "formule de Feynman–Kac stochastique" (Pardoux [68]) :

Proposition 2.5.6 *v désignant l'unique solution de l'équation (2.8) (au sens du théorème 2.5.5), on a l'égalité suivante $d\overset{\circ}{P} \times dx$ p.p. :*

$$v(t,x) = \overset{\circ}{E}_{tx} [Z_1^t / \mathcal{Y}] \ .$$

Preuve On va indiquer la démonstration proposée par Krylov–Rosovskii. Soit $\theta \in L^{\infty}(0,1;\mathbb{R}^N)$. On pose :

$$\rho_t = \exp\left(\int_t^1 (\theta_s, dY_s) - \frac{1}{2}\int_t^1 |\theta_s|^2\, ds\right)$$

ou encore, d'après la formule d'Itô rétrograde

$$\begin{cases} d\rho_t = -\rho_t(\theta_t, dY_t)\,, & 0 \le t \le 1\,, \\[2mm] \rho_1 = 1\,. \end{cases}$$

Posons $V(t) = \rho_t\, v(t)$. Soit $u \in C_c^{\infty}(\mathbb{R}^M)$. En appliquant à nouveau la formule d'Itô pour calculer la différentielle de $(V_t, u) = \rho_t\,(v(t), u)$, $(\cdot, \cdot)$ désigne le produit scalaire dans $L^2(\mathbb{R}^M)$,

$$d(V(t), u) + (LV(t), u)\, dt + (L^i V(t), u)\, dY_t^i + \theta_t^i\,(V(t), u)\, dY_t^i +$$
$$+ \theta_t^i\,(L^i V(t), u)\, dt = 0\,.$$

Ceci étant vrai $\forall u \in C_c^{\infty}(\mathbb{R}^M)$, il est facile d'en déduire que $\bar{V}(t) = \overset{\circ}{E}\, V(t)$ satisfait :

$$\begin{cases} \dfrac{d}{dt}\bar{V}(t) + L\bar{V}(t) + \theta_t^i\, L^i \bar{V}(t) = 0\,, & 0 \le t \le 1\,, \\[2mm] \bar{V}(1) = 1\,. \end{cases}$$

Il en résulte de la formule de Feynman–Kac "usuelle" que :

$$\bar{V}(t,x) = E_{tx}^{\theta}\left\{\exp\left[\int_t^1 (\theta_s, h(s, X_s))\, ds\right]\right\}$$

où P_{tx}^{θ} est la loi de probabilité sur $C([t,1];\mathbb{R}^M)$ de la solution de l'équation différentielle stochastique :

$$\begin{cases} dX_s = [b(s, X_s) + g(s, X_s)\,\theta_s]\, ds + f(s, X_s)\, dV_s + g(s, X_s)\, dW_s^{\theta}\,, & s \ge t\,, \\[2mm] X_t = x\,. \end{cases}$$

Il résulte du théorème de Girsanov que :

$$\left.\frac{dP_{tx}^{\theta}}{d\overset{\circ}{P}_{tx}}\right|_{\mathcal{F}_1^t} = \exp\left[\int_t^1 (\theta_s + h(s, X_s), dY_s) - \frac{1}{2}\int_t^1 |\theta_s + h(s, X_s)|^2\, ds\right]$$

$$= \rho_t\, Z_1^t \exp\left[-\int_t^1 (\theta_s, h(s, X_s))\, ds\right]$$

d'où l'on tire

$$\bar{V}(t,x) = \overset{\circ}{E}_{tx}\left[Z_1^t\, \rho_t\right]\,,$$

soit

$$\overset{\circ}{E}_{tx}\left[v(t,x)\,\rho_t(\theta)\right] = \overset{\circ}{E}_{tx}\left[Z_1^t\,\rho_t(\theta)\right]\,.$$

Or $\{\rho_t(\theta) \, ; \, \theta \in L^\infty(t,1;\mathbb{R}^N)\}$ est total dans $L^2(\Omega, \mathcal{Y}_1^t, \overset{\circ}{P}_{tx})$, et $v(t,x)$ est $\mathcal{Y}_1^t$ mesurable. On en déduit alors aisément que :

$$v(t,x) = \overset{\circ}{E}_{tx} \left[Z_1^t \, / \, \mathcal{Y}_1^t \right]$$

ce qui entraîne la proposition. □

La loi conditionnelle non normalisée du lissage à l'instant t s'obtient en résolvant l'équation progressive (Z) de 0 à t, et l'équation rétrograde (2.8) de 1 à t. On peut aussi établir une équation pour l'évolution de cette loi non normalisée, voir Pardoux [71].

Remarquons que l'on peut penser exprimer, à l'aide du calcul stochastique non adapté, la différentielle de μ_t. Mais il ne semble pas possible d'écrire une équation pour μ_t qui fasse intervenir μ_t seul, et non le couple $(\mu_t, v(t))$.

Notons $\Lambda_t = \mu_t(1)^{-1} \, \mu_t$. Alors, comme $\mu_1(1) = \sigma_t(v(t,\cdot)) = \sigma_1(1)$

$$\frac{d\Lambda_t}{d\Pi_t}(x) = \frac{\sigma_1(1)}{\sigma_t(1)} \, v(t,x) \; .$$

Sous des hypothèses de régularité ad hoc sur les coefficients, on peut établir, à l'aide du calcul stochastique non adapté, une équation pour

$$u(t,x) = \frac{\sigma_1(1)}{\sigma_t(1)} \, v(t,x)$$

qui s'écrit sous forme de Stratonovich :

$$du(t) + Au(t)\,dt + L_t^i u(t) \circ dY_t^i + u(t)\,\Pi_t(h^i) \circ dY_t^i =$$
$$= \left[L_t^i u(t) + \Pi_t(h_t^i) + \frac{1}{2} u(t)\,\Pi_t(L_t^i h^i) \right] dt$$

avec $A = L - \frac{1}{2}\sum_i (L^i)^2$. Remarquons que, contrairement au système des équations pour (σ, v), le système des équations pour (Π, u) est couplé : il faut résoudre l'équation (KS) pour Π jusqu'à l'instant final 1 avant de résoudre l'équation satisfaite par u.

Finalement, dans le cas où les coefficients dépendent du passé des observations, on s'attend à ce que $v(t) = d\mu_t/d\sigma_t$ satisfasse une EDPS rétrograde analogue à (2.8), mais cette fois on a besoin du calcul stochastique non adapté pour lui donner un sens. Nous allons écrire l'équation sous forme Stratonovich, sans donner les conditions sous lesquelles on peut l'établir, renvoyant à Ocone–Pardoux [67] pour les détails.

On suppose que pour tout $x \in \mathbb{R}^M$, les processus $g(t,Y,x)$ et $h(t,Y,x)$ possèdent une variation quadratique jointe avec Y_t, et on pose :

$$\bar{g}^{li}(t,Y,x) = \frac{d}{dt} < g^{li}(\cdot,Y,x), Y^i >_t , \quad \bar{g}^l = \sum_{i=1}^N \bar{g}^{li} ,$$

$$\bar{h}^i(t,Y,x) = \frac{d}{dt} < h^i(\cdot,Y,x), Y^i >_t , \quad \bar{h} = \sum_{i=1}^N \bar{h}^i .$$

On note alors $\bar{L}$ l'opérateur L^1 avec (g^1, h^1) remplacé par $(\bar{g}, \bar{h})$. Posons finalement :

$$\tilde{A}_{tY} = A_{tY} - \frac{1}{2} \bar{L}_{tY} \; .$$

L'équation satisfaite par v s'écrit alors :

$$(2.9) \quad \begin{cases} dv(t) + \tilde{A}_{tY}v(t)\,dt + L^i_{tY}v(t) \circ dY^i_t = 0 , & 0 \le t \le 1 , \\ \\ v(1) = 1 . \end{cases}$$

Notons qu'il n'existe pas à ce jour — à notre connaissance — de résultat d'existence ou unicité pour l'équation (2.9). Dans le cas où les coefficients ne dépendent que de Y à l'instant courant t, on peut établir et étudier l'équation (2.9) (ou sa version sous forme Itô) à l'aide de la théorie du grossissement d'une filtration (voir Pardoux [69]).

2.6 Application en statistique des processus : calcul de la vraisemblance

On se place ici pour simplifier dans le cadre de la section 2.2, en supposant que les dérives b et h dépendent d'un paramètre inconnu $\theta \in \Theta$, où Θ est un borélien de $\mathbb{R}^p$. On suppose donc que b ($resp.$ h) est une application bornée de $\Theta \times \mathbb{R}_+ \times C(\mathbb{R}_+; \mathbb{R}^N) \times \mathbb{R}^M$ à valeurs dans $\mathbb{R}^M$ ($resp.$ dans $\mathbb{R}^N$), qui est $\Sigma \otimes \mathcal{P}_2 \otimes \mathcal{B}_M / \mathcal{B}_N$ ($resp.$ $/ \mathcal{B}_M$) mesurable, où Σ désigne la trace de $\mathcal{B}_p$ sur Θ. f, g étant comme à la section 2.2, on suppose que pour chaque $\theta \in \Theta$ il existe une probabilité P_θ sur $(\Omega, \mathcal{F})$ et un $P_\theta - \mathcal{F}_t$ processus de Wiener standard $\{(V^\theta_t, W^\theta_t)'; t \ge 0\}$ à valeurs dans $\mathbb{R}^M \times \mathbb{R}^N$ tel que :

$$(2.10) \quad \begin{cases} dX_t = b_\theta(t, Y, X_t)dt + f(t, Y, X_t)dV^\theta_t + g(t, Y, X_t)dW^\theta_t , \\ \\ dY_t = h_\theta(t, Y, X_t)dt + dW^\theta_t . \end{cases}$$

On supposera en outre que

$$a(t,y,x) = (ff^* + gg^*)(t,y,x) \ge \alpha I, \ \forall (t,y,x) \in \mathbb{R}_+ \times C(\mathbb{R}_+; \mathbb{R}^N) \times \mathbb{R}^M$$

et que le problème de martingales associé à l'équation (2.10) est bien posé.

A l'instant t, on dispose de l'observation $\{Y_s; 0 \le s \le t\}$. Le modèle statistique correspondant à ce problème est : $(\Omega_2, \mathcal{Y}_t, P_\theta, \theta \in \Theta)$. A chaque $\theta \in \Theta$, on associe la "loi conditionnelle non normalisée" de X_t sachant $\mathcal{Y}_t$, σ^θ_t. $\{\sigma^\theta_t; t \ge 0\}$ satisfait une équation de Zakai paramétrée par θ. On a le :

Théorème 2.6.1 *Le modèle statistique* $(\Omega_2, \mathcal{Y}_t, P_\theta, \theta \in \Theta)$ *est dominé, et une fonction de vraisemblance est donnée par :*

$$\theta \rightarrow \sigma^\theta_t(1) .$$

Preuve On pose :

$$Z^\theta_t = \exp\left(\int_0^t (h_\theta(s, Y, X_s), dY_s) - \frac{1}{2}\int_0^t \mid h_\theta(s, Y, X_s) \mid^2 ds\right) ,$$

$$V^\theta_t = \exp\left(\int_0^t (a^{-1}\overline{b}_\theta(s, Y, X_s), dX_s) - \frac{1}{2}\int_0^t \mid a^{-1/2}\overline{b}_\theta(s, Y, X_s) \mid^2 ds\right)$$

où $\bar{b}_\theta = b_\theta - g h_\theta$. Il résulte des hypothèses faites ci–dessus que $\{(Z_t^\theta, V_t^\theta); \ t \geq 0\}$ est une P_θ martingale. On définit alors la mesure suivante sur $(\Omega, \mathcal{F}_t)$ (dont il est facile de voir qu'elle ne dépend pas de θ) :

$$\frac{dQ}{dP_\theta|_{\mathcal{F}_t}} = (Z_t^\theta V_t^\theta)^{-1} \ .$$

Alors Q est une mesure dominante, et

$$
\begin{aligned}
\frac{dP_\theta|_{\mathcal{Y}_t}}{dQ} &= E_Q(V_t^\theta Z_t^\theta \,|\, \mathcal{Y}_t) \\
&= \overset{\circ}{E}_\theta \,(Z_t^\theta \,|\, \mathcal{Y}_t) \\
&= \sigma_t^\theta(1) \ .
\end{aligned}
$$

En effet $\overset{\circ}{P}_\theta \,|_{\mathcal{Y}_t} = Q$, c'est la mesure de processus de Wiener standard sur $C([0,t]; \mathbb{R}^N)$, et pour tout $\Lambda \in \mathcal{Y}_t$,

$$
\begin{aligned}
E_Q(V_t^\theta Z_t^\theta \Lambda) &= \overset{\circ}{E}_\theta \,(Z_t^\theta \Lambda) \\
&= \overset{\circ}{E}_\theta \,(\overset{\circ}{E}_\theta \,(Z_t^\theta \,|\, \mathcal{Y}_t) \Lambda) \\
&= E_Q(\overset{\circ}{E}_\theta \,(Z_t^\theta \,|\, \mathcal{Y}_t) \Lambda)
\end{aligned}
$$

$\square$

On vient de voir que la solution de l'équation de Zakai permet de calculer une vraisemblance dans un problème de statistique de processus partiellement observé. Pour certains algorithmes d'estimation on a même besoin des équations du lissage et du calcul stochastique non adapté, voir Campillo, Le Gland [15].

Chapitre 3

Equations aux dérivées partielles stochastiques. Applications à l'équation de Zakai

3.1 Equations d'évolution déterministes dans les espaces de Hilbert.

Nous allons établir quelques résultats qui nous serviront dans la suite. Signalons tout de suite que la nécessité de considérer plusieurs espaces de Hilbert inclus les uns dans les autres vient de ce que les opérateurs aux dérivées partielles que nous considérerons (et qui ont déjà été introduits au chapitre 2) sont des opérateurs *non bornés*. On se donne deux espaces de Hilbert V et H, avec $V \subset H, V$ dense dans H avec injection continue. On identifie H avec son dual. Alors H s'identifie à un sous ensemble de V'. Autrement dit, on a le schéma :

$$V \subset H \subset V',$$

on notera respectivement $\| \cdot \|, | \cdot |$ et $\| \cdot \|_*$ les normes dans V, H et V', par $(\cdot, \cdot)$ le produit scalaire dans H, et par $< \cdot, \cdot >$ le produit de dualité entre V et V'. On se donne enfin $T > 0$.

Dans la suite, V et H seront des espaces de Sobolev : $V = H^{s+1}(\mathbb{R}^M)$, $H = H^s(\mathbb{R}^M)$, $s \in \mathbb{R}$; le choix canonique étant $s = 0$ (alors $H = L^2(\mathbb{R}^M)$). La définition de ces espaces de Sobolev sera donnée à la section 3.4.

Lemme 3.1.1 *Soit $t \to u(t)$ une fonction absolument continue de $[0,T]$ dans V', t.q. en outre $u \in L^2(0,T;V)$ et $\frac{du}{dt} \in L^2(0,T;V')$. Alors $u \in C([0,T]; H)$, $t \to |u(t)|^2$ est absolument continue et*

$$\frac{d}{dt}|u(t)|^2 = 2 < u(t), \frac{du}{dt}(t) > \quad p.p. \ dans \ [0,T].$$

Preuve Il est facile de prolonger u à $\mathbb{R}$ de telle sorte que u soit à support compact, $u \in L^2(\mathbb{R};V)$, $\frac{du}{dt} \in L^2(\mathbb{R};V')$. En régularisant, on approche u par une suite $u_n \in C^1(\mathbb{R};V)$

à support compact t.q. $u_n \to u$ dans $L^2(\mathbb{R}; V)$ et $\frac{du_n}{dt} \to \frac{du}{dt}$ dans $L^2(\mathbb{R}; V')$. En outre,

$$|u_n(t)|^2 = 2 \int_{-\infty}^t (u_n(s), \frac{du_n}{ds}(s)) ds$$

$$= 2 \int_{-\infty}^t < u_n(s), \frac{du_n}{ds}(s) > ds,$$

on en déduit aisément que $u_n \to u$ dans $C(\mathbb{R}; H)$, et le résultat. $\square$

On considère maintenant un opérateur $A \in \mathcal{L}(V, V')$ t.q. $\exists \lambda$ et $\gamma > 0$ avec :

(3.1) $$< Au, u > + \lambda |u|^2 \geq \gamma \|u\|^2, \forall u \in V.$$

Théorème 3.1.2 *Sous l'hypothèse (3.1), si $u_0 \in H$ et $f \in L^2(0, T; V')$, alors l'équation suivante possède une unique solution :*

(3.2)
$$\begin{cases} (i) \quad u \in L^2(0, T; V), \\[2mm] (ii) \quad \dfrac{du}{dt} + Au(t) = f(t) \ p.p. \ dans \ (0, T), \\[2mm] (iii) \quad u(0) = u_0. \end{cases}$$

Preuve (esquisse) Les hypothèses (3.2–i) + (3.2–ii) entraînent, d'après le Lemme 3.1.1, que $u \in C([0, T]; H)$, ce qui fait que la condition (3.2–iii) a un sens.

Supposons qu'il existe une solution u à (3.2). Alors, d'après le Lemme 3.1.1,

$$|u(t)|^2 + 2 \int_0^t < Au(s), u(s) > ds = |u_0|^2 + 2 \int_0^t < f(s), u(s) > ds,$$

on utilise maintenant (3.1) et Cauchy–Schwarz :

$$|u(t)|^2 + 2\gamma \int_0^t \|u(s)\|^2 ds \leq |u_0|^2 + \frac{1}{\gamma} \int_0^t \|f(s)\|_*^2 ds +$$

$$+ \gamma \int_0^t \|u(s)\|^2 ds + 2\lambda \int_0^t |u(s)|^2 ds.$$

On tire alors du Lemme de Gronwall :

(3.3) $$|u(t)|^2 \leq \left(|u_0|^2 + \frac{1}{\gamma} \int_0^T \|f(t)\|_*^2 dt \right) e^{2\lambda T},$$

(3.4) $$\int_0^T \|u(t)\|^2 dt \leq \left(|u_0|^2 + \frac{1}{\gamma} \int_0^T \|f(t)\|_*^2 dt \right) \frac{e^{2\lambda T}}{\gamma}.$$

L'unicité résulte immédiatement de ces inégalités. Pour établir l'existence, on approche l'équation (3.2) (par exemple en dimension finie par une méthode de Galerkin), et on établit une estimation uniforme du type (3.3)–(3.4) pour la suite correspondante. Il reste à montrer que toute limite d'une sous suite convergeant faiblement est solution de (3.2). $\square$

Remarque 3.1.3 Le résultat du théorème serait encore vrai avec A dépendant du temps, pourvu que $A \in L^\infty(0, T; \mathcal{L}(V, V'))$ et que (3.1) soit satisfaite avec des constantes λ et γ indépendantes de t.

$\square$

3.2 Equations d'évolution stochastiques dans les espaces de Hilbert

On reprend le cadre ci-dessus, et on se donne un processus de Wiener standard N-dimensionel $(\Omega, \mathcal{F}, \mathcal{F}_t, P, W_t)$. Etant donné X un espace de Hilbert réel séparable, on pose :

$$M^2(0, T; X) \triangleq L^2((0, T) \times \Omega, \mathcal{P}, dP \times dt; X),$$
$$M^2(X) \triangleq \bigcap_{T>0} M^2(0, T; X)$$

où $\mathcal{P}$ désigne la tribu progressive sur $\mathbb{R}_+ \times \Omega$. Si $\varphi \in M^2(X^N)$, on définit l'intégrale stochastique :

$$\int_0^t \varphi_s \cdot dW_s = \sum_{i=1}^N \int_0^t \varphi_s^i dW_s^i, \quad t \geq 0$$

qui est une martingale continue de carré intégrable à valeurs dans X.

Si $u \in X$, $\psi \in M^2(X)$, $\varphi \in M^2(X^N)$, alors le processus $\{u_t, \ t \geq 0\}$ défini par :

$$u_t = u + \int_0^t \psi_s \, ds + \int_0^t \varphi_s \cdot dW_s$$

est une semi-martingale continue à valeurs dans X, et on a la formule d'Itô suivante pour la norme dans X au carré :

$$|u_t|^2 = |u|^2 + 2 \int_0^t (u_s, \psi_s) \, ds + 2 \int_0^t (u_s, \varphi_s) \cdot dW_s + \sum_{i=1}^N \int_0^t |\varphi_s^i|^2 ds \, .$$

On va maintenant généraliser à la fois cette formule d'Itô et le Lemme 3.1.1 (on reprend le cadre $V \subset H \subset V'$ de la section 3.1) :

Lemme 3.2.1 *Supposons donnés* $u \in H$, $\psi \in M^2(V')$ *et* $\varphi \in M^2(H^N)$ *tels le processus* $\{u_t\}$ *à valeurs dans* V' *défini par :*

$$u_t = u + \int_0^r \psi_s \, ds + \int_0^t \varphi_s \cdot dW_s, \quad t \geq 0$$

vérifie $u \in M^2(V)$. *Alors* $u \in \bigcap_{T>0} L^2(\Omega; C([0, T]; H))$, *et :*

$$(3.5) \qquad |u_t|^2 = |u|^2 + 2 \int_0^t <u_s, \psi_s> ds + 2 \int_0^t (u_s, \varphi_s) \cdot dW_s + \sum_{i=1}^d \int_0^t |\varphi_s^i|^2 ds \, .$$

Le lemme est une conséquence de la :

Proposition 3.2.2 *Soit* $A \in \mathcal{L}(V, V')$ *qui satisfait (3.1), u, φ et ψ donnés comme au Lemme 3.2.1. Alors l'équation suivante a une solution unique :*

$$(3.6) \qquad \begin{cases} \{u_t\} \in M^2(0, T; V), \\[2mm] du_t + Au_t \, dt = \psi_t \, dt + \varphi_t \cdot dW_t, \quad 0 \leq t \leq T, \\[2mm] u_0 = u \end{cases}$$

qui satisfait en outre les conclusions du Lemme 3.2.1.

Preuve Remarquons que l'existence de A résulte de faits élémentaires sur les applications de dualité. Pour déduire le Lemme de la Proposition, il suffit de vérifier que le processus $\{u_t\}$ du Lemme est solution de l'équation (3.6) avec $\{\psi_t\}$ remplacé par $\{\psi_t + Au_t\}$. L'unicité de la solution est une conséquence du Théorème 3.1.2.

L'existence est facile à démontrer si l'on suppose en outre que $\varphi \in M^2(0, T; V^d)$. Car alors $\{u_t\}$ résout l'équation (3.6) si et seulement si $\{\bar{u}_t\}$ défini par

$$\bar{u}_t = u_t - \int_0^t \varphi_s \cdot dW_s$$

résout l'équation :

$$\begin{cases} \bar{u} \in L^2(0, T; V) \text{ p.s. }, \\[2mm] \dfrac{d\bar{u}_t}{dt} + A\bar{u}_t = \psi_t - A\left[\int_0^t \varphi_s \cdot dW_s\right], \quad 0 \le t \le T, \\[2mm] \bar{u}_0 = u, \end{cases}$$

à laquelle on peut appliquer le Théorème 3.1.2. En outre, toujours avec l'hypothèse supplémentaire ci-dessus, (3.5) se déduit du Lemme 3.1.1 à l'aide d'une intégration par parties. Finalement, on montre à l'aide de (3.5) et de l'inégalité de Burkholder que $u \in M^2(0, T; V) \cap L^2(\Omega; C([0, T]; H))$. Soit enfin $\{\varphi_n\} \subset M^2(0, T; V^d)$ t.q. $\varphi_n \to \varphi$ dans $M^2(0, T; H^d)$. On déduit alors de (3.5) que la suite $\{u_t^n\}$ correspondante est de Cauchy dans $M^2(0, T; V) \cap L^2(\Omega; C([0, T]; H))$, et que sa limite $\{u_t\}$ est solution de (3.6). $\quad\square$

Remarquons que l'on dispose d'une formule d'Itô plus générale que celle du Lemme 3.2.1 (qui est vraie lorsque V est un espace de Banach):

Corollaire 3.2.3 *Supposons satisfaites les hypothèses du Lemme 3.2.1. Soit $\Phi : H \to \mathbb{R}$ une application deux fois Fréchet différentiable qui satisfait :*

(i) Φ, Φ' et Φ'' sont localement bornées.

(ii) Φ et Φ' sont continues de H à valeurs dans $\mathbb{R}$ et H respectivement

(iii) $\forall Q \in \mathcal{L}^1(H)$, $u \to Tr[Q\Phi''(u)]$ est continue de H dans $\mathbb{R}$.

(iv) Si $u \in V$, $\Phi'(u) \in V$, et Φ' est continue de "V fort" dans "V faible".

(v) $\exists k$ t.q. $\|\Phi'(u)\| \le k(1 + \|u\|)$, $u \in V$.

Alors :

$$\Phi(u_t) = \Phi(u) + \int_0^t <\Phi'(u_s), \psi_s> ds + \int_0^t (\Phi'(u_s), \varphi_s) \cdot dW_s$$
$$+ \frac{1}{2} \int_0^t (\Phi''(u_s)\varphi_s^i, \varphi_s^i) ds .$$

Preuve Le corollaire se démontre exactement comme le Lemme 3.2.1, à l'aide de l'approximation utilisée à la Proposition 3.2.2. □

Donnons maintenant un résultat plus général d'existence et d'unicité de la solution d'une EDPS de type parabolique. On se donne maintenant, outre $A \in \mathcal{L}(V, V')$, $B \in \mathcal{L}(V; H^N)$, $u \in H$, $f \in M^2(0, T; V')$ et $g \in M^2(0, T; H^N)$. On suppose satisfaite l'hypothèse de "coercivité" (ou "ellipticité") suivante : $\exists \lambda, \gamma > 0$ t.q.

$$(3.7) \qquad 2 < Au, u > + \lambda |u|^2 \geq \gamma \|u\|^2 + \sum_{i=1}^{N} |B_i u|^2, \; u \in V .$$

On a alors le :

Théorème 3.2.4 *Sous les hypothèses ci-dessus (en particulier la condition (3.7)), l'équation :*

$$(3.8) \qquad \begin{cases} \{u_t\} \in M^2(0, T; V) , \\[2mm] du_t + (Au_t + f_t) \, dt = (Bu_t + g_t) \cdot dW_t , \; 0 \leq t \leq T , \\[2mm] u_0 = u , \end{cases}$$

a une solution unique.

Preuve Il est clair que toute solution de (3.8) satisfait les conclusions du Lemme 3.2.1, donc en particulier si $\{u_t\}$ est une solution,

$$(3.9) \qquad |u_t|^2 + 2 \int_0^t < Au_s + f_s, u_s > ds =$$

$$= |u|^2 + 2 \int_0^t (Bu_s + g_s, u_s) \cdot dW_s + \sum_{i=1}^{N} \int_0^t |B_i u_s + g_s^i|^2 ds$$

d'où l'on tire, grâce à (3.7) :

$$\begin{aligned} E(|u_t|^2) + \gamma E \int_0^t \|u_s\|^2 ds \; \leq \; & |u|^2 + \lambda E \int_0^t |u_s|^2 ds \\ & + \frac{\gamma}{2} E \int_0^t \|u_s\|^2 ds + \frac{3}{\gamma} E \int_0^t \|f_s\|_*^2 ds + \\ & + (1 + \frac{3N}{\gamma}) \sum_{i=1}^{N} E \int_0^t |g_s^i|^2 ds, \; 0 \leq t \leq T \end{aligned}$$

et en utilisant le lemme de Gronwall :

$$E(|u_t|^2) \; \leq \; e^{\lambda T} \left(|u|^2 + cE \int_0^T \|f_t\|_*^2 dt + \bar{c} \sum_{i=1}^{N} E \int_0^T |g_t^i|^2 dt \right) ,$$

$$E \int_0^T \|u_s\|^2 ds \; \leq \; \frac{2e^{\lambda T}}{\gamma} \left(|u|^2 + cE \int_0^T \|f_t\|_*^2 dt + \bar{c} \sum_{i=1}^{N} E \int_0^T |g_t^i|^2 dt \right) .$$

On tire alors de (3.9), à l'aide de l'inégalité de Burkholder pour $p = 1$:

$$E(\sup_{t \leq T} |u_t|^2) \leq \tilde{c} \left(|u|^2 + c \int_0^T E\|f_t\|^2 dt + \bar{c} \sum_{i=1}^N E \int_0^T |g_t^i|^2 dt \right).$$

La démonstration se termine alors comme au Théorème 3.1.2. ☐

Remarque 3.2.5 L'unicité est vraie même si $\gamma = 0$ dans (3.7). ☐

Remarque 3.2.6 Soit $A \in L^\infty((0,T) \times \Omega, \mathcal{P}, dt \times dP; \mathcal{L}(V,V'))$ et $B \in L^\infty((0,T) \times \Omega, \mathcal{P}, dt \times dP; \mathcal{L}(V;H^N))$ qui satisfont la condition (3.7) avec des constantes λ et γ indépendantes de (t,ω). Alors le Théorème 3.2.4 se généralise aisément à cette situation. ☐

3.3 Application à l'équation de Zakai

Nous revenons à la situation de la section 2.2. On choisit

$$(\Omega, \mathcal{F}, \mathcal{F}_t, P, W_t) = (C(\mathbb{R}_+; \mathbb{R}^N), \mathcal{Y}, \mathcal{Y}_t, Q, Y_t),$$

où Q est la mesure de Wiener sur $C(\mathbb{R}_+; \mathbb{R}^N)$. Supposons que pour tout $(t,y) \in \mathbb{R}_+ \times C(\mathbb{R}_+; \mathbb{R}^N)$, les applications $x \to a^{jl}(t,y,x)$ et $x \to g^{li}(t,y,x)$, $1 \leq i,j \leq M$, $1 \leq l \leq N$, sont des éléments de l'espace de Sobolev $W^{1,\infty}(\mathbb{R}^M)$. C'est à dire que l'on suppose que $a^{ij}(t,y,\cdot)$ est bornée et que ses dérivées premières au sens des distributions sont également des fonctions bornées ; on supposera que toutes les bornes en question sont indépendantes de (t,y). Rappelons que tous les autres coefficients qui interviendront ci-dessous seront également supposés bornés. On peut réécrire les opérateurs L_{ty}, L_{ty}^i sous la forme :

$$L_{ty} = \frac{1}{2} \frac{\partial}{\partial x^l} \left[a^{jl}(t,y,x) \frac{\partial}{\partial x^j} \right] + \bar{b}^j(t,y,x) \frac{\partial}{\partial x^j},$$

$$L_{ty}^i = \frac{\partial}{\partial x^l}[g^{li}(t,y,x) \cdot] - \frac{\partial g^{li}}{\partial x^l}(t,y,x) + h^i(t,y,x)$$

avec $\bar{b}^j(t,y,x) = b^j(t,y,x) - \frac{1}{2} \frac{\partial a^{jl}}{\partial x^l}(t,y,x)$. On choisit maintenant $V = H^1(\mathbb{R}^M)$, $H = L^2(\mathbb{R}^M)$ (c'est le choix le plus naturel du point de vue des EDP !), où $H^1(\mathbb{R}^M)$ est l'espace de Sobolev :

$$H^1(\mathbb{R}^M) = \left\{ u \in L^2(\mathbb{R}^M); \ \frac{\partial u}{\partial x^j} \in L^2(\mathbb{R}^M), \ 1 \leq j \leq M \right\}.$$

Alors V' s'identifie avec l'espace de distributions $H^{-1}(\mathbb{R}^M)$ (la définition de $H^s(\mathbb{R}^M)$, s réel quelconque, sera donnée dans la section suivante).

Il est facile de voir que si L^* désigne l'adjoint de l'opérateur L et L^{i*} celui de l'opérateur L^i, alors

$$L^* \in L^\infty((0,T) \times \Omega; \mathcal{L}(V,V')) \text{ et}$$
$$L^{i*} \in L^\infty((0,T) \times \Omega; \mathcal{L}(V,H)), \ 1 \leq i \leq N, \ \forall T > 0.$$

Soit $u \in H^1(\mathbb{R}^M)$. Alors :

$$(3.10) \qquad -2 < L_{ty}^* u, u > - \sum_{i=1}^{N} |L_{ty}^{i*} u|^2 =$$

$$= \int_{\mathbb{R}^M} [a^{jl}(t,y,x) - (gg^*)^{jl}(t,y,x)] \frac{\partial u}{\partial x^j}(x) \frac{\partial u}{\partial x^i}(x) dx$$

$$+ \int_{\mathbb{R}^M} \alpha^j(t,y,x) \frac{\partial u}{\partial x^j}(x) u(x) \, dx + \int_{\mathbb{R}^M} \beta(t,y,x) \, u^2(x) \, dx$$

avec $\alpha^j = -2\bar{b}^j - \frac{\partial g^{li}}{\partial x^i} g^{ji}$, $\beta = - \sum_{i=1}^{N} \left(\frac{\partial g^{ji}}{\partial x^j} \right)^2$. Notons que $a - gg^* = ff^*$. Il en résulte que si l'on pose $A = -L_{t,y}^*$, $B_i = L_{t,y}^{i*}$, l'hypothèse (3.7) est satisfaite si et seulement si il existe $\delta > 0$ t.q.

$$(3.11) \qquad (ff^*)(t,y,x) \geq \delta I, \quad \forall (t,y,x) \in \mathbb{R}_+ \times C(\mathbb{R}_+; \mathbb{R}^N) \times \mathbb{R}^M .$$

Notons en outre que l'équation (Z) se réécrit formellement :

$$(Z) \quad \begin{cases} d\sigma_t - L_{tY}^* \sigma_t \, dt - L_{tY}^{i*} \sigma_t \, dY_t^i = 0, & t \geq 0, \\ \\ \sigma_0 = \Pi_0, \end{cases}$$

où Π_0 est la loi de X_0. Si la condition (3.11) est satisfaite, et si σ_0 admet une densité $p(0, \cdot)$ appartenant à $L^2(\mathbb{R}^M)$, alors l'équation (Z) possède une solution $p \in L^2(0,T; V) \cap C([0,T]; H)$, $\forall T > 0$, où pour tout t, $p(t, \cdot)$ est "candidat" à être la densité de la mesure σ_t. L'affirmation $p(t,x) = \frac{d\sigma_t}{dx}(x)$ relève alors d'un théorème d'unicité, pour l'équation (Z). On va établir un tel théorème sans l'hypothèse (3.11) (et sans supposer que σ_t possède une densité), mais sous des hypothèses de régularité des coefficients.

3.4   Un résultat d'unicité pour l'équation de Zakai.

Reprenons la formule (3.10), et supposons maintenant que pour tout $(t,y) \in \mathbb{R}_+ \times C(\mathbb{R}_+; \mathbb{R}^N)$,

$$x \to \alpha^j(t,y,x)$$

appartient à $C_b^1(\mathbb{R}^M)$. Alors, si $u \in H^1(\mathbb{R}^M)$, on a la formule d'intégration par parties suivante (qui se vérifie en approchant u par des fonctions à support compact) :

$$\int_{\mathbb{R}^M} \alpha^j(t,y,x) \frac{\partial u}{\partial x^j}(x) u(x) \, dx = \frac{1}{2} \int_{\mathbb{R}^M} \alpha^j(t,y,x) \frac{\partial}{\partial x^j}(u^2)(x) \, dx$$

$$= -\frac{1}{2} \int_{\mathbb{R}^M} \frac{\partial \alpha^j}{\partial x^j}(t,y,x) u^2(x) \, dx .$$

Supposons que :

$$(3.12) \qquad \lambda = \sup_{(t,y,x)} \left\{ \frac{1}{2} \left| \frac{\partial \alpha^j}{\partial x^j}(t,y,x) \right| + |\beta(t,y,x)| \right\} < \infty,$$

alors (puisque $ff^* \geq 0$) :

$$(3.13) \qquad -2 < L_{ty} u, u > - \sum_{i=1}^{N} |L_{ty}^{i*} u|^2 + \lambda |u|^2 \geq 0, \quad \forall u \in H^1(\mathbb{R}^M) .$$

On déduit de la Remarque 3.2.5 que sous l'hypothèse (3.12) l'équation de Zakai possède au plus une solution dans $M^2(H^1(\mathbb{R}^M))$. Cependant, sans l'hypothèse d' "uniforme ellipticité" (3.11), il n'y a aucune raison pour que la loi conditionnelle non normalisée possède une densité dans $M^2(0, T; H^1(\mathbb{R}^M))$. Il est donc nécessaire d'établir un résultat d'unicité dans un espace plus gros.

Introduisons pour cela quelques nouvelles notions d'analyse fonctionnelle. Dans la suite s désigne un réel quelconque. On note $\mathcal{S}'(\mathbb{R}^M)$ l'espace des distributions tempérées. Pour $u \in \mathcal{S}'(\mathbb{R}^M)$, on note $\hat{u}$ sa transformée de Fourier. Si $u \in \mathcal{S}'(\mathbb{R}^M)$, et $s \in \mathbb{R}$, on définit $\Lambda_s u \in \mathcal{S}'(\mathbb{R}^M)$ par

$$\widehat{\Lambda_s u}(\xi) = (1 + |\xi|^2)^{\frac{s}{2}} \hat{u}(\xi),$$

et $H^s(\mathbb{R}^M) = \{u \in \mathcal{S}'(\mathbb{R}^M), \Lambda_s u \in L^2(\mathbb{R}^M)\}$. $H^s(\mathbb{R}^M)$, muni de la norme $\|f\|_s = \|\Lambda_s f\|_0$ ($\| \cdot \|_0$ désigne la norme usuelle de $L^2(\mathbb{R}^M)$), est un espace de Hilbert. Posons $V = H^{s+1}(\mathbb{R}^M), H = H^s(\mathbb{R}^M)$. Identifier H à son dual revient à identifier le dual V' de $H^{s+1}(\mathbb{R}^M)$ à $H^{s-1}(\mathbb{R}^M)$ à l'aide du produit de dualité :

$$
\begin{aligned}
< u, v >_s &= (\Lambda_{s+1} u, \Lambda_{s-1} v), u \in H^{s+1}(\mathbb{R}^M), v \in H^{s-1}(\mathbb{R}^M) \\
&= < \Lambda_s u, \Lambda_s v >_0 .
\end{aligned}
$$

La clé de notre résultat est la Proposition suivante, qui étend la remarque ci–dessus disant que (3.12) $\Rightarrow$ (3.13) :

Proposition 3.4.1 *Supposons que tous les coefficients du système différentiel stochastique (2.2) sont de classe C^∞ en x, et uniformément bornés ainsi que toutes leurs derivées. Alors, pour tout s réel,*

$$
\begin{aligned}
L, L^* &\in L^\infty(\mathbb{R}_+ \times \Omega_2; \mathcal{L}(H^{s+1}(\mathbb{R}^M), H^{s-1}(\mathbb{R}^M))) , \\
L^{i*} &\in L^\infty(\mathbb{R}_+ \times \Omega_2; \mathcal{L}(H^{s+1}(\mathbb{R}^M), H^s(\mathbb{R}^M)))
\end{aligned}
$$

et il existe un réel λ_s t.q. pour tout $(t, y) \in \mathbb{R}_+ \times C(\mathbb{R}_+; \mathbb{R}^N)$ et $u \in H^{s+1}(\mathbb{R}^M)$,

$$(3.14) \qquad -2 < L_{ty}^* u, \, u >_s - \sum_{i=1}^{N} \|L_{ty}^{i*} u\|_s^2 + \lambda_s \|u\|_s^2 \geq 0 .$$

Avant d'examiner la preuve de la Proposition 3.4.1, voyons comment on en déduit le résultat d'unicité cherché. On introduit la notation :

$$\mathcal{W}_T^s = M^2(0, T; H^s(\mathbb{R}^M)) \cap L^2(\Omega; C([0, T]; H^{s-1}(\mathbb{R}^M)))$$

Il résulte alors du Lemme 3.2.1 et de la Remarque 3.2.5, avec $V = H^{s+1}(\mathbb{R}^M), H = H^s(\mathbb{R}^M)$:

Corollaire 3.4.2 *Sous les hypothèses de la Proposition 3.4.1, pour tout $s \in \mathbb{R}$, l'équation (Z) possède au plus une solution dans $\cap_{T>0} \mathcal{W}_T^s$ dont la valeur en $t = 0$ soit un élément donné de $H^s(\mathbb{R}^M)$.* $\square$

Il reste à trouver s tel que $\Pi_0 \in H^s(\mathbb{R}^M)$ et $\sigma \in \mathcal{W}_T^s$. Notons que si $s < -\frac{M}{2}$,

$$
\begin{aligned}
\|\sigma_t\|_s^2 &= \int_{\mathbb{R}^M} (1 + |\xi|^2)^s |\hat{\sigma}_t(\xi)|^2 d\xi \\
&\leq \sup_{\xi \in \mathbb{R}^M} |\hat{\sigma}_t(\xi)|^2 \int_{\mathbb{R}^M} (1 + |\xi|^2)^s d\xi .
\end{aligned}
$$

Or $\hat{\sigma}_t(\xi) = \overset{\circ}{E}\left(e^{i\xi.X_t}Z_t/\mathcal{Y}_t\right)$

$$|\hat{\sigma}_t(\xi)| \leq \overset{\circ}{E}\left(Z_t/\mathcal{Y}_t\right)$$

Donc

$$\|\sigma\|^2_{M^2(0,T;H^s(\mathbb{R}^M))} \leq CT\,\overset{\circ}{E}\left(\sup_{t\leq T} Z_t^2\right),$$

s étant toujours fixé ($s < -\frac{M}{2}$), il résulte de cette estimation, du fait que σ satisfait l'équation (Z), de la première partie de la Proposition 3.4.1 et du Lemme 3.2.1 que $\sigma \in \cap_{T>0}\mathcal{W}_T^s$. On a donc démontré le :

Théorème 3.4.3 *Pour tout $s < -\frac{M}{2}$, sous les hypothèses de la Proposition 3.4.1 le processus "loi conditionnelle non normalisée" $\{\sigma_t,\ t \geq 0\}$ est l'unique solution de l'équation (Z) dans $\cap_{T>0}\mathcal{W}_T^s$.* $\qquad\square$

Il nous reste à donner des indications sur la :

Preuve de la Proposition 3.4.1 La première partie du résultat résulte de ce que $\frac{\partial}{\partial x^i}$ envoie $H^{s+1}(\mathbb{R}^M)$ dans $H^s(\mathbb{R}^M)$ et $\frac{\partial^2}{\partial x^i \partial x^j}$ envoie $H^{s+1}(\mathbb{R}^M)$ dans $H^{s-1}(\mathbb{R}^M)$, et d'autre part la multiplication par une fonction de $C_b^\infty(\mathbb{R}^M)$ est un opérateur borné dans chaque $H^s(\mathbb{R}^M)$.

a. *Cas s entier positif pair*

Ce n'est pas le cas qui nous intéresse, mais c'est le cas où tout est élémentaire. Dans ce cas Λ_s est un opérateur aux dérivées partielles :

$$\Lambda_s = (I - \Delta)^{\frac{s}{2}}.$$

On va utiliser ci-dessous la notation $[A, B] = AB - BA$. Remarquons que si A, B sont des opérateurs aux dérivées partielles à coefficients C_b^∞ d'ordre respectivement a et b, alors $[A, B]$ est un opérateur aux dérivées partielles d'ordre au plus $a + b - 1$. Il reste à établir (3.14). Soit $u \in H^{s+1}(\mathbb{R}^M)$.

$$2 < L_{ty}^* u,\, u >_s + \sum_{i=1}^N \|L_{ty}^{i*}u\|_s^2 =$$

$$= 2 < \Lambda_s L_{ty}u,\, \Lambda_s u >_0 + \sum_{i=1}^N |\Lambda_s L_{ty}^{i*}u|^2$$

$$= 2 < L_{ty}^*\Lambda_s u,\, \Lambda_s u >_0 + \sum_{i=1}^N |L_{ty}^{i*}\Lambda_s u|^2 + 2([\Lambda_s,\, L_{ty}^*]u,\, \Lambda_s u) +$$

$$+ \sum_{i=1}^N |[\Lambda_s,\, L_{ty}^{i*}]u|^2 + 2(L_{ty}^{i*}\Lambda_{su},\, [\Lambda_s,\, L_{ty}^{i*}]u)$$

$$\leq \lambda_s\|u\|_s^2 + 2([\Lambda_s,\, L_{ty}^*]u,\, \Lambda_s u) + 2(\Lambda_{ty}^i[\Lambda_s,\, L_{ty}^{i*}]u,\, \Lambda_s u).$$

On vient d'utiliser (3.13) et le fait que $[\Lambda_s,\, L_{ty}^{i*}]$ est un opérateur d'ordre $s + 1 - 1 = s$. On va utiliser ci-dessous le fait que Λ_s est autoadjoint, est une bijection de $H^{s+\alpha}(\mathbb{R}^M)$

sur $H^{\alpha}(\mathbb{R}^M)$ pour tout α réel, et que $\Lambda_{-s} = (\Lambda_s)^{-1}$. Supposons un instant que $u \in H^{2s+1}(\mathbb{R}^M)$.

$$
\begin{aligned}
([\Lambda_s, L^*]u, \Lambda_s u) &= (u, [L, \Lambda_s]\Lambda_s u) \\
&= (u, \Lambda_s[L, \Lambda_s]u) + (u, [[L, \Lambda_s], \Lambda_s]u) \\
&= (\Lambda_s u, [L, \Lambda_s]u) + (\Lambda_s u, \Lambda_{-s}[[L, \Lambda_s], \Lambda_s]u) .
\end{aligned}
$$

Par ailleurs,

$$([\Lambda_s, L^*]u, \Lambda_s u) = (\Lambda_s u, [-L^*, \Lambda_s]u) .$$

Donc

$$([\Lambda_s, L^*]u, \Lambda_s u) = \frac{1}{2}(\Lambda_s u, [L - L^*, \Lambda_s]u) + \frac{1}{2}(\Lambda_s u, \Lambda_{-s}[[L, \Lambda_s], \Lambda_s]u) .$$

Mais $L - L^*$ est un opérateur aux dérivées partielles d'ordre 1, donc $[L - L^*, \Lambda_s]$ est d'ordre s. En outre, $[[L, \Lambda_s], \Lambda_s]$ est d'ordre $2s$. Donc :

$$|\Lambda_{-s}[[L, \Lambda_s], \Lambda_s]u| \le c\|u\|_s .$$

Finalement,

$$2([\Lambda_s, L^*]u, \Lambda_s u) \le \gamma_s \|u\|_s^2$$

inégalité qui s'étend à tout $u \in H^{s+1}(\mathbb{R}^M)$. Enfin

$$(L^i[\Lambda_s, L^{i*}]u, \Lambda_s u) = (\Lambda_s u, \Lambda_{-s}[L^i, \Lambda_s]L^{i*}\Lambda_s u) .$$

Or $\Lambda_{-s}[L^i, \Lambda_s]$ est un opérateur borné, et $T^i = \Lambda_{-s}[L^i, \Lambda_s] L^{i*}$ est le produit d'un opérateur différentiel d'ordre 1 et d'un opérateur borné, donc $T^i + T^{i*}$ est un opérateur borné, d'où

$$
\begin{aligned}
(L^i[\Lambda_s, L^{i*}]u, \Lambda_s u) &= ((T^i + T^{i*})\Lambda_s u, \Lambda_s u) \\
&\le c \| u \|_s .
\end{aligned}
$$

b. *Cas général*

Dans le cas général, Λ_s n'est pas un opérateur aux dérivées partielles, mais un opérateur pseudo-différentiel d'ordre s. La même démonstration marche cependant, en utilisant des propriétés élémentaires des opérateurs pseudo-différentiels.

3.5 Un résultat de régularité pour l'équation de Zakai

Dans cette section, nous allons supposer satisfaites à la fois l'hypothèse de régularité des coefficients faite à la section précédente, et l'hypothèse de "forte ellipticité" (3.11). On obtient alors, au lieu de (3.13),

$$-2 < L_{ty}u, u > - \sum_{i=1}^{N} |L_{ty}^{i*}u|^2 + (\lambda + \delta)|u|^2 \ge \delta\|u\|_1^2, \ \forall u \in H^1(\mathbb{R}^M),$$

dont on déduit par le raisonnement de la Proposition 3.4.1 :

Proposition 3.5.1 *Pour tout réel s, il existe un réel λ_s tel que pour tout $(t,y) \in \mathbb{R}_+ \times C(\mathbb{R}_+; \mathbb{R}^M)$ et tout $u \in H^{s+1}(\mathbb{R}^M)$,*

$$-2 < L_{ty}^* u, u >_s + \lambda_s \|u\|_s^2 \geq \delta \|u\|_{s+1}^2 + \sum_{i=1}^N \left\| L_{ty}^{i*} u \right\|_s^2$$

□

Il résulte de cette proposition que la suite des opérateurs

$$-L_{ty}^*, \ -L_{ty}^{1*}, \ldots, \ -L_{ty}^{N*}$$

satisfait la condition (3.7) avec $V = H^{s+1}(\mathbb{R}^M)$, $H = H^s(\mathbb{R}^M)$, $V' = H^{s-1}(\mathbb{R}^M)$, pour tout s réel. $\{\sigma_t; t \geq 0\}$ désignant à nouveau l'unique solution de l'équation (Z) (au sens du Théorème 3.4.3), posons

$$\sigma_t^{(l)} = t^l \sigma_t, \quad t \geq 0, \quad l \in \mathbb{N}^*.$$

Alors $\sigma_t^{(l)}$ est solution de l'EDPS :

$$(3.15) \quad \begin{cases} d\sigma_t^{(l)} = L_{ty}^* \sigma_t^{(l)} dt + l\, \sigma_t^{(l-1)} dt + L_{ty}^{i*} \sigma_t^{(l)} dY_t^i, \\ \sigma_0^{(l)} = 0. \end{cases}$$

On a alors le

Théorème 3.5.2 $\sigma \in C(\mathbb{R}_+^*; H^l(\mathbb{R}^M))$, *pour tout $l \in \mathbb{N}$.*

qui est un simple corollaire de la :

Proposition 3.5.3 *Pour tout $l \in \mathbb{N}$, $s < -M/2$, $T > 0$:*

$$\sigma^{(l)} \in M^2(0, T; H^{s+2l}(\mathbb{R}^M)) \cap L^2(\Omega, C([0,T]; H^{s+2l-1}(\mathbb{R}^M)).$$

Preuve Par récurrence. Il suffit d'utiliser le Théorème 3.4.3, la Proposition 3.5.1 et Théorème 3.2.4. □

Remarque 3.5.4 Nous venons de donner en particulier une condition sous laquelle σ_t possède une densité de classe C_b^∞ (le Théorème d'injection de Sobolev, voir Adams [1], entraîne que $\cap_{l \in \mathbb{N}} H^l(\mathbb{R}^M) \subset C_b^\infty(\mathbb{R}^M)$), pour tout $t > 0$. On verra un résultat analogue sous une hypothèse plus faible que (3.11), mais avec une dépendance des coefficients beaucoup moins arbitraire en (t,y), à l'aide du calcul de Malliavin au chapitre 5. □

3.6 Commentaires bibliographiques

a. Les résultats sur les EDPS que nous avons présentés sont un cas très particuliers des résultats sur les EDPS paraboliques de Pardoux [68] et Krylov–Rosovskii [45] (voir aussi l'article de revue de Pardoux [72] et le livre de Rosovskii [80]). Les résultats contenus dans ces références sont plus généraux d'une part dans la mesure où les opérateurs A et $B_1, \ldots, B_N$ peuvent être non linéaires, à condition de satisfaire une hypothèse de monotonie du type :

$$2 < A(u) - A(v), u - v > + \lambda |u - v|^2 \geq \sum_{i=1}^N |B_i(u) - B_i(v)|^2$$

et d'autre part dans la mesure où le processus de Wiener directeur peut être de dimension infinie, à condition toutefois d'être de covariance nucléaire, ce qui exclut le cas d'un "bruit blanc spatio-temporel" considéré par exemple dans Walsh [85]. Notons en outre que Gyöngy [32] a considéré le cas d'EDPS paraboliques dirigées par une semi-martingale non nécessairement continue.

b. Le résultat d'unicité pour l'équation de Zakai que nous venons de présenter est dû à Chaleyat-Maurel, Michel, Pardoux [20]. L'intérêt de ce résultat est d'être applicable dans le cas où les coefficients dépendent de façon arbitraire du passé de l'observation. Malheureusement, il impose de supposer tous les coefficients bornés et réguliers en x. Dans le cas où les coefficients ne dépendent pas de l'observation (ou à la rigueur de Y_t seulement), deux autres techniques ont été mises au point pour établir l'unicité de la solution de l'équation de Zakai, dans la classe des processus à valeurs mesures. La première de ces méthodes, dûe indépendamment à Rozovskii [80] et Bensoussan [8] repose essentiellement sur la dualité, qui permet de déduire l'unicité de la solution d'une EDP de l'existence d'une solution à l'équation adjointe. En outre, plus la solution de l'équation adjointe est régulière, plus le résultat d'unicité est vrai dans une large classe de processus. Enfin, grâce à la technique utilisée ci-dessus dans la preuve de la Proposition 2.5.6, il suffit d'étudier une équation "adjointe" déterministe. Signalons que le même type de technique de dualité peut être utilisé pour les problèmes de martingales (voir Ethier, Kurtz [26]). La deuxième méthode, dûe à Kurtz, Ocone [50], utilise la notion de "problème de martingale filtré", et permet d'étudier l'unicité aussi bien de l'équation de Zakai ou de Kushner-Stratonovich. Si $\{A_t, \ t \geq 0\}$ désigne le générateur infinitésimal du processus de Markov $\{(X_t, Y_t), \ t \geq 0\}$,

$$\Pi_t \varphi(\cdot, Y_t) - \int_0^t \Pi_s A_s \varphi(\cdot, Y_s) ds$$

est une $\mathcal{Y}_t$-martingale, pour toute fonction $\varphi \in C_c^2(\mathbb{R}^M \times \mathbb{R}^N)$, qui satisfait en outre :

$$E[\Pi_0 \varphi(\cdot, Y_0)] = E[\varphi(X_0, Y_0)] .$$

Notons que si ce "problème de martingale filtré" possède une seule solution, alors la loi du couple (Π, Y) est entièrement caractérisée, et donc aussi Π comme fonction de Y. Soit $\{\mu_t\}$ une solution de l'équation de Kushner-Stratonovich. Alors le couple (μ, Y) est une solution du problème de martingale filtré. De l'unicité du problème de martingale filtré résulte alors le fait que $\mu = \Pi$. Le résultat de Kurtz-Ocone s'applique en particulier au modèle suivant :

$$\begin{cases} dX_t = f(X_t)dt + g(X_t)dV_t \,, \\[2mm] dY_t = h(X_t)dt + GdV_t + \bar{G}dW_t \end{cases}$$

avec f, g globalement lipschitziens, h continue telle $E \int_0^T |h(X_t)|^2 dt$, $T > 0$, $\bar{G}\bar{G}^* > 0$, $\{V_t\}$ et $\{W_t\}$ sont des Wieners standard mutuellement indépendants et indépendants de X_0 .

Chapitre 4

Continuité du filtre par rapport à l'observation

4.1   Introduction

Nous avons défini au chapitre 2 deux processus à valeurs mesures $\{\Pi_t\}$ et $\{\sigma_t\}$. Ces processus, ainsi que les processus $\{\Pi_t(\varphi)\}$, $\{\sigma_t(\varphi)\}$ ($\varphi \in C_b(\mathbb{R}^M)$) sont des fonctions de l'observation, i.e. sont définis sur l'espace de probabilité $(\Omega_2, \mathcal{Y}, P^Y)$. Pour l'instant, ils sont définis soit comme une projection optionnelle (ou à t fixé comme une espérance conditionnelle) soit comme la solution d'une EDP stochastique. Dans tous les cas, ils ne sont définis que P^Y p.s. Or en pratique on voudrait évaluer la valeur qu'ils prennent en *une* trajectoire du processus observé. Pour qu'une telle évaluation ait un sens, il faudrait disposer d'une version continue de

$$y \to \sigma_t(\varphi, y) \ .$$

On va voir que ceci est possible lorsque le signal et le bruit d'observation sont indépendants, la fonction h étant régulière et bornée. On indiquera ensuite une généralisation de ce résultat dans le cas où le processus d'observation est scalaire. Enfin on donnera un résultat de continuité plus faible qui est vrai beaucoup plus généralement.

4.2   Le cas où signal et bruit d'observation sont indépendants.

On reprend le modèle (2.2) du Chapitre 2, en supposant que $g \equiv 0$ et h ne dépend pas de l'observation, i.e. :

$$(4.1) \quad \begin{cases} X_t = X_0 + \displaystyle\int_0^t b(s, Y, X_s)\, ds + \int_0^t f(s, Y, X_s)\, dV_s \ , \\[2mm] Y_t = \displaystyle\int_0^t h(s, X_s)\, ds + W_t \end{cases}$$

et on suppose en outre que $h \in C^{1,2}(\mathbb{R}_+ \times \mathbb{R}^M)$, $h, A^1 h, \ldots, A^M h$ et $\frac{\partial h}{\partial t} + Lh$ sont bornés sur $[0, t] \times C([0, t]; \mathbb{R}^N) \times \mathbb{R}^M$ quelque soit t, et que $y \to (b(s, y, x), f(s, y, x))$ est localement lipschitzienne, uniformément par rapport à $(s, x) \in [0, t] \times \mathbb{R}^M$, quelque soit t.

On remarque alors que

$$
\begin{aligned}
Z_t &= \exp\left[\int_0^t (h(s, X_s), dY_s) - \frac{1}{2}\int_0^1 | h(s, X_s) |^2 \, ds\right] \\
&= \exp\left[(h(t, X_t), Y_t) - \int_0^t \left(\left(\frac{\partial h}{\partial s} + L_{sY} h\right)(s, X_s), Y_s\right) ds \right. \\
&\quad \left. - \int_0^t (A_{sY}^l h(s, X_s), Y_s) dV_s^l - \frac{1}{2}\int_0^t | h(s, X_s) |^2 \, ds\right].
\end{aligned}
$$

Pour $y \in C(\mathbb{R}_+; \mathbb{R}^N)$, posons :

$$
\begin{aligned}
Z_t(y) &= \exp\left[(h(t, X_t), y(t)) - \int_0^t \left(\left(\frac{\partial h}{\partial s} + L_{sy} h\right)(s, X_s), y(s)\right) ds \right. \\
&\quad \left. - \int_0^t (A_{sy}^l h(s, X_s), y(s)) dV_s^l - \frac{1}{2}\int_0^t | h(s, X_s) |^2 \, ds\right].
\end{aligned}
$$

On définit alors deux collections de mesures sur $\mathbb{R}^M$, indexées par $(t, y) \in \mathbb{R}_+ \times C(\mathbb{R}_+; \mathbb{R}^N)$:

$$
\begin{aligned}
\sigma_t(y, \varphi) &= E(\varphi(X_t) Z_t(y)), \quad \varphi \in C_b(\mathbb{R}^M), \\
\Pi_t(y, \varphi) &= \sigma_t(y, 1)^{-1} \sigma_t(y, \varphi), \quad \varphi \in C_b(\mathbb{R}^M).
\end{aligned}
$$

On constate alors que les processus $\sigma_t(\varphi)$ et $\sigma_t(Y, \varphi)$ (resp $\Pi_t(\varphi)$ et $\Pi_t(Y, \varphi)$) sont indistinguables. On déduit des hypothèses faites sur h et du théorème de convergence dominée :

Proposition 4.2.1 *Pour tout $t > 0$ l'application $y \to \sigma_t(y, \cdot)$ (resp $\Pi_t(y, \cdot)$) est continue de $C([0, t]; \mathbb{R}^N)$ dans $\mathcal{M}_+(\mathbb{R}^M)$ muni de la convergence étroite. En outre si $\varphi \in C_b(\mathbb{R}^M)$, l'application $y \to \sigma_t(y, \varphi)$ (resp $\Pi_t(y, \varphi)$) est localement lipschitzienne de $C([0, t]; \mathbb{R}^M)$ dans $\mathbb{R}$.* □

Il y a une autre façon de définir $\sigma_t(y, \cdot)$. Supposons maintenant que tous les coefficients sont bornés. Réécrivons l'équation de Zakai dans le cas particulier considéré dans cette section.

$$
(Z) \qquad \sigma_t = \sigma_0 + \int_0^t L_{sY}^* \sigma_s \, ds + \int_0^t h_i(s, \cdot) \sigma_s dY_s^i.
$$

(Z) est une EDS dans l'espace $\mathcal{M}_+(\mathbb{R}^M)$ des mesures finies sur $\mathbb{R}^M$. Les coefficients de diffusion de cette équation sont les applications linéaires de $\mathcal{M}_+(\mathbb{R}^M)$ dans lui-même :

$$
\mu \to H_{t,i}(\mu); \ t \geq 0, \ 1 \leq i \leq N
$$

définies par :

$$
H_{t,i}(\mu)(\varphi) = \mu(h_i(t, \cdot)\varphi).
$$

Ces applications $\{H_{t,i};\ t \geq 0,\ 1 \leq i \leq N\}$ commutent entre elles. On va donc pouvoir appliquer à l'équation (Z) la "transformation de Doss-Sussmann", qui permet de résoudre (Z) trajectoire par trajectoire, et de construire une version de la solution qui est continue en Y, voir Doss [23], Sussmann [82]. En fait dans notre cas la transformation de Doss-Sussmann est explicite. Définissons un nouveau processus à valeurs mesure $\{\overline{\sigma}_t\}$ par :

$$\frac{d\overline{\sigma}_t}{d\sigma_t}(x) = \exp(-h_i(t,x)Y_t^i)\ .$$

$\{\overline{\sigma}_t\}$ satisfait une version continue de l'équation de Zakai, communément appelée "équation de Zakai robuste " .

Posons

$$\overline{L}_{sy} = L_{sy} - \frac{1}{2}\sum_{i=1}^{N} h_i^2(s,\cdot)\ ,$$

$$c(s,y,x) = \left(\frac{\partial h_i}{\partial s}(s,x) + L_{sy}h_i(s,x)\right)y^i(s) - \frac{1}{2}\sum_{l=1}^{M}\mid A_{sy}^l h_i(s,x)y^i(s)\mid^2\ .$$

On a alors le :

Théorème 4.2.2 *Pour tout* $\varphi \in C_b^2(\mathbb{R}^M)$,

$$(ZR) \qquad \overline{\sigma}_t(\varphi) = \sigma_0(\varphi) + \int_0^t \overline{\sigma}_s(\overline{L}_{sY})\,ds - \int_0^t \overline{\sigma}_s(Y_s^j A_{sY}^l h_j(s,\cdot)A_{sY}^l\varphi + c(s,Y,\cdot)\varphi)\,ds\ .$$

Preuve On reprend la démonstration du Théorème 2.2.3

$$Z_t\varphi(X_t) = \varphi(X_0) + \int_0^t Z_s L_{sY}\varphi(X_s)\,ds$$

$$+ \int_0^t Z_s A_{sY}^l\varphi(X_s)dV_s^l + \int_0^t Z_s h_i(s,X_s)\varphi(X_s)dY_s^i\ ,$$

$$e^{-h_i(t,X_t)Y_t^i} = 1 - \int_0^t e^{-h_i(s,X_s)Y_s^i}c(s,Y,X_s)\,ds$$

$$- \int_0^t e^{-h_i(s,X_s)Y_s^i}Y_s^j A_{sY}^l h_j(s,X_s)dV_s^l - \int_0^t e^{-h_i(s,X_s)Y_s^i}h_j(s,X_s)dY_s^j$$

$$+ \frac{1}{2}\int_0^t e^{-h_i(s,X_s)Y_s^i}\left(\sum_{j=1}^{N} h_j^2(s,X_s)\right)ds\ .$$

Donc, à nouveau par la formule d'Itô :

$$e^{-h_i(t,X_t)Y_t^i}Z_t\varphi(X_t) = \varphi(X_0) + \int_0^t e^{-h_i(s,X_s)Y_s^i}Z_s\overline{L}_{sY}\varphi(X_s)\,ds$$

$$- \int_0^t e^{-h_i(s,X_s)Y_s^i}Z_s[Y_s^j A_{sY}^l h_j(s,X_s)A_{sY}^l\varphi(X_s) + c(s,Y,X_s)]\,ds$$

$$+ \int_0^t e^{-h_i(s,X_s)Y_s^i}Z_s[A_{sY}^l\varphi(X_s) - Y_s^j(A_{sY}^l h_j(s,X_s))\varphi(X_s)]dV_s^l\ .$$

Il reste à appliquer $\overset{o}{E}{}^{y}(\cdot)$ aux deux membres de la dernière égalité. □

On peut maintenant réécrire l'équation (ZR) pour chaque trajectoire fixée du processus d'observation, ce qui donne une EDP *déterministe* paramétrée par $y \in C(\mathbb{R}_+; \mathbb{R}^N)$. Sous les hypothèses de la Proposition 3.4.1, il existe un unique

$$\overline{\sigma}_{\cdot,y} \in L^2_{loc}(\mathbb{R}_+; H^s(\mathbb{R}^M)) \cap C(\mathbb{R}_+; H^{s-1}(\mathbb{R}^M))$$

$(s < -M/2)$ solution de :

$$\overline{\sigma}_{t,y} = \sigma_0 + \int_0^t \overline{L}_{sy}\overline{\sigma}_{sy}\,ds - \int_0^t y^j(s)(A^l_{sy})^*[\overline{\sigma}_{s,y}(A^l_{sy}h_j(s,\cdot))]\,ds$$
$$- \int_0^t \overline{\sigma}_{s,y}(c(s,y,\cdot))\,ds .$$

On peut alors redéfinir $\sigma_t(y,\cdot)$ par :

$$\sigma_t(y,\varphi) = \overline{\sigma}_{t,y}(e^{h_i(t,\cdot)v^i_t}\varphi) .$$

Notons que, à l'aide des techniques du Théorème 3.4.3 appliquées à l'équation ci-dessus, on peut établir le :

Théorème 4.2.3 *On suppose satisfaites les hypothèses du début de cette section, et en outre que tous les coefficients sont de classe C^∞ en x, les coefficients et toutes leurs dérivées étant bornés. Alors pour tout $T > 0$, $s < -M/2$, l'application*

$$y \to \sigma_\cdot(y,\cdot)$$

est localement lipschitzienne de

$$C([0,T]; \mathbb{R}^N) \text{ dans } L^2(0,T; H^s(\mathbb{R}^M)) \cap C([0,T]; H^{s-1}(\mathbb{R}^M))$$

□

4.3 Extension du résultat de continuité.

Les résultats de la section 4.2 ont été obtenus sous deux hypothèses contraignantes : la bornitude de h, et l'indépendance du signal et du bruit d'observation ($g \equiv 0$).

Notons que l'intégration par parties que nous avons faite dans l'argument de l'exponentielle fait apparaître le terme :

$$-\frac{\partial h}{\partial s} - Lh - \frac{1}{2}h^2 .$$

Donc si h^2 domine $|\frac{\partial h}{\partial s} + Lh|$, une partie au moins de l'exponentielle sera bornée.

Cette remarque permet de traiter (toujours avec $g \equiv 0$) en particulier le cas où b et h sont à croissance au plus linéaire, et f bornée, voir en particulier Baras, Blankenship, Hopkins [3] et Pardoux [69].

En ce qui concerne l'hypothèse $g \equiv 0$, elle est cruciale pour que les coefficients de diffusion de l'équation de Zakai commutent entre eux. Cette restriction ne devrait donc pas être nécessaire lorsque l'observation est scalaire. C'est aussi le principal cas où l'on sait obtenir un résultat de continuité pour une large classe de fonctionnelles h non nécessairement à croissance linéaire.

La continuité du filtre dans le cas d'une observation scalaire a été étudiée dans le cas h non bornée par Sussmann [82] et dans le cas $g \not\equiv 0$ par Davis, Spathopoulos [22]. Florchinger [30] a combiné ces deux résultats. Enonçons le résultat de Florchinger. On reprend le problème de filtrage de la section 2.2, en supposant que les coefficients ne dépendent ni du temps, ni de l'observation, et que $N = 1$. On écrit le système sous forme Stratonovich :

$$(4.2) \quad \begin{cases} X_t = X_0 + \displaystyle\int_0^t b(X_s)\,ds + \int_0^t f_j(X_s) \circ dV_s^j + \int_0^t g(X_s) \circ dY_s\,, \\[2mm] Y_t = \displaystyle\int_0^t h(X_s)\,ds + W_t\,. \end{cases}$$

On suppose que b est de classe C^1, à croissance au plus linéaire, $f_1, \ldots, f_M$, g de classe C_b^3, h de classe C^3 à croissance au plus exponentielle ainsi que ses dérivées, et $b + hg$ est à croissance au plus linéaire. On note U_j l'opérateur aux dérivées partielles

$$f_j^l(x)\frac{\partial}{\partial x^l} \qquad (1 \le j \le M)$$

et $\overline{U}$ l'opérateur

$$g^l(x)\frac{\partial}{\partial x^l}\,.$$

On note enfin $\phi_t(x)$ le flot associé au champ de vecteurs $\overline{U}$, et on suppose vérifiée l'hypothèse :

$$(H) \quad \begin{cases} \forall r > 0,\ \varepsilon > 0,\ \exists K_\varepsilon \quad t.q.: \\[2mm] |\,\overline{U}h\,| + \displaystyle\sup_{s \le r} |\,L(h \circ \phi_s)\,| + \sum_{j=1}^M \sup_{s \le r} |\,U_j(h \circ \phi_s)\,| \le \varepsilon h^2 + K_\varepsilon \end{cases}$$

Dans le cas $\phi_s = I$, $\forall s$, cette hypothèse est celle de Sussmann. On a le :

Théorème 4.3.1 *Sous les hypothèses ci-dessus, en particulier l'hypothèse (H), et si en outre la mesure σ_0 intègre $\exp(k \mid x \mid)$ pour tout $k > 0$, alors quelque soit $\varphi \in C^1(\mathbb{R}^M)$, $t > 0$, $\sigma_t(\varphi)$ et $\Pi_t(\varphi)$ possèdent des versions localement lipschitziennes par rapport à Y.*

$\square$

4.4   Un résultat de continuité dans le cas général

Il est bien connu que pour une EDS en dimension finie, la solution n'est pas nécessairement une fonction continue du Wiener directeur, lorsque les champs de vecteurs associés aux coefficients de diffusion ne commutent pas. Au vu de l'équation de Zakai, on ne peut donc pas espérer que $\sigma_t(\varphi)$ soit en général continu par rapport à Y, pour la norme $\sup_{0 \le s \le t} |Y_s|$.

On a cependant, sous des hypothèses très faibles, le résultat suivant, qui a été conjecturé par Sussmann, et démontré par Chaleyat-Maurel, Michel [19]. On reprend le problème de filtrage de la section précédente, mais cette fois avec $\{Y_t\}$ de dimension quelconque N :

$$(4.3) \quad \begin{cases} X_t = X_0 + \int_0^t b(X_s)\,ds + \int_0^t f_j(X_s) \circ dV_s^j + \int_0^t g_i(X_s) \circ dY_s^i \,, \\ \\ Y_t = \int_0^t h(X_s)\,ds + W_t \end{cases}$$

et on suppose que tous les coefficients sont de classe C_b^∞.

Soit $u \in L_{loc}^2(\mathbb{R}_+; \mathbb{R}^N)$ (u peut être interprété comme une "commande"). On définit les processus fonction de u :

$$X_t^u = X_0 + \int_0^t b(X_s^u)\,ds + \int_0^t f_j(X_s^u) \circ dV_s^j + \int_0^t g_i(X_s^u)u^i(s)\,ds \,,$$

$$Z_t^u = \exp\left\{ \int_0^t [h_i(X_s^u)u^i(s) - \frac{1}{2}\overline{U}_i h_i(X_s^u) - \frac{1}{2}\sum_{i=1}^N h_i^2(X_s^u)]\,ds \right\}$$

où $\overline{U}_i$ désigne l'opérateur aux dérivées partielles $g_i^j \frac{\partial}{\partial x^j}$. Notons que X_t^u et Z_t^u sont obtenus à partir de X_t et Z_t en remplaçant "odY_s^i" par "$u^i(s)ds$". A toute commande u, on associe pour chaque $t > 0$ la mesure σ_t^u définie par :

$$\sigma_t^u(\varphi) = E(\varphi(X_t^u)Z_t^u) \,, \quad \varphi \in C_b(\mathbb{R}^M) \,.$$

On a alors le :

Théorème 4.4.1 *Pour tout* $u \in L^2(0, T; \mathbb{R}^N), \varepsilon > 0$ *et* φ *fonction lipschitzienne de* $\mathbb{R}^M$ *dans* $\mathbb{R}$,

$$P\left(\sup_{0 \le t \le T} |\sigma_t(\varphi) - \sigma_t^u(\varphi)| > \varepsilon \,\Big/\, \sup_{0 \le t \le T} |Y_t - \int_0^t u(s)\,ds| \le \delta \right) \to 0, \quad \text{quand } \delta \to 0$$

□

Chapitre 5

Deux applications du calcul de Malliavin au filtrage non linéaire

5.1 Introduction

Le calcul de Malliavin s'appuie sur un calcul différentiel sur l'espace de Wiener pour établir par une méthode probabiliste des résultats d'existence de densité régulière pour la loi de la solution d'une EDS. En fait, la méthodologie de Malliavin se décompose en deux parties :

a Un critère général d'existence d'une densité régulière pour la loi d'un vecteur aléatoire, en terme de la non dégénérescence de la "matrice de covariance de Malliavin".

b La preuve que dans le cas d'une EDS la condition de rang d'Hörmander (au point de départ de l'EDS) entraîne la non-dégénérescence de la matrice de covariance de Malliavin.

Le but de ce chapitre est de présenter les idées de Malliavin, et leur application au filtrage. Nous essaierons d'être assez complet, mais en nous limitant au problème de l'existence d'une densité, ce qui évite l'essentiel des "larmes" liées aux estimations nécessaires pour établir la régularité de la densité.

Notons que ces dernières années ont vu la floraison d'articles qui "simplifient" le calcul de Malliavin (ou son application au filtrage). Nous ne prétendons pas faire œuvre novatrice de ce point de vue, tout en espérant que le lecteur trouvera dans ce chapitre un exposé clair des idées essentielles de Malliavin et de leurs deux applications au filtrage.

5.2 Le calcul de Malliavin. Application aux EDS

Dans cette section, on suppose que $\Omega = C([0,1]; \mathbb{R}^k)$ (dans les sections suivantes, $[0,1]$ deviendra $[0,t]!$), $\mathcal{F}$ est la tribu borélienne de Ω, P la mesure de Wiener, et $W_t(\omega) = \omega(t), 0 \le t \le 1$. Pour $0 < t < 1$, on notera $\mathcal{F}_t = \sigma\{W_s; \ 0 \le s \le t\}$ $\mathcal{F}^t = \sigma\{W_s - W_t; \ t \le s \le 1\}$, et $\mathcal{P}$ la tribu sur $[0,1] \times \Omega$ des ensembles $\mathcal{F}_t$–progressivement mesurables. On pose $H = L^2(0,1; \mathbb{R}^k)$, et on désigne par S la classe des v.a.r. "élémentaires" de la forme :

$$(5.1) \qquad F = f(W(h_1), \ldots, W(h_n))$$

où $n \in \mathbb{N}, f \in C_b^\infty(\mathbb{R}^n), h_j \in H$ et $W(h_j) = \sum_{i=1}^k \int_0^1 h_j^i(t)dW_t^i$ désigne l'intégrale de Wiener de $h_j, j = 1, \ldots, n$. Notons que $\mathbf{S}$ est dense dans $L^2(\Omega)$. Pour $F \in \mathbf{S}$ de la forme (5.1), on définit le processus k-dimensionel $\{D_t F; 0 \leq t \leq 1\}$ par :

$$D_t F = \sum_{j=1}^n \frac{\partial f}{\partial x^j}(W(h_1), \ldots, W(h_n))h_j(t) .$$

Posons la :

Définition 5.2.1 *On appelle intégrale de Skorohod l'opérateur $\delta = D^*$; i.e. δ est l'opérateur non borné de $L^2(\Omega; H)$ dans $L^2(\Omega)$ défini par :*

(i) Domδ est l'ensemble des u dans $L^2(\Omega; H)$ qui sont tels qu'il existe $c > 0$ avec :

$$|E \int_0^1 (u_t, D_t F)dt| \leq c\|F\|_2, \forall F \in \mathbf{S}$$

(ii) Si $u \in$ Domδ, $\delta(u)$ est l'élément de $L^2(\Omega)$ (dont l'existence est assurée par le théorème de Riesz) qui satisfait :

$$(5.2) \qquad E(\delta(u)F) = E \int_0^1 (u_t, D_t F)dt, F \in \mathbf{S}$$

(5.2) est la "formule d'intégration par parties", qui fait intervenir l'intégrale d'Itô dans le cas adapté :

Proposition 5.2.2 $L^2(\Omega \times (0,1), \mathcal{P}, dP \times dt; \mathbb{R}^k) \subset$ Domδ, et si $u \in L^2(\Omega \times (0,1), \mathcal{P}, dP \times dt; \mathbb{R}^k)$, $\delta(u)$ est l'intégrale d'Itô $\int_0^1 (u_t, dW_t)$.

Preuve   Soit $u \in L^2(\Omega \times (0,1), \mathcal{P}, dP \times dt; \mathbb{R}^k)$. Il suffit de montrer que $\forall F \in \mathbf{S}$,

$$(5.3) \qquad E[F \int_0^1 (u_t, dW_t)] = E \int_0^1 (u_t, D_t F)dt$$

En approchant u par des processus en escalier, on se ramène à établir (5.3) avec u de la forme $\sum_1^p \xi_j k_j$, avec $\xi_j \in L^2(\Omega, \mathcal{F}_{s_j}, P)$, $k_j \in H$ tel que sup$(k_j) \subset [s_j, 1]$. Il suffit donc de calculer (en oubliant l'indice j) :

$$\begin{aligned} E\left[F \int_0^1 \xi k(t)dW_t\right] &= E\left[F\xi \int_s^1 k(t)dW_t\right] \\ &= E\left[E^{\mathcal{F}^s}(F\xi) \int_0^1 k(t)dW_t\right] \\ &= E\left[G \int_0^1 k(t)dW_t\right] \end{aligned}$$

où $G \in \mathbf{S} \cap L^2(\Omega, \mathcal{F}^s, P)$, $G = g(W(k_1), \ldots, W(k_n))$. Chaque k_i admet une décomposition orthogonale dans H :

$$k_i = \alpha_i k + \overline{k_i}, < k, \overline{k_i} > = 0; 1 \leq i \leq n .$$

Il existe $\overline{g} \in C_b^\infty(\mathbb{R}^{n+1})$ t.q. :

$$G = \overline{g}(W(k); W(\overline{k}_1), \ldots, W(\overline{k}_n))$$

Notons :

$$\phi(x) = E[\overline{g}(x; W(\overline{k}_1), \ldots, W(\overline{k}_n))]$$

On a alors :

$$
\begin{aligned}
E\left[F \int_0^1 \xi k(t) dW_t\right] &= E[\phi(W(k)) W(k)] \\
&= E[\phi'(W(k)) \parallel k \parallel^2] \\
&= E \int_0^1 D_t \phi(W(k) k(t) dt
\end{aligned}
$$

L'avant–dernière égalité résulte d'une intégration par parties dans :

$$(\sqrt{2\pi}\|k\|)^{-1} \int_{\mathbb{R}} x\phi(x) \exp\left(-x^2/2\|k\|^2\right) dx$$

On remarque finalement que :

$$
\begin{aligned}
E \int_0^1 D_t \phi(W(k)) k(t) dt &= E \int_0^1 D_t G k(t) dt \\
&= E \int_0^1 D_t F \xi k(t) dt
\end{aligned}
$$

$\square$

Il y a un autre cas où l'on a une formule "explicite" pour l'intégrale de Skorohod :

Proposition 5.2.3 *Si* $h \in H, F \in \mathbf{S}$, *alors* $hF \in Dom\delta$ *et* :

$$\delta(hF) = F\delta(h) - \int_0^1 h(t) D_t F dt .$$

Cette Proposition est une conséquence de la définition de δ et du Lemme suivant (qui est immédiat) :

Lemme 5.2.4 *Si* $F, G \in \mathbf{S}$, *alors* $FG \in \mathbf{S}$ *et* $D_t(FG) = F D_t G + G D_t F$.

Il résulte de la Proposition 5.2.3 que $H \otimes \mathbf{S} \subset Dom\delta$, d'où $Dom\delta$ est dense dans $L^2(\Omega, H)$. On en déduit alors aisément que D est fermable. Définissons sur $\mathbf{S}$ la norme :

$$\|F\|_{1,2} = \|F\|_2 + \|\|DF\|_H\|_2$$

(où $\| \cdot \|_2$ désigne la norme dans $L^2(\Omega)$). Alors la fermeture de D (que nous noterons encore D par abus de notation) est une application linéaire continue de $\mathbb{D}^{1,2} \triangleq \overline{\mathbf{S}}^{\|\cdot\|_{1,2}}$ dans $L^2(\Omega; H)$. On vérifie aisément que la Proposition 5.2.3 reste vraie avec $F \in \mathbb{D}^{1,2}$, et le Lemme 5.2.4 sous l'hypothèse $F, G, FG \in \mathbb{D}^{1,2}$. Notons en outre que si F est $\sigma(W_s; 0 \le s \le t)$ mesurable, $D_r F = 0, r > t$. Nous aurons besoin ci-dessous des résultats techniques suivants. Le premier se démontre en approchant les intégrales par des "sommes de Darboux".

Lemme 5.2.5 *Si* $u \in M^2(0,1;\mathbb{R}) \cap L^2(0,1;\mathbb{D}^{1,2})$, $\int_0^1 u_t dt$, $\int_0^1 u_t dW_t^i \in \mathbb{D}^{1,2}$, $1 \leq i \leq k$, *et*

$$D_t \int_0^1 u_r dr = \int_t^1 D_t u_r dr$$

$$D_t^j \int_0^1 u_r dW_r^i = \delta_{ij} u_t + \int_t^1 D_t^j u_r dW_r^i$$

Notons que si $u \in Dom\delta$, $E[\delta(u)] = 0$ (choisir $F \equiv 1$ dans la Définition 5.2.1). On a en outre :

Proposition 5.2.6 $L^2(0,1;(\mathbb{D}^{1,2})^k) \subset Dom\delta$ *et pour* $u \in L^2(0,1;(\mathbb{D}^{1,2})^k)$,

$$E[\delta(u)^2] = E \int_0^1 u_t^2 dt + E \int_0^1 \int_0^1 D_s^i u_t^j D_t^j u_s^i ds dt$$

Preuve On utilise le Lemme 5.2.3, et on calcule $E[\delta(hF)\delta(kG)]$, avec $h, k \in H$, $F, G \in$ **S**. Le résultat s'en déduit pour $u = \sum_1^n h_l F_l$; $h_l \in H$, $F_l \in \mathbb{D}^{1,2}$, $1 \leq l \leq n$. Il reste à approcher $u \in L^2(0,1;(\mathbb{D}^{1,2})^k)$ par des u de cette forme. □

Le second outil dont nous aurons besoin est le résultat suivant, adapté de Stroock [81] :

Lemme 5.2.7 *Soit* μ *une mesure finie sur* $(\mathbb{R}^d, \mathcal{B}_d)$. *Supposons qu'il existe des mesures signées* $\mu_j, 1 \leq j \leq d$, *t.q. :*

$$\int_{\mathbb{R}^d} \frac{\partial f}{\partial x^j}(x)\mu(dx) = \int_{\mathbb{R}^d} f(x)\mu_j(dx), f \in C_c^\infty(\mathbb{R}^d), 1 \leq j \leq d.$$

Alors μ *admet une densité par rapport à la mesure de Lebesgue.*

Preuve Posons

$$f(t,x) = (2\pi t)^{\frac{-d}{2}} \exp(-|x|^2/2t)$$

$$g(t,x) = (f(t,\cdot) * \mu)(x)$$

Alors $f'_t = \frac{1}{2}\Delta_x f$,

$$\frac{\partial g}{\partial t}(t,x) = \frac{1}{2}\sum_1^d (\frac{\partial f}{\partial x^j}(t,\cdot) * \mu_j)(x) , \quad t > 0 , \quad x \in \mathbb{R}^d$$

Mais, par intégration par parties, si $\lambda > 0$,

$$e^{-\lambda s} g(s,x) = \lambda \int_s^\infty e^{-\lambda t} g(t,x) dt - \int_s^\infty e^{-\lambda t} \frac{\partial g}{\partial t}(t,x) dt$$

Faisant tendre $s \rightarrow 0$, on obtient :

$$\mu = \lambda \int_0^\infty e^{-\lambda t} g(t,\cdot) dt - \frac{1}{2}\sum_j \int_0^\infty e^{-\lambda t} \frac{\partial f}{\partial x^j}(t,\cdot) * \mu_j dt$$

Posons

$$F_\lambda(x) = \int_0^\infty e^{-\lambda t} f(t,x)dt.$$

On a :

$$\mu = \lambda F_\lambda * \mu - \frac{1}{2} \sum_j \frac{\partial F_\lambda}{\partial x^j} * \mu_j$$

Or

$$F_\lambda, \frac{\partial F_\lambda}{\partial x^1}, \ldots, \frac{\partial F_\lambda}{\partial x^d} \in L^1(\mathbb{R}^d).$$

La convolée d'une fonction intégrable et d'une mesure signée est une fonction intégrable. Donc μ, en tant que distribution, est une fonction intégrable, i.e. en tant que mesure admet une densité. □

On peut maintenant établir le critère général de Malliavin :

Théorème 5.2.8 *Soit $X = (X_1, \ldots, X_d)' \in (\mathbb{D}^{1,2})^d$. Supposons en outre qu'il existe $u_1, \ldots, u_d \in Dom\delta$ t.q. si $A(u) = (< DX_i, u_j >)_{i,j}$,*

- *$A(u)_{ij} \in \mathbb{D}^{1,2}; i,j = 1, \ldots, d$*
- *$d\acute{e}t\,(A(u)) \neq 0 \quad p.s.$*

Alors la loi de X est absolument continue par rapport à la mesure de Lebesgue.

Preuve Pour $\varepsilon > 0$, soit $\varphi_\varepsilon, \psi_\varepsilon \in C_b^\infty(\mathbb{R}_+; [0,1])$ t.q. :

$$\varphi_\varepsilon(x) = \begin{cases} 1, & x \leq 1/\varepsilon \\ 0, & x \geq 1 + 1/\varepsilon \end{cases}$$

$$\psi_\varepsilon(x) = \begin{cases} 1, & x \geq \varepsilon \\ 0, & x \leq \varepsilon/2 \end{cases}$$

Posons $\xi_\varepsilon = \varphi_\varepsilon(TrAA^*)\psi_\varepsilon(|d\acute{e}tA|)$. Notons que l'hypothèse du théorème entraîne $\xi_\varepsilon \to 1$ p.s., quand $\varepsilon \to 0$. On pose $B^\varepsilon = \xi_\varepsilon A^{-1}$. On peut alors vérifier que $B^\varepsilon \in L^\infty(\Omega) \cap \mathbb{D}^{1,2}$. Soit $f \in C_c^\infty(\mathbb{R}^d)$. On rappelle que l'on utilise la convention de sommation sur indice répété.

$$E[f(X)\delta(u_j)B_{jl}^\varepsilon] = E\left[\frac{\partial f}{\partial x^i}(X) < DX^i, u_j > B_{jl}^\varepsilon\right] + E[f(X) < DB_{jl}^\varepsilon, u_j >],$$

où $< \cdot, \cdot >$ désigne le produit scalaire dans H. Donc, si l'on pose :

$$\rho_l = \delta(u_j)B_{jl}^\varepsilon - < DB_{jl}^\varepsilon, u_j >, \rho = (\rho_1, \ldots, \rho_d),$$

on a :

$$E[f(X)\rho] = E([\nabla f(X)]'AB^\varepsilon) = E([\nabla f(X)]'\xi_\varepsilon)$$

Posons $\mu_\varepsilon = (\xi_\varepsilon \cdot P)X^{-1}; \mu_\varepsilon^i = (\rho_i \cdot P)X^{-1}, 1 \leq j \leq d$. On vient de montrer que :

$$\int \frac{\partial f}{\partial x^i}(x)\mu_\varepsilon(dx) = \int f(x)\mu_\varepsilon^i(dx), 1 \leq i \leq d$$

Donc d'après le Lemme 5.2.7, ceci entraîne que μ_ε est absolument continue par rapport à la mesure de Lebesgue, i.e. pour tout $N \in \mathcal{B}_d$ de mesure de Lebesgue nulle,

$$E(\xi_\varepsilon 1_N(X)) = 0$$

et par convergence dominée :

$$P(X \in N) = 0.$$

$\square$

Corollaire 5.2.9 *Si $X = (X_1, \ldots, X_d)' \in (\mathbb{D}^{1,2})^d$, $DX_i \in Dom\delta$, $1 \leq i \leq d$ et de plus $< DX_i, DX_j >\in \mathbb{D}^{1,2}$, $i,j = 1, \ldots, d$, alors la condition*

$$[(< DX_i, DX_j >)_{ij}] > 0\, p.s.$$

assure l'absolue continuité de la loi de X.

Remarque 5.2.10 La matrice $(< DX_i, DX_j >)_{ij}$ s'appelle la matrice de covariance de Malliavin. Bouleau et Hirsch [13] ont montré que la conclusion du Théorème est vraie sous les seules hypothèses :

$$X \in (\mathbb{D}^{1,2})^d, \quad [(< DX_i, DX_j >)_{ij}] > 0\ p.s. .$$

$\square$

Nous pouvons maintenant appliquer le critère général aux EDS. Soit $b \in C^\infty(\mathbb{R}^d; \mathbb{R}^d)$, dont toutes les dérivées sont supposées bornées, $\sigma_1, \ldots, \sigma_d \in C_b^\infty(\mathbb{R}^d; \mathbb{R}^d)$ et $x_0 \in \mathbb{R}^d$. Soit $\{X_t; t \geq 0\}$ l'unique solution de l'EDS au sens de Stratonovich :

$$(5.4) \qquad X_t = x_0 + \int_0^t b(X_s)ds + \int_0^t \sigma_i(X_s) \circ dW_s^i, 0 \leq t \leq 1$$

Proposition 5.2.11 *Soit $0 \leq t \leq 1$. $X_t \in (\mathbb{D}^{1,2})^d$ et pour $0 \leq r \leq t$,*

$$D_r^j X_t = \sigma_j(X_r) + \int_r^t b'(X_s)D_r^j X_s ds + \int_r^t \sigma_i'(X_s)D_r^j X_s \circ dW_s^i$$

Preuve On considère l'approximation de Picard de l'EDS (5.4) écrite au sens d'Itô :

$$(5.5) \qquad X_t^{n+1} = x_0 + \int_0^t \overline{b}(X_s^n)ds + \int_0^t \sigma_i(X_s^n)dW_s^i, t \geq 0, n \in \mathbb{N}$$

On montre par récurrence, à l'aide du Lemme 5.2.5, que

$$X^n \in M^2(0, T; \mathbb{R}^d) \cap L^2(0, T; (\mathbb{D}^{1,2})^d), \quad \forall T > 0$$

et que :

$$(5.6) \qquad D_r^j X_t^{n+1} = \sigma_j(X_r^n)\ + \int_r^t \overline{b}'(X_s^n)D_r^j X_s^n ds +$$

$$+ \int_r^t \sigma_i'(X_s^n)D_r^j X_s^n dW_s^i; t \geq r; j = 1, \ldots, k$$

Le théorème de convergence de la méthode de Picard montre que le couple d'équations (5.5)–(5.6) converge vers le couple formé de (5.4) et de :

$$Z_t^{j,r} = \sigma_j(X_r) + \int_r^t \bar{b}'(X_s)Z_s^{j,r}ds + \int_r^t \sigma_i'(X_s)Z_s^{j,r}dW_s^i, t \geq r$$

Il résulte de ce que D est fermé : $X_t \in \mathbb{D}^{1,2}$, et $D_r^j X_t = Z_t^{j,r}$. $\qquad\qquad\square$

Soit $\{\Phi_t, t \geq 0\}$ le processus à valeurs matrices $d \times d$, unique solution de l'EDS :

$$\Phi_t = I + \int_0^t \bar{b}'(X_s)\Phi_s ds + \int_0^t \sigma_i'(X_s)\Phi_s dW_s^i, 0 \leq t \leq 1$$

On montre aisément que $\det\Phi_t \neq 0 \quad \forall t$, p.s. en exhibant l'équation pour Φ_t^{-1}. Alors pour $0 \leq r \leq t$,

$$\Phi_t\Phi_r^{-1} = I + \int_r^t \bar{b}'(X_s)\Phi_s\Phi_r^{-1}ds + \int_r^t \sigma_i'(X_s)\Phi_s\Phi_r^{-1}dW_s^i$$

En multipliant l'égalité matricielle ci-dessus à droite par le vecteur $\sigma_j(X_r)$, on déduit du théorème d'unicité de la solution d'une EDS à coefficients localement lipschitziens :

$$D_r^j X_t = \Phi_t\Phi_r^{-1}\sigma_j(X_r), \quad 0 \leq r \leq t \leq 1$$

D'où la matrice de covariance de Malliavin de X_t s'écrit :

$$A_t = \Phi_t[\int_0^t \Phi_r^{-1}\sigma_j(X_r)\sigma_j^*(X_r)(\Phi_r^{-1})^* dr]\Phi_t^*$$

En itérant l'argumentation de la Proposition 5.2.11 et à l'aide de la Proposition 5.2.6, on vérifie que les conditions de régularité du Corollaire 5.2.9 sont satisfaites. Il reste à trouver une condition qui assure que $A_t > 0$ p.s. Notons qu'il est équivalent de montrer que :

$$B_t = \int_0^t \Phi_r^{-1}\sigma_j(X_r)\sigma_j^*(X_r)(\Phi_r^{-1})^* dr > 0 \ p.s.$$

Pour $0 \leq r \leq 1, 0 \leq t \leq 1$, posons

$$\begin{aligned}
\mathcal{U}_s &= \text{ e.v.}\{\Phi_s^{-1}\sigma_j(X_s), 1 \leq j \leq k\}, \\
\mathcal{V}_t &= \text{ e.v.}\{\cup_{0 \leq s \leq t}\mathcal{U}_s\}, \\
\mathcal{V}_t^+ &= \cap_{s>t}\mathcal{V}_s.
\end{aligned}$$

D'après la loi $0-1$, $\mathcal{V}_0^+$ est p.s. égal à un sous e.v. fixe de $\mathbb{R}^d$. Pour montrer que $B_t > 0$ p.s., $\forall 0 < t \leq 1$, il suffit de montrer que $\mathcal{V}_0^+ = \mathbb{R}^d$. On a donc la :

Proposition 5.2.12 *Une condition suffisante pour que la loi de X_t possède une densité, $\forall t > 0$, et que $\mathcal{V}_0^+ = \mathbb{R}^d$.*

On notera $U_0, U_1, \ldots, U_d$ les opérateurs aux dérivées partielles du premier ordre :

$$U_0(x) = b^j(x)\frac{\partial}{\partial x^j}; U_i(x) = \sigma_i^j(x)\frac{\partial}{\partial x^j}, 1 \leq i \leq d.$$

Considérons les algèbres de Lie de champs de vecteurs (ou d'opérateurs aux dérivées partielles du premier ordre à coefficients C^∞) suivantes (pour le crochet de Lie $[C,D] = DC - CD$) :

$$\mathcal{A} = A.L.\{U_0, U_1, \ldots, U_k\},$$
$$\mathcal{B} = A.L.\{U_1, \ldots, U_k\},$$
$$\mathcal{I} = \text{idéal engendré par } \mathcal{B} \text{ dans } \mathcal{A}$$

(i.e. $\mathcal{I}$ contient les $U_i, 1 \le i \le k$, leurs crochets et les crochets de U_0 avec les $U_i, 1 \le i \le k,\ldots$, mais pas U_0 lui-même). Si $x \in \mathbb{R}^d$, on note $\mathcal{I}(x)$ l'espace vectoriel (de dimension $\le d$) des opérateurs aux dérivées partielles à coefficients constants obtenus en prenant les opérateurs de $\mathcal{I}$, avec leurs coefficients fixés à la valeur qu'ils prennent au point x.

Théorème 5.2.13 *Si $dim\mathcal{I}(x_0) = d$, alors $\mathcal{V}_0^+ = \mathbb{R}^d$, et donc la loi de X_t possède une densité, $\forall t > 0$.*

Preuve Soit q un vecteur de $\mathbb{R}^d$, t.q. $q \in (\mathcal{V}_0^+)^\perp$. On va montrer que $q = 0$. On pose :

$$\tau = inf\{0 \le t \le 1, \mathcal{V}_t \neq \mathcal{V}_0^+\}.$$

Alors $\tau > 0$ p.s. et pour tout $t \in [0, \tau]$, $(q, \Phi_t^{-1}\sigma_j(X_t)) = 0, 1 \le j \le k$. Or

$$\sigma_j(X_t) = \sigma_j(x_0) + \int_0^t \sigma_j'(X_s)b(X_s)ds + \int_0^t \sigma_j'(X_s)\sigma_i(X_s) \circ dW_s^i,$$
$$\Phi_t^{-1} = I - \int_0^t \Phi_s^{-1}b'(X_s)ds - \int_0^t \Phi_s^{-1}\sigma_i'(X_s) \circ dW_s^i.$$

Il résulte donc de la formule de Stratonovich (en identifiant un champ de vecteurs avec ses coefficients dans la base $(\frac{\partial}{\partial x^1}, \ldots, \frac{\partial}{\partial x^d})$) :

$$(\Phi_t^{-1}\sigma_j(X_t), q) = (U_j(x_0), q) + \int_0^t (\Phi_s^{-1}[U_j, U_0](X_s), q)ds$$
$$+ \int_0^t (\Phi_s^{-1}[U_j, U_i](X_s), q) \circ dW_s^i.$$

En outre $q \in (\mathcal{V}_0^+)^\perp, (U_j(x_0), q) = 0$. Comme la variation quadratique de $(\Phi_t^{-1}\sigma_j(X_t), q)$ est nulle sur $[0, \tau], 1 \le j \le k$, pour $t \in [0, \tau]$,

(5.7) $$(\Phi_t^{-1}[U_j, U_i](X_t), q) = 0, 1 \le i, j \le k.$$

Et finalement puisque $(\Phi_t^{-1}\sigma_j(X_t), q)$ est nul, pour $t \in [0, \tau]$,

(5.8) $$(\Phi_t^{-1}[U_j, U_0](X_t), q) = 0, 1 \le j \le k.$$

En particulier,

$$([U_j, U_i](x_0), q) = 0, \quad 0 \le i \le k, \quad 1 \le j \le k.$$

En repartant de (5.7), (5.8), on montre ensuite que q est orthogonal aux vecteurs

$$[[U_j, U_i], U_\ell](x_0), \quad 0 \le i, \quad \ell \le k, \quad 1 \le j \le k.$$

En itérant l'argument, on montre que $q \perp \mathcal{I}(x_0)$, ce qui d'après l'hypothèse du théorème entraîne $q = 0$. $\square$

Remarque 5.2.14 L'hypothèse du théorème est la plus faible possible, au sens où si $dim\mathcal{I}(x) < d$ pour tout $x \in \mathbb{R}^d$, alors la solution de l'EDS reste confinée dans une sous variété de dimension $< d$.

Remarquons que sous l'hypothèse globale $dim\mathcal{I}(x) = d, \forall x \in \mathbb{R}^d$, le résultat est une conséquence du célèbre "théorème de la somme des carrés" d'Hörmander [38], puisque si l'on pose

$$\bar{U}_0 = (\frac{\partial}{\partial t}, U_0), \bar{U}_1 = (0, U_0), \ldots, \bar{U}_k = (0, U_k), \ \bar{Q} = A.L.\{\bar{U}_0, \bar{U}_1, \ldots, \bar{U}_k\} ,$$

$$dim\mathcal{I}(x) = d, \forall x \quad \Leftrightarrow \quad dim\bar{Q}(t,x) = d + 1, \forall (t,x)$$
$$\Rightarrow \quad -\frac{\partial}{\partial t} + L^* \text{ est hypoelliptique}$$

(cf. l'équation de Fokker–Planck pour l'évolution de la loi de $\{X_t\}$).

Enfin, pour le système contrôlé :

$$(5.9) \qquad \frac{d\varphi_t(x,u)}{dt} = b(\varphi_t(x,u)) + \sigma_i(\varphi_t(x,u)u^i(t) , \qquad \varphi_0(x,u) = x ,$$

la condition $dim\mathcal{I}(x) = d, \forall x \in \mathbb{R}$ entraîne que l'ensemble :

$$A(t, x_0) = \overline{\{\varphi_t(x_0, u); u \in L^1(0, t; \mathbb{R}^k)\}}$$

est d'intérieur non vide (voir Lobry [56], Isidori [40]). Or cet ensemble est précisément, d'après un théorème de Stroock–Varadhan (cf. Ikeda–Watanabe [39]) le support de la loi de X_t, qui est bien d'intérieur non vide lorsque cette loi est absolument continue. $\qquad \square$

Le but du reste de ce chapitre est de donner deux applications très différentes des idées de Malliavin au filtrage. La première, partant de l'idée que la loi conditionnelle de X_t sachant $\mathcal{Y}_t$ est la loi d'une diffusion (conditionnelle) ou encore que l'équation de Zakai est une EDP (stochastique), établit l'existence d'une densité (régulière) à cette loi conditionnelle. La seconde, partant de l'idée que l'équation de Zakai est une EDS (aux dérivées partielles), montre que sous une hypothèse ad hoc, quand $\{Y_t\}$ varie, la solution "remplit un espace de dimension infinie". Ce deuxième résultat a des conséquences importantes quand à la non existence de "filtres de dimension finie ".

5.3 Existence d'une densité pour la loi conditionnelle du filtrage

On reprend le modèle du 2.2, mais avec des coefficients ne dépendant que de Y à l'instant courant et indépendants de t, b remplacé par $b + gh$, et écrit sous forme Stratonovich. **Attention** il y a un "double" changement de notation par rapport au Chapitre 2 !

$$(5.10) \qquad \begin{cases} X_t = x_0 + \int_0^t (b + gh)(X_s, Y_s)ds + \int_0^t f_j(X_s, Y_s) \circ dV_s^j + \\ \qquad + \int_0^t g_i(X_s, Y_s) \circ dW_s^i , \\ \\ Y_t = \int_0^t h(X_s, Y_s)ds + W_t \end{cases}$$

avec $b, f_1, \ldots, f_M, g_1, \ldots, g_N, h$ de $\mathbb{R}^M \times \mathbb{R}^N$ à valeurs dans $\mathbb{R}^M, \mathbb{R}^M, \mathbb{R}^M$ et $\mathbb{R}^N$ respectivement, de classe C_b^∞. On définit les champs de vecteurs sur $\mathbb{R}^M \times \mathbb{R}^N$:

$$U_0(x,y) = b^\ell(x,y)\frac{\partial}{\partial x^\ell},$$

$$U_j(x,y) = f_j^\ell(x,y)\frac{\partial}{\partial x^\ell}, 1 \le j \le M,$$

$$\bar{U}_i(x,y) = g_i^\ell(x,y)\frac{\partial}{\partial x^\ell} + \frac{\partial}{\partial y^i}, 1 \le i \le N.$$

Afin de "deviner" la bonne condition de rang qu'il nous faudra faire ici, considérons tout d'abord le cas (sans grand intérêt !) $h \equiv 0$. Formellement, la première équation de (5.10) s'écrit alors :

$$(5.11) \qquad X_t = x_0 + \int_0^t [b(X_s, Y_s) + g_i(X_s, Y_s)\frac{dY_s^i}{ds}]ds + \int_0^t f_j(X_s, Y_s) \circ dV_s^j.$$

Faisons comme si les trajectoires de $\{Y_t\}$ étaient régulières. La loi conditionnelle de X_t sachant $\mathcal{Y}_t$ est la loi de la solution de (5.11), obtenue en fixant $\{Y_s; 0 \le s \le t\}$. La différence avec la situation de la section précédente est que les coefficients dépendent du temps (par l'intermédiaire de Y !). D'après la Remarque 5.2.14 (analogie avec le Théorème d'Hörmander), il est facile de se convaincre que les champs de vecteurs qui interviennent ici sont les $U_j (1 \le j \le M)$ ainsi que les crochets $(1 \le j \le M)$:

$$\left[\frac{\partial}{\partial t} + (b^\ell(\cdot, Y_t) + g_i^\ell(Y_t)\frac{dY_t^i}{dt})\frac{\partial}{\partial x^\ell}, f_j^\ell(\cdot, Y_t)\frac{\partial}{\partial x^\ell}\right](x)$$

$$= [U_0, U_j](x, Y_t) + [\bar{U}_i, U_j](x, Y_t)\frac{dY_t^i}{dt}$$

(on a utilisé $\frac{\partial}{\partial t} f_j^\ell(x, Y_t) = \frac{\partial f_j^\ell}{\partial y^i}(x, Y_t)\frac{dY_t^i}{dt}$) et les "crochets itérés". Comme les $\frac{dY_t^i}{dt}$ sont très erratiques, on peut admettre qu'apparaissent les champs de vecteurs :

$$[U_0, U_j], [\bar{U}_1, U_j], \ldots, [\bar{U}_N, U_j]; 1 \le j \le M.$$

Ceci signifie que la condition de rang devrait porter sur l'idéal $\bar{\mathcal{I}}$ engendré par

$$\bar{\mathcal{B}} = A.L.\{U_1, \ldots, U_M\}$$

dans

$$\bar{\mathcal{A}} = A.L.\{U_0, \bar{U}_1, \ldots, \bar{U}_N, U_1, \ldots, U_M\}.$$

Remarquons que la dimension de $\bar{\mathcal{I}}(x_0, 0)$ est au plus M. On va maintenant voir effectivement que la condition

$$dim\bar{\mathcal{I}}(x_0, 0) = M$$

entraîne l'existence d'une densité pour la mesure Π_t- ou équivalemment pour la mesure σ_t (en fait même d'une densité C^∞, mais nous n'examinerons pas la question de la régularité de la densité). Nous allons maintenant donner deux démonstrations de l'existence d'une densité. La première consiste à se ramener au calcul de Malliavin usuel, la seconde utilise un "calcul de Malliavin partiel". Seule la seconde approche s'étend aisément pour démontrer la régularité de la densité.

5.3.1 L'absolue continuité de σ_t par le calcul de Malliavin

Notons que, avec les notations des chapitres précédents,

$$\begin{aligned} \sigma_t(\varphi) &= \overset{\circ}{E}\left[\varphi(X_t)Z_t/\mathcal{Y}\right] \\ &= \overset{\circ}{E}^{\,y}\left[\varphi(X_t)\,\overset{\circ}{E}^{\,y}\,(Z_t/X_t)\right]. \end{aligned}$$

Définissons la mesure aléatoire $\bar{\sigma}_t$ par :

$$\frac{d\bar{\sigma}_t}{d\sigma_t}(x) = (\overset{\circ}{E}^{\,y}\,(Z_t/X_t = x))^{-1}.$$

Il est clair que σ_t est absolument continue ssi $\bar{\sigma}_t$ l'est. Mais $\bar{\sigma}_t$ n'est autre que la loi de X_t sous $\overset{\circ}{P}$, $\{Y_s;\ 0 \le s \le t\}$ étant fixé. Rappelons que

$$(5.12) \qquad X_t = x_0 + \int_0^t b(X_s, Y_s)ds \ + \ \int_0^t f_j(X_s, Y_s) \circ dV_s^j$$
$$+ \ \int_0^t g_i(X_s, Y_s) \circ dY_s^i.$$

Comment peut-on fixer la trajectoire de $\{Y_s\}$ dans (5.12) ? Grâce à la théorie des flots (comme on l'a déjà vu au chapitre 4, et de la façon suivante. Soit $\{\psi_t^Y(x); t \ge 0, x \in \mathbb{R}^M\}$ le flot associé à l'EDS :

$$(5.13) \qquad \tilde{X}_t = x + \int_0^t g_i(\tilde{X}_s, Y_s) \circ dY_s^i.$$

Notons que par définition du flot, on a choisi une version de $\psi_t^Y(x)$ t.q.

$$(t, x) \to \psi_t^Y(x)$$

soit de classe $C^{0,\infty}(\mathbb{R}_+ \times \mathbb{R}^M; \mathbb{R}^M)$ pour tout $Y \in C(\mathbb{R}_+; \mathbb{R}^N)$ et que pour tout $(t, Y) \in \mathbb{R}_+ \times C(\mathbb{R}_+; \mathbb{R}^N)$,

$$x \to \psi_t^Y(x)$$

est un difféomorphisme. On en déduit que la matrice :

$$\nabla \psi_t^Y(x) = \left(\frac{\partial(\psi_t^Y)^i(x)}{\partial x^j}\right)_{i,j}$$

est inversible pour tout $(t, Y, x) \in \mathbb{R}_+ \times C(\mathbb{R}_+; \mathbb{R}^N) \times \mathbb{R}^M$. Avec la notation :

$$\psi_t^{Y*}b(x, Y_t) = (\nabla \psi_t^Y(x))^{-1}b(\psi_t^Y(x), Y_t)$$

et

$$\psi_t^{Y*}f_j(x, Y_t) = (\nabla \psi_t^Y(x))^{-1}f_j(\psi_t^Y(x), Y_t), 1 \le j \le M$$

il résulte de la formule de Stratonovich généralisée (voir Bismut [9], Kunita [49]) que $X_t = \psi_t^Y(\overline{X}_t)$, $t \ge 0$ où $\{\overline{X}_t,\ t \ge 0\}$ est l'unique solution de l'EDS :

$$\overline{X}_t = x_0 + \int_0^t \psi_s^{Y*}b(\overline{X}_s, Y_s)ds + \int_0^t \psi_s^{Y*}f_j(\overline{X}_s, Y_s) \circ dV_s^j$$

On peut maintenant fixer la trajectoire du processus $\{Y_t\}$ dans cette équation. Etant donné $y \in C(\mathbb{R}_+; \mathbb{R}^N)$, on pose $X_t^y = \psi_t^y(\overline{X}_t^y)$, $t \geq 0$, où $\{\overline{X}_t^y\}$ est la solution de l'EDS :

$$(5.14) \qquad \overline{X}_t^y = x_0 + \int_0^t \psi_s^{y*} b(\overline{X}_s^y, y_s) ds + \int_0^t \psi_s^{y*} f_j(\overline{X}_s^y, y_s) \circ dV_s^j, \ t \geq 0 \ .$$

Il nous reste à appliquer la technique de la section précédente. On ne peut pas se contenter d'appliquer le Théorème 5.2.13, en particulier parce que les coefficients dépendent de t par l'intermédiaire de y – on verra pourtant que tout marche ici comme à la section 5.2. Il y a cependant une autre difficulté : ni les coefficients de l'équation (5.14), ni (et c'est plus grave) leurs dérivées en x ne sont bornées. On est obligé de "localiser" la démarche de la section 5.2. Ceci peut se faire aisément de la façon suivante. Soit $n \in \mathbb{N}$. Supposons que l'on remplace les coefficients g_i de l'équation (5.13) par de nouveaux coefficients de classe C_b^∞ qui coïncident avec les g_i sur $\{|x| \leq n\} \times \mathbb{R}^N$, et sont nuls sur $\{|x| \geq n+1\} \times \mathbb{R}^N$. Les nouveaux coefficients correspondants de l'équation (5.14) sont bornés ainsi que leurs dérivées, et le "nouveau" X_t^y coïncide avec l'ancien sur $\Omega_t^n = \{sup_{0 \leq r \leq s \leq t} |\psi_r^y(\overline{X}_s^y)| \leq n\}$. Or puisque $\cup_n \Omega_t^n = \Omega$, si pour tout n l'image de $1_{\Omega_t^n} \cdot P$ par X_t^y est absolument continue, il en est de même de l'image de P par X_t^y. Il suffit donc d'établir le résultat avec les g_i nuls pour x en dehors d'un compact. On peut donc supposer les coefficients de l'équation (5.14) bornés ainsi que leurs dérivées. Les premières conditions du Corollaire 5.2.9 se vérifient alors comme à la section précédente, et il reste à établir l'inversibilité de la matrice de covariance de Malliavin, qui est équivalente à l'inversibilité d'une matrice de la forme :

$$B_t^y = \int_0^t \Phi_s^{y-1} f_j(X_s^y, y_s) f_j^*(X_s^y, y_s) (\Phi_s^{y-1})^* ds \ .$$

Théorème 5.3.1 *Si $dim\overline{\mathcal{I}}(x_0, 0) = M$, alors $B_t^Y > 0$ pour tout $t > 0$, p.s., et pour tout $t > 0, \sigma_t$ est p.s. absolument continue par rapport à la mesure de Lebesgue sur $\mathbb{R}^M$.*

Preuve L'argument est identique à celui du Théorème 5.2.13. Indiquons seulement le "développement" de $(\Phi_t^{-1} f_j(X_t, Y_t), q)$:

$$\begin{aligned}
(\Phi_t^{-1} f_j(X_t, Y_t), q) = & (U_j(x_0, 0), q) + \\
& + \int_0^t (\Phi_s^{-1}[U_j, U_0](X_s, Y_s), q) ds \\
& + \int_0^t (\Phi_s^{-1}[U_j, U_\ell](X_s, Y_s), q) \circ dW_s^\ell \\
& + \int_0^t (\Phi_s^{-1}[U_j, \bar{U}_i](X_s, Y_s), q) \circ dY_s^i
\end{aligned}$$

$\square$

5.3.2 L'absolue continuité de σ_t par le calcul de Malliavin partiel

On reprend le modèle (5.11). Puisque sous $\overset{\circ}{P}$ $\{V_t\}$ et $\{Y_t\}$ sont deux processus de Wiener indépendants, il sera commode de supposer que ce sont les processus canoniques sur

$\Omega = \Omega_1 \times \Omega_2, \Omega_1 = C_0(\mathbb{R}_+; \mathbb{R}^M), \Omega_2 = C_0(\mathbb{R}_+; \mathbb{R}^N)$; i.e.. $V_t(\omega) = \omega_1(t), Y_t(\omega) = \omega_2(t)$.
Soit P^V et P^Y respectivement la mesure de Wiener sur Ω_1 et Ω_2. $\overset{\circ}{P} = P_V \times P_Y$ et

$$\sigma_t(\varphi) = \overset{\circ}{E}{}^{y}\,[\varphi(X_t)Z_t] = E_V[\varphi(X_t)Z_t]\ .$$

Dans toute la suite, **S** désignera l'espace des v.a. simples définies sur Ω_1, et D l'opérateur de dérivation sur Ω_1 (i.e. "dans la direction de $\{V_t\}$"). On définit $I\!\!D^{1,2}$ comme le complété de $\mathbf{S} \otimes L^2(\Omega_2, \mathcal{F}_2, P_Y)$ pour la norme :

$$\|F\|_{1,2} = \|F\|_2 + \|\|DF\|_H\|_2$$

avec $H = L^2(0, T; \mathbb{R}^M)$, $\|\cdot\|_2$ norme de $L^2(\Omega, \mathcal{F}, \overset{\circ}{P})$. Il est alors clair qu'un élément de $L^2(\Omega_2, \mathcal{F}_2, P_Y)$ s'identifie à un élément de $I\!\!D^{1,2}$. On démontre comme à la section 5.2 que $X_t \in I\!\!D^{1,2}$. Il résulte alors de la Définition 5.2.1 et de la Proposition 5.2.2 (à nouveau, $< \cdot, \cdot >$ désigne le produit scalaire dans H) :

Proposition 5.3.2 *Si* $\varphi \in C_b^\infty(\mathbb{R}^M)$, $H \in \mathbf{S}$, $G \in L^4(\Omega_2, \mathcal{F}_2, P_Y)$,

$$\overset{\circ}{E}\left[\frac{\partial \varphi}{\partial x^i}(X_t) < DX_t^i, u > HZ_tG\right] =$$
$$= \overset{\circ}{E}\left[\{H\int_0^t(u_s, dV_s) - < DH + HD(Log\,Z), u >\}\varphi(X_t)Z_tG\right]$$

et donc :

$$(5.15) \qquad E_V\left[\frac{\partial \varphi}{\partial x^i}(X_t) < DX_t^i, u > HZ_t\right] =$$
$$= E_V\left[\{H\int_0^t(u_s, dV_s) - < DH + HD(Log\,Z), u >\}\varphi(X_t)Z_t\right]\ .$$

(5.15) est la "formule d'intégration par parties" du "calcul de Malliavin partiel". C'est l'analogue de la formule qui sert de point de départ à la preuve du Théorème 5.2.8. Il est alors clair que l'on peut démontrer, par un raisonnement analogue :

Proposition 5.3.3 *Une condition suffisante pour que* σ_t *soit absolument continue est que la "matrice de covariance partielle de Malliavin"*

$$(< DX_t^i, DX_t^j >)_{i,j}$$

soit p.s. non dégénérée.

On peut maintenant établir le :

Théorème 5.3.4 *Si* $dim[\bar{\mathcal{I}}(x_0, 0)] = M$, *alors la matrice de covariance partielle de Malliavin de* X_t *est p.s. non dégénérée, et* σ_t *est absolument continue par rapport à la mesure de Lebesgue de* $\mathbb{R}^M$.

Preuve Désignons par $\{\Phi_t\}$ le processus à valeurs matrice $M \times M$ solution de l'EDS :

$$\Phi_t = I + \int_0^t b_1'(X_s, Y_s)\Phi_s ds \; + \; \int_0^t f_{j,1}'(X_s, Y_s)\Phi_s \circ dV_s^j +$$
$$+ \; \int_0^t g_{i,1}'(X_s, Y_s)\Phi_s \circ dY_s^i$$

où b_1' désigne la matrice $M \times M$ des dérivées des coordonnées de b par rapport aux coordonnées de la première variable x. On a, pour $0 \le s \le t$, $D_s^j X_t = \Phi_t \Phi_s^{-1} f_j(X_s, Y_s)$, et comme à la section 5.2 on étudie la nondégénérescence de la matrice :

$$B_t = \int_0^t \Phi_s^{-1} f_j(X_s, Y_s) f_j^*(X_s, Y_s)(\Phi_s^{-1})^* ds .$$

On raisonne comme au Théorème 5.2.13. Il nous faut donc développer par Itô–Stratonovich $(\Phi_t^{-1} f_j(X_t, Y_t), q)$.

$$\begin{aligned}
f_j(X_t, Y_t) &= f_j(x_0, 0) + \int_0^t f_{j,1}'(X_s, Y_s)b(X_s, Y_s)ds \\
&\quad + \int_0^t f_{j,1}'(X_s, Y_s)f_\ell(X_s, Y_s) \circ dV_s^\ell \\
&\quad + \int_0^t f_{j,1}'(X_s, Y_s)g_i(X_s, Y_s) \circ dY_s^i + \int_0^t f_{j,2}'(X_s, Y_s)^i \circ dY_s^i , \\
\Phi_t^{-1} &= I - \int_0^t \Phi_s^{-1} b_1'(X_s, Y_s)ds \\
&\quad - \int_0^t \Phi_s^{-1} f_{\ell,1}'(X_s, Y_s) \circ dV_s^\ell \\
&\quad - \int_0^t \Phi_s^{-1} g_{i,1}'(X_s, Y_s) \circ dY_s^i
\end{aligned}$$

ce qui donne :

$$\begin{aligned}
(\Phi_t^{-1} f_j(X_t, Y_t), q) &= (U_j(x_0, 0), q) + \int_0^t (\Phi_s^{-1}[U_j, U_0](X_s, Y_s), q)ds \\
&\quad + \int_0^t (\Phi_s^{-1}[U_j, U_\ell](X_s, Y_s), q) \circ dV_s^\ell \\
&\quad + \int_0^t (\Phi_s^{-1}[U_j, \bar{U}_i](X_s, Y_s), q) \circ dY_s^i .
\end{aligned}$$

La preuve se termine alors comme au Théorème 5.2.13. □

5.4 Application du calcul de Malliavin à l'équation de Zakai : non existence de filtres de dimension finie

On va considérer dans cette section un modèle de filtrage simplifié par rapport aux sections précédentes :

(5.16)
$$\begin{cases}
X_t = x_0 + \displaystyle\int_0^t b(X_s)ds + \int_0^t f_j(X_s)dV_s^j , \\[2mm]
Y_t = \displaystyle\int_0^t h(X_s)ds + W_t .
\end{cases}$$

Des hypothèses précises sur les coefficients seront formulées ci-dessous, qui entraîneront en particulier l'existence d'une densité pour la mesure $\sigma_t(t > 0)$:

$$p(t,x) = \frac{d\sigma_t}{dx}(x)$$

qui satisfait l'équation de Zakai (sous forme Stratonovich) :

$$(5.17) \quad \begin{cases} d_t p(t,x) = Ap(t,x)dt + h^i(x)p(t,x) \circ dY_t^i, \ t > 0, \\ p(0,\cdot) = \delta_{x_0}(\cdot). \end{cases}$$

Notons que, les coefficients de (5.16) étant réguliers, l'opérateur $A = L^* - \frac{1}{2}\sum_{i=1}^N (h^i)^2$ se met sous la forme :

$$A = \frac{1}{2}a^{j,\ell}(x)\frac{\partial^2}{\partial x^j \partial x^\ell} + \tilde{b}^j(x)\frac{\partial}{\partial x^j} + c(x)$$

avec

$$a = f_j f_j^*, \tilde{b} = -b + \frac{\partial a^j}{\partial x^j}, c = \frac{1}{2}\frac{\partial^2 a^{j\ell}}{\partial x^j \partial x^\ell} - \frac{\partial b^j}{\partial x^j} - \frac{1}{2}\sum_{i=1}^N (h^i)^2.$$

Notre but est d'appliquer le calcul de Malliavin à l'EDPS (5.17). Notons cependant que la solution de (5.17) prend ses valeurs dans un espace de Hilbert, dans lequel il n'existe pas de mesure de Lebesgue. Quel peut donc être l'énoncé d'un résultat de type "calcul de Malliavin" pour l'équation (5.17)? Cet énoncé dit que sous des conditions techniques et une hypothèse ad hoc de rang d'algèbre de Lie, toute projection orthogonale de $p(t,\cdot)$ sur un sous e.v. de dimension finie d de $L^2(\mathbb{R}^M)$ possède une densité par rapport à la mesure de Lebesgue sur $\mathbb{R}^d$. Pour énoncer précisément le résultat, introduisons "l'algèbre de Lie formelle" Λ d'opérateurs engendrée par A et les opérateurs de multiplication par $h_1(\cdot),\ldots,h_N(\cdot)$, pour le crochet de Lie $[B,C] = CB - BC$ (i.e. Λ est une algèbre pour le crochet $[\cdot,\cdot]$). Les éléments de Λ sont des opérateurs aux dérivées partielles à coefficients variables. On notera $\Lambda(x_0)$ l'espace vectoriel d'opérateurs aux dérivées partielles à coefficients constants constitué des opérateurs de Λ dont les coefficients sont fixés à leur valeur prise en x_0. On notera en outre $C_b^\omega(\mathbb{R}^M)$ l'espace des applications de $\mathbb{R}^M$ dans $\mathbb{R}$ analytiques en chaque point, bornées ainsi que toutes leurs dérivées.

Théorème 5.4.1 *Supposons satisfaites les trois hypothèses :*

(i) $a^{j\ell}, b^j, h^i \in C_b^\omega(\mathbb{R}^M); 1 \leq j, \ell \leq M, 1 \leq i \leq N,$

(ii) $a(x) \geq \epsilon I$, *pour* $\epsilon > 0$ *et tout* $x \in \mathbb{R}^M,$

(iii) $\Lambda(x_0)$ *contient tous les opérateurs aux dérivées partielles en x de tous ordres, à coefficients constants.*

Alors pour tout $t > 0$, tout $n \in \mathbb{N}$, et toute suite linéairement indépendante

$$\{\varphi_1,\ldots,\varphi_n\} \subset L^2(\mathbb{R}^M),$$

la loi de probabilité de

$$\Phi_t^n = ((p(t,\cdot),\varphi_1),\ldots,(p(t,\cdot),\varphi_n))$$

admet une densité par rapport à la mesure de Lebesgue sur $\mathbb{R}^n$.

Avant de démontrer ce théorème, discutons-en les conséquences. Etant donnée une suite $\{\varphi_i, i \geq 1\} \subset L^2(\mathbb{R}^d) \cap C_b(\mathbb{R}^d)$, on pose les définitions suivantes (σ_t et Π_t sont les lois conditionnelles non normalisées et normalisées définies au Chapitre 2) :

Définition 5.4.2 *Etant donné $t > 0$ fixé, on dit que la collection de statistiques*

$$\{\sigma_t(\varphi_i); i \geq 1\}$$

admet une statistique exhaustive régulière β en dimension finie r si β est une application $\mathcal{Y}_t-$ mesurable de Ω_2 dans $\mathbb{R}^r$ t.q. pour tout $i \geq 1$, il existe $\theta_i \in C^1(\mathbb{R}^r; \mathbb{R})$ t.q.

$$\sigma_t(\varphi_i) = \theta_i(\beta) .$$

Notons qu'une statistique exhaustive réalise une factorisation de l' application :

$$Y \quad \longrightarrow \quad \{\sigma_{t,Y}(\varphi_i); \ i \geq 1\}$$
$$\searrow \qquad \nearrow$$
$$\beta(Y)$$

On déduit alors du Théorème 5.4.1 le :

Corollaire 5.4.3 *Sous les hypothèses du Théorème 5.4.1, il n'existe pas de suite infinie linéairement indépendante $\{\varphi_i, i \geq 1\} \subset L^2(\mathbb{R}^M)$ t.q. $\{\sigma_t(\varphi_i), i \geq 1\}$ admette une statistique exhaustive régulière en dimension finie pour un $t > 0$. Le même énoncé vaut pour Π_t.*

Preuve La première affirmation découle aisément du Théorème (raisonner par l'absurde). La deuxième conclusion s'en déduit (à nouveau par l'absurde), en utilisant l'identité :

$$\sigma_t = \Pi_t \exp\left(\int_0^t \Pi_s(h^i) dY_s^i - \frac{1}{2}\int_0^t |\Pi_s(h)|^2 ds\right)$$

$\square$

Remarque 5.4.4 Le lien entre non existence de statistique exhaustive régulière et récursive (cf. Définition 5.5.1 dans l'Appendice ci–dessous) en dimension finie et des propriétés de l'algèbre de Lie Λ a été conjecturé par Brockett [14] et Mitter [62], par analogie avec la théorie géométrique du contrôle. On trouvera dans l'Appendice à la fin de ce chapitre un résultat dans ce sens. L'originalité de l'approche par le calcul de Malliavin, dûe à Ocone [65] (voir aussi Ocone, Pardoux [66]) est d'obtenir la non existence d'une statistique exhaustive de dimension finie à t fixé. Malheureusement, elle nécessite, en l'état actuel de la théorie, les hypothèses très restrictives (i) et (ii) du Théorème 5.4.1. $\square$

La fin de cette section va être consacrée à la preuve du Théorème 5.4.1. On va appliquer le calcul de Malliavin sur $(\Omega_2, \mathcal{Y}, P_Y)$. D désignera ci-dessous l'opérateur de dérivation sur Ω_2 (dans la direction de $\{Y_t\}$). On appliquera D soit à des v.a. définies sur $(\Omega_2, \mathcal{Y}, P_Y)$, soit à des v.a. définies sur $(\Omega, \mathcal{F}, \overset{\circ}{P}) = (\Omega_1 \times \Omega_2, \mathcal{F}, P_V \times P_Y)$, comme on l'avait fait avec l'autre opérateur de dérivation à la section précédente (bien que nous conservions les mêmes notations, l'opérateur D et les espaces $\mathbb{S}$ et H ne sont plus du tout les mêmes !). Le Théorème 5.4.1 sera une conséquence du Corollaire 5.2.9 (ou plutôt de la Remarque 5.2.10) si l'on démontre que pour tout $t > 0$:

(i) $\sigma_t(\varphi_i) = (p(t,\cdot), \varphi_i) \in I\!\!D^{1,2}, i = 1, \ldots, n$,

(ii) $< D\sigma_t(\varphi_i), D\sigma_t(\varphi_j) >_{i,j} > 0$ p.s.

Notons que :

$$\sigma_t(\varphi_i) = E_V[\varphi_i(X_t)Z_t] \ .$$

Formellement (rappelons que X_t ne dépend pas de Y) :

(5.18) $$D_s\sigma_t(\varphi_i) = E_V[\varphi_i(X_t)Z_t h(X_s)] \ .$$

Pour montrer (i) et justifier (5.18), on remarque tout d'abord que (par intégration par parties de l'intégrale stochastique, comme au Chapitre 4) Z_t possède une version qui, pour chaque trajectoire de Y fixée, est une fonction bornée de X, et ensuite que $\varphi_i(X_t)$ est intégrable, car X_t possède une densité qui appartient à $L^2(I\!\!R^M)$. Enfin, si l'on approche l'intégrale stochastique par des sommes de Darboux et l'exponentielle par sa série de Taylor tronquée, $\sigma_t(\varphi_i)$ est approchée par une suite de polynômes en les accroissements de Y, donc par une suite d'éléments de $I\!\!D^{1,2}$. Il reste à montrer que l'on a une suite de Cauchy dans $I\!\!D^{1,2}$ et à identifier la limite des dérivées. Etudions maintenant (ii), qui est équivalente à : $\exists \mathcal{N} \in \mathcal{Y}$ t.q. $P_Y(\mathcal{N}) = 0$ et $\forall y \notin \mathcal{N}, \forall \xi \in I\!\!R^n, < D\sigma_t(\varphi_i), D\sigma_t(\varphi_j) > (y)\xi_i\xi_j =< D\sigma_t(\varphi), D\sigma_t(\varphi) > (y) \neq 0$ avec $\varphi = \xi_i\varphi_i$. Comme la suite $(\varphi_1, \ldots, \varphi_n)$ est linéairement indépendante, (ii) est une conséquence de :

$$(ii)' \quad \begin{cases} \exists \mathcal{N} \in \mathcal{Y} \text{ t.q. } P^Y(\mathcal{N}) = 0 \text{ et } \forall y \notin \mathcal{N}, \varphi \in L^2(I\!\!R^M) \setminus \{0\}, \\ \\ \|D\sigma_t(\varphi)(y)\|_H \neq 0 \ . \end{cases}$$

Afin d'établir (ii)', commençons par donner une autre expression pour $D_s\sigma_t(\varphi)$.

$$D_s\sigma_t(\varphi) = E^V[h(X_s)Z_s E^V_{sX_s}(\varphi(X_t)Z^s_t)] \ .$$

Posons, pour $0 \leq s \leq t$ et $x \in I\!\!R^M$,

$$v_\varphi(s,x) = E^V_{sx}[\varphi(X_t)Z^t_s] \ .$$

Une extension facile de la Proposition 2.5.6 permet d'établir que v est l'unique solution dans $M^2_r(0,t; H^1(I\!\!R^M))$ de l'EDPS rétrograde :

$$\begin{cases} \dfrac{\partial v_\varphi}{\partial_s}(s,x) + A^*v_\varphi(s,x) + h_i v(s,x) \circ dY^i_s = 0, \quad 0 < s < t, x \in I\!\!R^M , \\ \\ v_\varphi(t,x) = \varphi(x) \ . \end{cases}$$

Une extension immédiate du Théorème 2.5.1 nous permet de conclure que :

$$D_s\sigma_t(\varphi) = (p(s,\cdot), h(\cdot)v_\varphi(s,\cdot)) \ .$$

Notons que d'après le Théorème 3.5.2,

$$p \in C(]0,t]; \ H^\ell(I\!\!R^M)) , \quad \text{et } v_\varphi \in C([0,t[; \ H^\ell(I\!\!R^M)) \text{ p.s.}$$

pour tout $\ell \in I\!\!N$. L'étape cruciale dans la démonstration du Théorème 5.4.1 est contenue dans la :

Proposition 5.4.5 *Il existe un ensemble* $\mathcal{N} \in \mathcal{Y}$ *t.q.* $P_Y(\mathcal{N}) = 0$ *et pour* $y \notin \mathcal{N}$, $\varphi \in L^2(\mathbb{R}^M) \setminus \{0\}$,

$$\int_0^t \mid (p(s,\cdot), h(\cdot)v_\varphi(s,\cdot)) \mid^2 (y)ds = 0$$

entraîne

$$(v_\varphi(s,\cdot), Cp(s,\cdot))(y) = 0 , \quad 0 < s \le t$$

pour tout $C \in \Lambda$.

Admettons un instant cette Proposition. Soit $y \in \Omega_2$ tel que :

$$\begin{aligned}(v_\varphi(s,\cdot), Cp(s,\cdot))(y) &= (C^* v_\varphi(s,\cdot), p(s,\cdot))(y) \\ &= 0, \quad 0 < s \le t, \quad C \in \Lambda .\end{aligned}$$

Pour tout $C \in \Lambda, \exists \ell \in \mathbb{N}$ tel que $C^* \in \mathcal{L}(H^{r+\ell}(\mathbb{R}^M), H^r(\mathbb{R}^M)), r \in \mathbb{N}$. Mais quand $s \downarrow 0, p(s,\cdot) \to \delta_{x_0}$ dans $H^{-r}(\mathbb{R}^M)$, pour $r > M/2$. Donc

$$C^* v_\varphi(0, x_0) = 0 , \quad \forall C \in \Lambda$$

ou encore :

$$C^* v_\varphi(0, x_0) = 0 , \quad \forall C \in \Lambda(x_0)$$

La propriété (iii) de $\Lambda(x_0)$ est vraie aussi bien pour $\{C^*; C \in \Lambda(x_0)\}$. On en déduit que $v(0, \cdot)$ et toutes ses dérivées partielles d'ordre quelconque sont nulles au point x_0. Remarquons (cf. Chapitre 4) que

$$u_\varphi(s, x) \triangleq e^{(h(s,x), y(s))} v_\varphi(s, x)$$

satisfait une EDP parabolique paramétrée par $\{Y_s\}$, à laquelle, grâce aux hypothèses (i) et (ii) du Théorème, on peut appliquer le Théorème 6.2. p. 221 d'Eidel'man [24], qui entraîne que $u_\varphi(0, \cdot) \in C_b^\omega$. Donc $u_\varphi(0, x) = 0, x \in \mathbb{R}^M$. D'après le résultat d'unicité rétrograde de Bardos-Tartar [4], ceci est en contradiction avec $\varphi \not\equiv 0$. Il ne nous reste plus qu'à établir la :

Preuve de la Proposition 5.4.5 Indiquons tout d'abord formellement l'argument de la démonstration. Si

$$\begin{aligned}(v_\varphi(s,\cdot), h_i(\cdot)p(s,\cdot)) &= 0, \quad 0 < s < t, \quad 1 \le i \le N, \\ d(v_\varphi(s,\cdot), h_i(\cdot)p(s,\cdot)) &= (v_\varphi(s,\cdot), [A, h_i] p(s,\cdot)) ds \\ &= 0, \quad 1 \le i \le N .\end{aligned}$$

En itérant l'argument, si $C_i = [A, h_i]$

$$\begin{aligned}d(v_\varphi(s,\cdot), C_i p(s,\cdot)) &= (v_\varphi(s,\cdot), [A, C_i]p(s,\cdot)) ds + \\ &\quad + (v_\varphi(s,\cdot), [h_j, C_i]p(s,\cdot)) \circ dY_s^j .\end{aligned}$$

On en déduit que le coefficient de ds et ceux des dY_s^j sont nuls, et par récurrence on obtient le résultat. Il y a cependant deux points qui demandent justification :

1. Nous avons utilisé le calcul stochastique de Stratonovich, alors que $p(s, \cdot)$ est $\mathcal{Y}_s$ adapté, et $v_\varphi(s, \cdot)$ est $\mathcal{Y}_t^s = \sigma\{Y_r - Y_s; s \leq r \leq t\}$ adapté, donc le couple n'est pas adapté.

2. L'énoncé de la Proposition 5.4.5 donne un résultat vrai en dehors d'un ensemble des P^Y mesure nulle $\mathcal{N}$ qui est indépendant de φ.

Le premier point peut être résolu par le calcul stochastique non adapté. On va cependant donner un argument plus simple, qui permettra aussi de résoudre le point 2. Remarquons que si l'on pose, comme au chapitre 4,

$$
\begin{aligned}
u_\varphi(t, x) &= v_\varphi(t, x) \exp[h_i(x)Y_t^i] , \\
q(t, x) &= p(t, x) \exp[-h_i(x)Y_t^i] ,
\end{aligned}
$$

on remarque que pour tout $\ell \in \mathbb{N}$,

$$
\begin{aligned}
u_\varphi &\in C^1([0, t]; H^\ell(\mathbb{R}^M)) , \\
q &\in C^1([0, t]; H^\ell(\mathbb{R}^M)) .
\end{aligned}
$$

Donc pour tout $C \in \Lambda$, il existe $n \in \mathbb{N}$ t.q.

$$
\begin{aligned}
(v_\varphi(s, \cdot), Cp(s, \cdot)) &= (u_\varphi(s, \cdot), e^{-h_i(\cdot)Y_s^i} C[e^{h_i(\cdot)Y_s^i} q(s, \cdot)]) \\
&= \sum_{\ell=1}^n (Y_s^i)^\ell \xi_{i\ell}(s)
\end{aligned}
$$

avec des processus $\{\xi_{i\ell}(s), s \in (0, t)\}$ non nécessairement adaptés, mais à trajectoires dans $C^1(0, t)$. On en déduit que les intégrales stochastiques du type :

$$
\int_r^s (v_\varphi(\theta, \cdot), Cp(\theta, \cdot)) \circ dY_\theta^i , \ 0 < r < s < t
$$

peuvent être définies par intégration par parties, et toutes les formules d'Itô–Stratonovich utilisées ont un sens pour chaque trajectoire de Y. Il reste à montrer que lorsque la somme d'une intégrale de Lebesgue et d'intégrales de Stratonovich du type ci-dessus est nulle, alors chaque intégrand est nul pour toute trajectoire de Y en dehors d'un ensemble de mesure nulle universel $\mathcal{N}$. Ceci résulte du Lemme suivant, pour la démonstration duquel nous renvoyons à l'appendice d'Ocone [65] :

Lemme 5.4.6 *Il existe un ensemble de P^Y mesure nulle $\mathcal{N} \in \mathcal{Y}$ tel que pour tout $1 \leq i \leq N$ et $m \in \mathbb{N}$, pour tous $\rho_{k_j}, 1 \leq j \leq N$, processus à trajectoires de classe C^1 (non nécessairement adaptés), si*

$$
\psi(s) = \sum_{\ell=1}^m \int_0^s \rho_{k_j}(s)(Y_s^j)^\ell \circ dY_s^i, 0 \leq s \leq t
$$

alors, pour tout $0 \leq s \leq t$,

$$
\sum_{r=1}^\infty [\psi(\tfrac{r+1}{2^n} \wedge s) - \psi(\tfrac{r}{2^n} \wedge s)]^2 \to \int_0^s |\sum_{\ell=1}^m \rho_{k_j}(s)(Y_s^j)^\ell|^2 ds
$$

sur $\mathcal{N}^C$.

5.5   Appendice : Non existence de filtres de dimention finie sans le calcul de Malliavin

Dans cette section, nous allons présenter un résultat qui s'apparente au Corollaire 5.4.3. Les hypothèses seront beaucoup plus faibles, mais la conclusion aussi.

On considère ici le problème de filtrage suivant :

$$(5.19) \quad \begin{cases} X_t = x_0 + \int_0^t b(X_s)ds + \int_0^t f_j(X_s)dV_s^j + \int_0^t g_i(X_s)dY_s^i, \\ \\ Y_t = \int_0^t h(X_s)ds + W_t \end{cases}$$

où les notations sont celles des sections précédentes, et on suppose ici que b^j, a^{jl} et h^i sont des élément de $C^\infty(\mathbb{R}^M)$, $i, j = 1, \ldots M$, $i = 1, \ldots, N$. On suppose en outre que soit l'hypothèse $(H.3)$ soit l'hypothèse $(H.4)$ de la section 2.1 est satisfaite. On peut alors définir une "loi de probabilité conditionnelle non normalisée" $\{\sigma_t, \ t \geq 0\}$ qui satisfait l'équation de Zakai.

$$(Z) \quad \sigma_t(\varphi) = \varphi(x_0) + \int_0^t \sigma_s(L\varphi)ds + \int_0^t \sigma_s(L^i\varphi)dY_s^i, \ \varphi \in C_c^\infty(\mathbb{R}^M) \ ,$$

on constate que pour $\varphi \in C_c^\infty(\mathbb{R}^M)$, $L^i\varphi \in C_c^\infty(\mathbb{R}^M)$, et $\{\sigma_t(L^i\varphi), \ t \geq 0\}$ est une semi–martingale réelle dont la variation quatratique jointe avec $\{Y_t^i\}$ est donnée par :

$$< \sigma.(L^i\varphi), Y^i >_t = \int_0^t \sigma_s((L^i)^2\varphi)ds \ .$$

On peut donc réécrire l'équation de Zakai au sens de Stratonovich :

$$\sigma_t(\varphi) = \varphi(x_0) + \int_0^t \sigma(L^0\varphi)ds + \int_0^t \sigma(L^i\varphi) \circ dY_s^i, \ \varphi \in C_c^\infty(\mathbb{R}^M)$$

avec $L^0 = L - \frac{1}{2} \sum_{i=1}^N (L^i)^2$.

Posons maintenant la :

Définition 5.5.1 *On dira que le problème de filtrage 5.19 admet une statistique exhaustive récursive et régulière en dimension finie s'il existe :*

(i) $r \in \mathbb{N}$, des champs de vecteurs $U_0, U_1, \ldots, U_\infty$ de classe C^∞ sur $\mathbb{R}^r$, $z_0 \in \mathbb{R}^r$ t.q. l'EDS

$$\xi_t = z_0 + \int_0^t U_0(\xi_s)ds + \int_0^t U_i(\xi_0)dY_s^i$$

admette une solution non explosive,

(ii) une application $\theta = \mathbb{R}^r \to \mathcal{M}_+(\mathbb{R}^M)$ telle que

$$z \to < \theta(z), \ \varphi >$$

soit de classe C^3, pour tout $\varphi \in C_c^\infty(\mathbb{R}^M)$ et que pour $1 \leq i \leq N$, il existe $\theta_i \in \mathcal{D}'(\mathbb{R}^M)$ t.q.

$$\frac{\partial}{\partial z^i} < \theta(\cdot), \ \varphi > (z_0) = < \theta_i, \ \varphi > ,$$

tels que $\sigma_t = \theta(\xi_t)$, $t \geq 0$, *p.s.* □

On désignera ci–dessous par $\sum$ l'"algèbre de Lie formelle" d'opérateurs aux dérivées partielles engendrée par $L^0, L^1, \ldots, L^N$, et par $\mathcal{A}$ l'algèbre de Lie de champs de vecteurs sur $\mathbb{R}^r$ engendrée par $U_0, U_1, \ldots, U_N$. $\sum(x_0)$ désignera l'espace vectoriel d'opérateurs aux dérivées partielles à coefficients constants obtenue en "gelant les coefficients des éléments de $\sum$ au point x_0". On a alors le :

Théorème 5.5.2 *Une condition nécessaire pour que le problème de filtrage (5.19) possède une statistique exhaustive récursive et régulière en dimension finie est que la dimension de $\sum(x_0)$ soit finie.*

Preuve Soit $\varphi \in C_c^\infty(\mathbb{R}^M)$. On a l'identité :

$$\sigma_t(\varphi) = < \theta(\xi_t), \ \varphi >, \ t \geq 0, \text{ p.s.}$$

En développant le membre de gauche de cette identité à l'aide de l'équation de Zakai, et d'autre part le membre de droite par la formule d'Itô, on obtient :

$$\varphi(x_0) + \int_0^t \sigma_s(L^0\varphi)ds + \int_0^t \sigma_s(L^i\varphi) \circ dY_s^i =$$

$$= < \theta(\xi_0), \ \varphi > + \int_0^t U_0 < \theta(\cdot), \ \varphi > (\xi_s)ds +$$

$$+ \int_0^t U_i < \theta(\cdot), \ \varphi > (\xi_s) \circ dY_s^i$$

En égalant les termes martingales et les termes à variation finie de l'identité ci–dessus, on obtient :

$$\sigma_t(L^0\varphi) = U_0 < \theta(\cdot), \ \varphi > (\xi_t), \ t \geq 0 \text{ p.s.}$$
$$\sigma_t(L^i\varphi) = U_i < \theta(\cdot), \ \varphi > (\xi_t), \ 1 \leq i \leq N, \ t \geq 0, \text{ p.s.}$$

D'après les hypothèses ci–dessus, $L^i\varphi \in C_c^\infty(\mathbb{R}^M)$, $0 \leq i \leq N$, $z \to U_i < \theta(z), \ \varphi >$ est de classe C^2, $0 \leq i \leq N$. On peut donc itérer le raisonnement ci–dessus, d'où pour $0 \leq i \leq N$,

$$L^i\varphi(x_0) + \int_0^t \sigma_s(L^0 L^i\varphi)ds + \int_0^t \sigma_s(L^j L^i\varphi) \circ dY_s^j =$$

$$= U_i < \theta(\cdot), \ \varphi > (\xi_0) + \int_0^t U_0 U_i < \theta(\cdot), \ \varphi > (\xi_s)ds$$

$$+ \int_0^t U_j U_i < \theta(\cdot), \ \varphi > (\xi_s) \circ dY_s^j$$

d'où cette fois :

$$\sigma_t(L^i L^j\varphi) = U_i U_j < \theta(\cdot), \ \varphi > (\xi_t); \ 0 \leq i, \ j \leq N, \ t \geq 0, \text{ p.s.}$$

et par différence :

$$\sigma_t([L^i, L^j]\varphi) = [U_i, U_j] < \theta(\cdot), \ \varphi > (\xi_t); \ 0 \leq i, \ j \leq N, \ t \geq 0, \text{ p.s.}$$

En itérant l'argument ci–dessus, on obtient la même identité avec des crochets de tous ordres. L'égalité de tous les crochets à l'instant $t = 0$ s'écrit de façon symbolique :

$$\sum \varphi(x_0) = <\mathcal{A}\theta(z_0), \varphi>,$$

et ceci pour tout $\varphi \in C_c^\infty(\mathbb{R}^M)$. Mais $\mathcal{A}\theta(z_0)$ est un e.v. de dimension au plus r, engendré par $\theta_1, \ldots, \theta_r$. Donc il existe $A_1, \ldots, A_r \in \sum(x_0)$ t.q. pour tout $\varphi \in C_c^\infty(\mathbb{R}^M)$ et tout $A \in \sum(x_0)$,

$$A\varphi(x_0) \in \text{ e.v. } \{A_1\varphi(x_0), \ldots, A_r\varphi(x_0)\}.$$

Ceci entraîne que dim $\sum(x_0) \leq r$. □

5.6   Commentaires bibliographiques

a.   Notre traitement du calcul de Malliavin est inspiré de Bismut [10] et de Zakai [88]. Pour les estimations supplémentaires permettant de conclure à la régularité de la densité, nous renvoyons à Norris [64].

b.   Les résultats d'existence et de régularité de la loi conditionnelle par le calcul de Malliavin ont été obtenus par Bismut, Michel [11]. On pourra aussi consulter Kusuoka, Stroock [52]. Des résultats analogues peuvent être obtenus en adaptant la preuve du théorème d'Hörmander au cas des EDPS, voir Chaleyat–Maurel, Michel [18] et Michel [59].

c.   Le résultat de non existence de filtre de dimension finie par le calcul de Malliavin est dû à Ocone [65] et à Oconc, Pardoux [66]. Le résultat de non existence d'une statistique exhaustive et récursive en dimension finie que nous avons présenté est dû à Lévine [54]. On trouvera des idées analogues dans Hijab [37], Chaleyat–Maurel, Michel [17], Hazewinkel, Marcus [34] et Hazewinkel, Marcus, Sussmann [35]. Voir aussi l'article de revue de Marcus [58].

Chapitre 6

Filtres de dimension finie et filtres de dimension finie approchés

6.1   Introduction

Nous avons décrit jusqu'ici beaucoup de résultats concernant le filtrage. Cependant, en ce qui concerne le calcul effectif d'un "filtre" (i.e. de la loi conditionnelle Π_t, ou de sa version non normalisée σ_t), nous ne sommes pas encore très avancés. En effet, $\{\sigma_t\}$ est solution d'une EDPS parabolique, dont la variable spatiale varie dans $\mathbb{R}^M$. Dans le cas de l'Exemple 1.1.1, $M = 6$. Or la résolution numérique d'une EDP parabolique dont la variable d'espace est en dimension 6 est difficile, sans compter que l'on voudrait souvent pouvoir faire les calculs en temps réel, et sur des machines pas trop grosses.

Le but de ce chapitre est triple. Nous voulons tout d'abord présenter les filtres de dimension finie dans les cas où ils existent, puis présenter des filtres de dimension finie approchés, dans le cas d'un grand rapport signal/bruit (faible bruit d'observation), et enfin donner un exemple d'algorithme de filtrage non linéaire par une méthode d'analyse numérique de l'équation de Zakai.

6.2   Le problème de filtrage linéaire gaussien : le filtre de Kalman-Bucy

Considérons le système différentiel stochastique :

$$(6.1) \quad \begin{cases} X_t = X_0 + \displaystyle\int_0^t (B(s)X_s + b(s))ds + \int_0^t F(s)dV_s + \int_0^t G(s)dY_s \,, \\[2mm] Y_t = \displaystyle\int_0^t H(s)X_s ds + W_t \,, \end{cases}$$

où B, $F \in L^\infty(\mathbb{R}_+; \mathbb{R}^{M \times M})$, $b \in L^\infty(\mathbb{R}_+; \mathbb{R}^M)$, $G \in L^\infty(\mathbb{R}_+; \mathbb{R}^{M \times N})$, $H \in L^\infty(\mathbb{R}_+; \mathbb{R}^{N \times M})$; X_0 est un v.a. gaussien de loi $N(\overline{X}_0, R_0)$ indépendant du Wiener standard $M+N$ dimensionel $\{(V_t, W_t)'\}$. Alors le couple $\{(X_t, Y_t)\}$ est un processus gaussien, et on en déduit aisément que la loi conditionnelle de X_t sachant $\mathcal{Y}_t$ est une loi gaussienne $N(\hat{X}_t, R_t)$, où $\hat{X}_t$ seul dépend des observations.

Notons que les hypothèses de la section 2.3 sont satisfaites, et l'équation de Kushner-Stratonovich est satisfaite dans cette situation. On écrit l'équation (KS) pour le système (6.1) avec $\varphi(x) = x^i$, $\varphi(x) = x^i x^j$, $1 \leq i$, $j \leq M$. Sachant que la loi Π_t est une loi de Gauss, on en déduit alors aisément les équations du filtre de Kalman-Bucy pour le couple $\{(\hat{X}_t, R_t)\}$ (il suffit d'exprimer les moments d'ordre supérieur de Π_t en fonction des deux premiers) :

$$(KB) \begin{cases} d\hat{X}_t = (B(t)\hat{X}_t + b(t))dt + G(t)dY_t + R_t H^*(t) [dY_t - H(t)\hat{X}_t \, dt] \, , \\[2mm] \hat{X}_0 = E(X_0) \, , \\[2mm] \dfrac{dR_t}{dt} = B(t)R_t + R_t B^*(t) + F(t)F^*(t) - R_t H^*(t)H(t)R_t \, , \\[2mm] R_0 = Cov(X_0) \, . \end{cases}$$

On a établi le :

Théorème 6.2.1 *Pour tout $t \geq 0$, la loi conditionnelle de X_t, sachant $\mathcal{Y}_t$, est la loi $N(\hat{X}_t, R_t)$, où $(\hat{X}_t, R_t)$ est l'unique solution du système d'équations (KB).*

Une autre façon de déduire les équations (KB) de l'équation (KS) consiste à écrire que Π_t (du moins si elle est non dégénérée) est de la forme :

$$\Pi_t(dx) = (2\pi)^{-M/2} (\det R_t)^{-1/2} \exp\left[-\frac{1}{2}(R_t^{-1}(x - \hat{X}_t), \, x - \hat{X}_t)\right] dx \, .$$

Comment peut-on établir les équations (KB) sans passer par la théorie du filtrage non linéaire ? Il y a au moins trois méthodes. La première utilise le calcul stochastique et est proche de l'une des méthodes de dérivation de l'équation de Zakai ou de Kushner-Stratonovich (soit ce que nous avons fait au Chapitre 2, soit une méthode basée sur l'innovation comme celle de Fujisaki, Kallianpur, Kunita [31]) ; on exploite en outre le caractère gaussien de la loi conditionnelle. Ce n'est pas très différent de ce que nous avons fait.

La seconde consiste à approcher la loi conditionnelle Π_t par la loi conditionnelle Π_t^n de X_t, sachant

$$\mathcal{Y}_t^n = \sigma\{Y_{\frac{k}{n}t}, \ 0 \leq k \leq n\} \cdot \Pi_t^n = N(\hat{X}_t^n, R_t^n),$$

et les équations qui donnent $(\hat{X}_t^n, R_t^n)$ sont celles du filtre de Kalman en temps discret. L'écriture de ces dernières est un simple exercice sur le conditionnement dans le cas gaussien.

Une troisième méthode consiste à réécrire le système (6.1) sous forme "bruit blanc", à ramener le problème de filtrage linéaire à celui de la recherche de l'état optimal dans un problème de contrôle "linéaire quadratique", et à utiliser les résultats de la théorie du contrôle. Nous allons écrire le problème de contrôle. Pour l'exposé du cadre mathématique correspondant, nous renvoyons le lecteur à Bensoussan [6].

L'équation d'état du système contrôlé est :

$$\begin{cases} \dfrac{dX_s}{ds} &= B(s)X_s + b(s) + F(s)\zeta_s + G(s)y(s) \,, \\[2mm] X_0 &= \overline{X}_0 + \xi \,, \end{cases}$$

où $(\xi; \ \zeta_s, \ 0 \le s \le t)$ est le contrôle, qui est un élément arbitraire de $\mathbb{R}^M \times L^2(0, t; \mathbb{R}^M)$, et $(y(s); \ 0 \le s \le t)$ est l'observation $\left(y(t) = \frac{dY_t}{dt}\right)$. La fonction coût à minimiser est :

$$J_t(\xi, \zeta) = (R_0^{-1}\xi, \xi) + \int_0^t \mid \zeta_s \mid^2 ds + \int_0^t \mid y(s) - H(s)X_s(\xi, \zeta) \mid^2 ds \,,$$

$\hat{X}_t$ est alors l'état optimal à l'instant t dans ce problème de contrôle.

6.3   Généralisations du filtre de Kalman–Bucy

Le cas linéaire gaussien est le cas le plus simple où un filtre de dimension finie existe. On sait (voir la section 5.4) que cette situation est "exceptionnelle". Elle est cependant réalisée en tout cas chaque fois que la loi conditionnelle Π_t est une loi gaussienne, ce qui peut être vrai sans que le couple $\{(X_t, Y_t)\}$ soit gaussien. C'est le cas du modèle suivant :

$$(6.2) \quad \begin{cases} X_t &= X_0 + \displaystyle\int_0^t [B(s, Y)\, X_s + b(s, Y)]ds + \int_0^t F(s, Y)dV_s \\[2mm] &\quad + \displaystyle\int_0^t [G_i(s, Y)X_s + g_i(s, Y)]dY_s^i \,, \\[3mm] Y_t &= \displaystyle\int_0^t [H(s, Y)X_s + h(s, Y)]ds + W_t \,, \end{cases}$$

où $B, F, G_1, \ldots, G_N$ sont progressivement mesurables de $\mathbb{R}_+ \times \Omega_2$ à valeurs dans $\mathbb{R}^{M \times M}$, ainsi que $H, b, g_1, \ldots, g_N$ à valeurs dans $\mathbb{R}^{M \times N}$ et dans $\mathbb{R}^M$, toutes ces fonctions étant localement bornées. On suppose en outre que $X_0 \simeq N(\overline{X}_0, R_0)$ est indépendant du Wiener standard $M + N$ dimensionel $\{(V_t, W_t)\}$, et que l'hypothèse $(H.4)$ du chapitre 2 est satisfaite.

On a alors le (H_j désigne la j^{eme} ligne de la matrice H) :

Théorème 6.3.1 *Etant donné le modèle de filtrage (6.2), pour tout $t \ge 0$ la loi conditionnelle Π_t est la loi $N(\hat{X}_t, R_t)$, où $\{(\hat{X}_t, R_t), \ t \ge 0\}$ est l'unique solution forte du système différentiel :*

$$\begin{aligned} d\hat{X}_t &= [B(t, Y)\hat{X}_t + b(t, Y) + G_j(t, Y)R_t H_j^*(t, Y)]dt + \\ &\quad + [G_i(t, Y)\hat{X}_t + g_i(t, Y)]dY_t^i + \\ &\quad + R_t H^*(t, Y)[dY_t - (H(t, Y)\hat{X}_t + h(t, Y))dt] \,, \\ \hat{X}_0 &= \overline{X}_0 = E(X_0) \,, \end{aligned}$$

$$
\begin{aligned}
dR_t &= [B(t,Y)R_t + R_t B^*(t,Y) + F(t,Y)F^*(t,Y) + \\
&\quad + G_i(t,Y)R_t G_i^*(t,Y) - R_t H^*(t,Y)H(t,Y)R_t]dt + \\
&\quad + [G_i(t,Y)R_t + R_t G_i^*(t,Y)]dY_t^i , \\
R_0 &= Cov(X_0).
\end{aligned}
$$

Preuve Sans restreindre la généralité, on suppose que $h \equiv 0$. Il suffit de montrer que Π_t est une loi de Gauss, la suite de la démonstration étant analogue à celle de la section précédente. Nous allons seulement esquisser l'argumentation, renvoyant pour les détails à Haussmann, Pardoux [33].

Il suffit de montrer qu'il existe des v.a. $\mathcal{Y}_t$ mesurables k, α et β (de dimension respectives $1, M$ et M^2) t.q.

$$
\overset{\circ}{E}{}^{\mathcal{Y}} [Z_t \exp(iu^*X_t)] = k \exp[iu^*\alpha - u^*\beta u] \text{ p.s.}
$$

pour tout $u \in \mathbb{R}^M$.

$\{\Phi_t,\ t \geq 0\}$ désignant le processus à valeurs matrices $M \times M$ solution de :

$$
\Phi_t = I + \int_0^t B(s,Y)\Phi_s ds + \int_0^t G_i(s,Y)\Phi_s dY_s^i,
$$

on a :

$$
\begin{aligned}
X_t &= \Phi_t \left[X_0 + \int_0^t \Phi_s^{-1}F(s,Y)dV_s \right] \\
&\quad + \Phi_t \left[\int_0^t \Phi_s^{-1}b(s,Y)ds + \int_0^t \Phi_s^{-1}g_i(s,Y)dY_s^i \right] = \eta_t + \gamma_t
\end{aligned}
$$

où η_t est un v.a. dont la loi conditionnelle sous $\overset{\circ}{P}$ sachant $\mathcal{Y}$ est une loi de Gauss connue et γ_t est $\mathcal{Y}_t$ mesurable.

$$
\overset{\circ}{E}{}^{\mathcal{Y}} (e^{iu^*X_t}Z_t) = \xi_1 \ \overset{\circ}{E}{}^{\mathcal{Y}} (\xi_2),
$$

avec

$$
\begin{aligned}
\xi_1 &= \exp\left\{ iu^*\gamma_t + \int_0^t (H(s,Y)\gamma_s, dY_s) - \frac{1}{2}\int_0^t |H(s,Y)\gamma_s|^2\, ds \right\}, \\
\xi_2 &= \exp\left\{ iu^*\eta_t + \int_0^t (H(s,Y)\eta_s, dY_s) - \frac{1}{2}\int_0^t (H^*H(s,Y)\eta_s, \eta_s + 2\gamma_s)ds \right\} \\
&= \exp\left\{ iu^*\eta_t + \left(\int_0^t \Phi_s^*H^*(s,Y)dY_s, X_0 \right) + \int_s^t \left(\int_s^t \Phi_\theta^* dY_\theta, \Phi_s^{-1}F(s,Y)dV_s \right) \right. \\
&\quad \left. - \int_0^t (H^*H(s,Y)\gamma_s, \eta_s)ds - \frac{1}{2}\int_0^t |H(s,Y)\eta_s|^2\, ds \right\} \\
&= \exp(\xi - \|\zeta\|^2),
\end{aligned}
$$

où (ξ, ζ) est un v.a. à valeurs dans $\mathbb{C} \times L^2(0,t;\mathbb{R}^N)$, dont la loi conditionnelle sous $\overset{\circ}{P}$ sachant $\mathcal{Y}_t$ est une loi gaussienne connue ; $\|\cdot\|$ désigne la norme dans $L^2(0,t;\mathbb{R}^N)$. Un calcul explicite permet de conclure. □

On peut encore compliquer le modèle (6.2) de deux façons, tout en conservant un filtre de dimension finie : d'une part, on peut ajouter une dérive non linéaire "à la Beneš" dans l'équation de $\{X_t\}$, et d'autre part la loi de X_0 peut être arbitraire.

Nous renvoyons le lecteur intéressé aux travaux de Beneš [5], Makowski [57] et Haussmann, Pardoux [33].

6.4   Le filtre de Kalman-Bucy étendu

Considérons le problème de filtrage :

$$
(6.3) \quad
\begin{cases}
X_t = X_0 + \displaystyle\int_0^t b(X_s)ds + \int_0^t f(X_s)dV_s + \int_0^t g(X_s)dW_s \,, \\[2mm]
Y_t = \displaystyle\int_0^t h(X_s)ds + W_t \,,
\end{cases}
$$

où, pour simplifier, on a supposé que les coefficients ne dépendent ni de t ni de Y, et sont de classe C^1 en x. Supposons tout d'abord que X_0 est presque connu (variance faible) et que f et g sont "petits". Alors $\{X_t,\ t \geq 0\}$ est "proche" de la "trajectoire nominale" :

$$
\frac{d\overline{X}_t}{dt} = b(\overline{X}_t),\ t \geq 0 \,, \quad \overline{X}_0 = X_0
$$

donc

$$
\begin{aligned}
b(X_t) &\simeq b(\overline{X}_t) + b'(\overline{X}_t)(X_t - \overline{X}_t) \,, \\
f(X_t) &\simeq f(\overline{X}_t) \,, \\
g(X_t) &\simeq g(\overline{X}_t) \,, \\
h(X_t) &\simeq h(\overline{X}_t) + h'(\overline{X}_t)(X_t - \overline{X}_t) \,,
\end{aligned}
$$

d'où

$$
(6.4) \quad
\begin{cases}
X_t \simeq X_0 + \displaystyle\int_0^t [b'(\overline{X}_s)(X_s - \overline{X}_s) + b(\overline{X}_s)]ds \\[2mm]
\qquad\qquad + \displaystyle\int_0^t f(\overline{X}_s)dV_s + \int_0^t g(\overline{X}_s)dW_s \,, \\[2mm]
Y_t \simeq \displaystyle\int_0^t [h'(\overline{X}_s)(X_s - \overline{X}_s) + h(\overline{X}_s)]ds + W_t \,.
\end{cases}
$$

Les signes "$\simeq$" n'ont aucun sens mathématique précis. Si on les remplace par des égalités, on obtient un système différentiel stochastique linéarisé autour de la "trajectoire nominale" $\{\overline{X}_t,\ t \geq 0\}$ qui est une fonction connue de t. On obtient un problème de filtrage linéaire gaussien, dont la solution est donnée par le filtre de Kalman-Bucy, que nous appellons "filtre de Kalman linéarisé" :

$$(6.5) \quad \begin{cases} d\hat{X}_t = [(b' - gh')(\overline{X}_t)\hat{X}_t + (b - gh)(\overline{X}_t) - (b' - gh')(\overline{X}_t)\overline{X}_t]dt \\ \qquad\qquad + g(\overline{X}_t)dY_t \\ \qquad\qquad + R_t h'^*(\overline{X}_t)[dY_t - (h'(\overline{X}_t)\hat{X}_t + h(\overline{X}_t) - h'(\overline{X}_t)\overline{X}_t)dt] \,, \\ \hat{X}_0 = E(X_0) \,, \\ \dfrac{dR_t}{dt} = (b' - gh')(\overline{X}_t)R_t + R_t(b' - gh')^*(\overline{X}_t) + \\ \qquad\qquad + f(\overline{X}_t)f^*(\overline{X}_t) - R_t h'^*(\overline{X}_t)h'(\overline{X}_t)R_t \,, \\ R_0 = Cov(X_0) \,. \end{cases}$$

Nous allons maintenant voir ce qu'est le "filtre de Kalman étendu". Nous revenons au problème de filtrage (6.3). Supposons que l'on dispose à chaque instant t d'un "estimateur" M_t de X_t (i.e. M_t est une v.a. $\mathcal{Y}_t$ mesurable). Si M_t est un "bon estimateur", on peut penser linéariser le problème (6.3) autour de l'estimée $\{M_t\}$, ce qui donne :

$$(6.6) \quad \begin{cases} X_t \simeq X_0 + \displaystyle\int_0^t [b'(M_s)(X_s - M_s) + b(M_s)]ds \\ \qquad\qquad + \displaystyle\int_0^t f(M_s)dV_s + \int_0^t g(M_s)dW_s \,, \\ Y_t \simeq \displaystyle\int_0^t [h'(M_s)(X_s - M_s) + h(M_s)]ds + W_t \,. \end{cases}$$

Si l'on remplace les signes "$\simeq$" par des égalités, on obtient un problème de filtrage qui possède un filtre de Kalman généralisé (au sens de la section 6.3), dont la solution est le couple $(\hat{X}_t, R_t)$. On obtient ainsi une application $\{M_t\} \to \{\hat{X}_t\}$. Le "filtre de Kalman étendu" produit un point fixe de cette application , i.e. c'est le filtre de Kalman généralisé correspondant à (6.6) où l'on a choisit $M_t = \hat{X}_t$.

$$(6.7) \quad \begin{cases} d\hat{X}_t = (b - gh)(\hat{X}_t)dt + g(\hat{X}_t)dY_t + R_t h'^*(\hat{X}_t)[dY_t - h(\hat{X}_t)dt] \,, \\ \hat{X}_0 = E(X_0) \,, \\ \dfrac{dR_t}{dt} = (b' - gh')(\hat{X}_t)R_t + R_t(b' - gh')^*(\hat{X}_t) + f(\hat{X}_t)f^*(\hat{X}_t) \\ \qquad\qquad - R_t h'^*(\hat{X}_t)h'(\hat{X}_t)R_t \,, \\ R_0 = Cov(X_0) \,. \end{cases}$$

Bien entendu, les démarches conduisant aux filtres de Kalman linéarisé et étendu ne sont pas justifiées. Mais les algorithmes correspondants sont très utilisés en pratique. Les résultats sont souvent bons, et parfois catastrophiques. On se doute que le filtre de Kalman étendu a des chances d'être efficace lorsque l'estimateur est "bon", et que les fonctions que l'on linéarise ne sont "pas trop non linéaires". Le résultat peut être très mauvais dès que ces conditions ne sont pas satisfaites. En particulier, une mauvaise connaissance de la condition initiale peut suffire à rendre la méthode inefficace.

Exemple 6.4.1 Considérons le problème de filtrage suivant $(M = N = 1)$:

$$dX_t = f\,dV_t\,,$$
$$dY_t = h(X_t)\,dt + dW_t\,,$$

avec $X_0 \simeq N(\overline{X}_0, R_0)$. On suppose que $h \in C^1(\mathbb{R})$ et possède un unique minimum en $x = 0$, $xh'(x) > 0$ pour $x \neq 0$, et $lim_{x\to\pm\infty}h(x) = +\infty, | h(x) | \leq c(1+ | x |)$.

Pour cet exemple, les équations du filtre de Kalman étendu peuvent s'écrire sous la forme :

$$(6.8) \quad \begin{cases} d\hat{X}_t = R_t h'(\hat{X}_t)(h(X_t) - h(\hat{X}_t))dt + R_t h'(\hat{X}_t)dW_t\,, \\ \dfrac{dR_t}{dt} = f^2 - h'(\hat{X}_t)^2 R_t^2\,. \end{cases}$$

A tout $x \in \mathbb{R}$, on associe $\varphi(x) \in \mathbb{R}$ tel que :

$$x\varphi(x) < 0\,,$$
$$h(\varphi(x)) = h(x)\,.$$

Si $X_t \cdot \hat{X}_t < 0$, la dérive dans la première équation tend à ce que $\hat{X}_t$ "suive" $\varphi(X_t)$. Donc dans cette situation le filtre de Kalman étendu n'a pas tendance à corriger les fautes de signe. D'ailleurs, le signe de $\hat{X}_t$ reste p.s. constant pour $t \geq 0$. Pourtant, si h n'est "pas trop proche d'une fonction paire" et si le rapport signal/bruit est grand, il est possible de retrouver le signe de X_t avec une faible probabilité d'erreur (voir Fleming, Pardoux [27]). □

On trouvera d'autres exemples où le filtre de Kalman étendu est inefficace dans Picard [77].

6.5 Filtres approchés dans le cas d'un grand rapport signal/bruit

On va maintenant, en suivant les travaux de Picard [73], (voir aussi Bobrovsky, Zakai [12], Katzur, Bobrovsky, Schuss [43], Bensoussan [7], Picard [74], [76], [77]) étudier une situation où le filtre de Kalman étendu et d'autres filtres obtenus par linéarisation sont efficaces.

On considèrera un cas unidimensionel $(M = N = 1)$, avec $g \equiv 0$, et un "petit bruit d'observation" ou ce qui revient au même un "grand rapport signal/bruit". Plus précisément, on considère le modèle :

$$\begin{cases} X_t = X_0 + \int_0^t b(X_s)ds + \int_0^t f(X_s)dV_s \,, \\[2mm] Y_t = \int_0^t h(X_s)ds + \varepsilon W_t \,. \end{cases}$$

On va supposer ci-dessous que h est injective, ce qui fait que le problème est trivial pour $\varepsilon = 0$: dans ce cas, la loi conditionnelle Π_t est la mesure de Dirac $\delta_{h^{-1}(\frac{dY_t}{dt})}$. Notons que dans ce cas la "mémoire" du filtre est "nulle" (on peut oublier les anciennes observations). On va voir que pour "ε petit", la variance de la loi conditionnelle est "petite" et sous certaines hypothèses, la mémoire du filtre est "courte".

Introduisons tout d'abord les hypothèses dont nous aurons besoin.

$$(H.1) \qquad b, f \in C^2(\mathbb{R}) \,; \ h \in C^3(\mathbb{R}) \,.$$

On suppose en outre que les dérivées de b, f, h sont bornées, et que f est bornée.

$$(H.2) \qquad \exists \alpha > 0 \ \text{t.q.} \ f(x) \geq \alpha, \ \forall x \in \mathbb{R} \,,$$

$$(H.3) \qquad h'b \ \text{et} \ f^{-1}b \ \text{sont lipschitziennes} \,,$$

$$(H.4) \qquad \exists \delta > 0 \ \text{t.q.} \ | h(x) - h(y) | \geq \delta \mid x - y \mid, \ \forall \, x,y \in \mathbb{R} \,.$$

On suppose en outre que tous les moments de X_0 sont finis. Dans la suite, on notera $\| \cdot \|_p$ la norme dans $L^p(\Omega, \mathcal{F}, P)$, $p \geq 1$. Considérons le "filtre sous-optimal" suivant :

$$M_t = m_0 + \int_0^t b(M_s)ds + \frac{1}{\varepsilon} \int_0^t K_s(dY_s - h(M_s)ds)$$

où $m_0 \in \mathbb{R}$ est arbitraire et $\{K_t, \ t \geq 0\}$ est un processus $\mathcal{Y}_t$–progressivement mesurable et borné tel que :

$$(H.5) \qquad \exists \beta > 0 \ \text{t.q.} \ K_t(\omega) \, \text{signe} \, (h') \geq \beta \,, \quad \forall (t,\omega) \,.$$

Notons que d'après $(H.4)$ $\quad | h'(x) | \geq \delta \ \forall x$, et en particulier le signe de h' est constant.

Proposition 6.5.1 *Supposons $(H.1)$, $(H.4)$ et $(H.5)$ satisfaites. Alors pour tous $t_0 > 0, p \geq 1$,*

$$\sup_{t \geq t_0} \| X_t - M_t \|_p = 0(\sqrt{\varepsilon}) \,.$$

Corollaire 6.5.2 *Pour tout $t_0 > 0$, $p \geq 1$,*

$$\sup_{t \geq t_0} \| M_t - E(X_t/\mathcal{Y}_t) \|_p = 0(\sqrt{\varepsilon}) \,,$$

$$\sup_{t \geq t_0} \| X_t - E(X_t/\mathcal{Y}_t) \|_p = 0(\sqrt{\varepsilon}) \,.$$

Preuve de la Proposition Il résulte de la formule d'Itô :

$$d(h(X_t) - h(M_t)) = -\frac{h'(M_t)K_t}{\varepsilon}(h(X_t) - h(M_t))dt$$
$$+ (Lh(X_t) - \tilde{L}_t h(M_t))dt + h'f(X_t)dV_t - h'(M_t)K_t dW_t$$

où $\tilde{L}_t = b(m)\frac{\partial}{\partial m} + \frac{1}{2}K_t^2 \frac{\partial^2}{\partial m^2}$.

$$h(X_t) - h(M_t) = \exp\left(-\frac{1}{\varepsilon}\int_0^t h'(M_s)K_s ds\right)[h(X_0) - h(m_0)]$$
$$+ \int_0^t e^{-\frac{1}{\varepsilon}\int_s^t h'(M_r)K_r dr}[Lh(X_s) - \tilde{L}h(M_s)]ds$$
$$+ \int_0^t e^{-\frac{1}{\varepsilon}\int_s^t h'(M_r)K_r dr} h'f(X_s)dV_s$$
$$+ \int_0^t e^{-\frac{1}{\varepsilon}\int_s^t h'(M_r)K_r dr} h'(M_s)K_s dW_s .$$

Puisque $h'(M_t)K_t \geq \beta\,\delta > 0$, pour $t \geq t_0$, le premier terme est d'ordre $e^{-c/\varepsilon}$, le second d'ordre ε et les deux derniers d'ordre $\sqrt{\varepsilon}$. □

Le résultat principal de cette section est que la première partie du Corollaire peut être améliorée sous l'hypothèse $(H.2)$, si l'on choisit convenablement K_t. On suppose dans toute la suite pour fixer les idées que $h' > 0$.

Théorème 6.5.3 *Supposons les hypothèses $(H.1)$, $(H.2)$, $(H.3)$ et $(H.4)$ satisfaites. Alors, si $K_t = f(M_t)$, pour tout $t_0 > 0$, $p \geq 1$,*

$$\sup_{t \geq t_0} \| E(X_t/\mathcal{Y}_t) - M_t \|_p = 0(\varepsilon) .$$

Preuve La démonstration, assez longue, va se faire en plusieurs étapes.

a. Deux changements de lois de probabilité.

On effectue tout d'abord le changement de probabilité usuel :

$$\left.\frac{d\overset{\circ}{P}}{dP}\right|_{\mathcal{F}_t} = Z_t^{-1}$$

où $Z_t = \exp\left(\frac{1}{\varepsilon^2}\int_0^t h(X_s)dY_s - \frac{1}{2\varepsilon^2}\int_0^t h^2(X_s)ds\right)$, puis le changement de probabilité supplémentaire :

$$\left.\frac{d\tilde{P}}{d\overset{\circ}{P}}\right|_{\mathcal{F}_t} = \Lambda_t^{-1}$$

où $\Lambda_t = \exp\left(-\frac{1}{\varepsilon}\int_0^t (h(X_s) - h(M_s))dV_s + \frac{1}{2\varepsilon^2}\int_0^t (h(X_s) - h(M_s))^2 ds\right)$. Sous $\tilde{P}$, $\tilde{V}_t = V_t - \frac{1}{\varepsilon}\int_0^t (h(X_s) - h(M_s))ds$ et $Y_{t/\varepsilon}$ sont des processus de Wiener standard indépendants, et :

$$dX_t = \frac{1}{\varepsilon} f(X_t)(h(X_t) - h(M_t))dt + b(X_t)dt + f(X_t)d\tilde{V}_t .$$

Si $\Gamma \in L^1(\Omega, \mathcal{F}_t, P)$,

$$E(\Gamma/\mathcal{Y}_t) = \frac{\tilde{E}(\Gamma Z_t \Lambda_t / \mathcal{Y}_t)}{\tilde{E}(Z_t \Lambda_t / \mathcal{Y}_t)} .$$

b. Calcul de $Z_t \Lambda_t$

$$Z_t \Lambda_t = \exp\left\{ -\frac{1}{\varepsilon} \int_0^t (h(X_s) - h(M_s))dV_s + \frac{1}{\varepsilon^2} \int_0^t h(X_s)(dY_s - h(M_s)ds) \right.$$
$$\left. + \frac{1}{2\varepsilon^2} \int_0^t h^2(M_s)ds \right\} .$$

On note $\gamma = h'f$, $F(x,m) = \gamma^{-1}(m)[h(x) - h(m)]^2$, $A_x = f^2(x)\frac{\partial^2}{\partial x^2}$, $A_m = f^2(m)\frac{\partial^2}{\partial m^2}$. Il résulte de la formule d'Itô :

$$F(X_t, M_t)$$
$$= F(X_0, M_0) + 2\int_0^t (h(X_s) - h(M_s))\gamma^{-1}(M_s)h'(X_s)dX_s$$
$$\quad -2\int_0^t (h(X_s) - h(M_s))f^{-1}(M_s)dM_s + \int_0^t (h(X_s) - $$
$$\quad -h(M_s))^2 \frac{\partial \gamma^{-1}}{\partial m}(M_s)dM_s + \int_0^t (A_x F + A_m F)(X_s, M_s)ds$$
$$= F(X_0, M_0) - 2\int_0^t (h(X_s) - h(M_s))(\gamma^{-1}(X_s) - \gamma^{-1}(M_s))h'(X_s)dX_s$$
$$\quad +2\int_0^t (h(X_s) - h(M_s))dV_s - \frac{2}{\varepsilon}\int_0^t [h(X_s) - h(M_s)][dY_s - h(M_s)ds]$$
$$\quad +2\int_0^t (h(X_s) - h(M_s))(f^{-1}b(X_s) - f^{-1}b(M_s))ds$$
$$\quad +\int_0^t (h(X_s) - h(M_s))^2 \frac{\partial \gamma^{-1}}{\partial m}(M_s)dM_s + \int_0^t (A_x F + A_m F)(X_s, M_s)ds ,$$

$$Z_t \Lambda_t = \exp\left\{ -\frac{1}{2\varepsilon}(F(X_t, M_t) - F(X_0, M_0)) + \frac{1}{\varepsilon}\int_0^t \psi_1(X_s, M_s)ds \right.$$
$$\quad +\frac{1}{\varepsilon}\int_0^t \psi_2(X_s, M_s)dM_s - \frac{1}{\varepsilon}\int_0^t \psi_3(X_s, M_s)dX_s$$
$$\quad \left. +\frac{1}{\varepsilon^2}\int_0^t h(M_s)dY_s - \frac{1}{2\varepsilon^2}\int_0^t h^2(M_s)ds \right\}$$

avec

$$\psi_1(x,m) = (h(x) - h(m))(f^{-1}b(x) - f^{-1}b(m)) + \frac{1}{2}(A_x F + A_m F)(x,m) ,$$
$$\psi_2(x,m) = \frac{1}{2}(h(x) - h(m))^2 \frac{\partial \gamma^{-1}}{\partial m}(m) ,$$
$$\psi_3(x,m) = h'(x)((fh'(x))^{-1} - (fh'(m))^{-1})(h(x) - h(m)) .$$

c. Dérivation sur l'espace du processus de Wiener

$D(resp. \ \tilde{D})$ désignera l'opérateur de dérivation dans la direction de $V(resp. \ \tilde{V})$ (définis de façon analogue à ce qui a été fait à la section 5.3.2). Alors pour $0 \le s \le t$,

$$D_s X_t = \zeta_{st} f(X_s)$$

où $\{\zeta_{st}, \ t \ge s\}$ est la solution de l'EDS :

$$\zeta_{st} = 1 + \frac{1}{\varepsilon} \int_s^t f'(X_r)(h(X_r) - h(M_r))\zeta_{sr} dr +$$
$$+ \frac{1}{\varepsilon} \int_s^t fh'(X_r)\zeta_{sr} dr + \int_s^t b'(X_r)\zeta_{sr} dr + \int_s^t f'(X_r)\zeta_{sr} d\tilde{V}_r \ .$$

Dans la suite, si g est une fonction de (x, m), on notera g' pour $\frac{\partial g}{\partial x}$. Il résulte du calcul fait ci-dessus :

$$\tilde{D}_s \, Log(Z_t \Lambda_t) = -\frac{1}{2\varepsilon} F'(X_t, M_t)\tilde{D}_s X_t + \frac{1}{\varepsilon} \int_s^t \psi_1'(X_r, M_r)\tilde{D}_s X_r dr$$
$$+ \frac{1}{\varepsilon} \int_s^t \psi_2'(X_r, M_r)\tilde{D}_s X_r dM_r - \frac{1}{\varepsilon} \int_s^t \psi_3'(X_r, M_r)\tilde{D}_s X_r dX_r$$
$$- \frac{1}{\varepsilon} \int_s^t \psi_3(X_r, M_r) dr(\tilde{D}_s X_r) \ ,$$

$$\frac{1}{2} F'(X_t, M_t) = -\frac{\varepsilon}{t} \int_0^t (\tilde{D}_s X_t)^{-1} \tilde{D}_s \, Log(Z_t \Lambda_t) ds$$
$$+ \frac{1}{t} \int_0^t ds \int_s^t \psi_1'(X_r, M_r)(\tilde{D}_s X_t)^{-1} \tilde{D}_s X_r dr$$
$$+ \frac{1}{t} \int_0^t ds \int_s^t \psi_2'(X_r, M_r)(\tilde{D}_s X_t)^{-1} \tilde{D}_s X_r dM_r$$
$$- \frac{1}{t} \int_0^t ds(\tilde{D}_s X_t)^{-1} \int_s^t \psi_3'(X_r, M_r)\tilde{D}_s X_r dX_r$$
$$- \frac{1}{t} \int_0^t ds(\tilde{D}_s X_t)^{-1} \int_s^t \psi_3(X_r, M_r) dr(\tilde{D}_s X_r) \ .$$

d. Estimation de $E(X_t/\mathcal{Y}_t) - M_t$

Puisque $\frac{1}{2} F'(x, m) = (h(x) - h(m))(h'(m)f(m))^{-1} h'(x)$

$$E\left[\frac{1}{2} F'(X_t, M_t)/\mathcal{Y}_t\right]$$
$$= f^{-1}(M_t) E[h(X_t) - h(M_t)/\mathcal{Y}_t]$$
$$+ (h'(M_t)f(M_t))^{-1} E[(h(X_t) - h(M_t))(h'(X_t) - h'(M_t))/\mathcal{Y}_t]$$
$$= f^{-1}(M_t) E[h(X_t) - h(M_t)/\mathcal{Y}_t] + 0(\varepsilon)$$

d'après la Proposition 6.5.1. Il résulte alors des hypothèses $(H.2)$ et $(H.4)$ que le Théorème sera établi si l'on montre que pour tous $t_0 > 0$, $p \ge 1$,

$$\sup_{t \ge t_0} \| \ E[\frac{1}{2} F'(X_t, M_t)/\mathcal{Y}_t] \ \|_p = 0(\varepsilon)$$

ce qui, au vu de la dernière égalité de la partie b de la démonstration résulte de :

(i) $\displaystyle\sup_{t \geq t_0} \frac{1}{t} \parallel E\left(\int_0^t (\tilde{D}_s X_t)^{-1} \tilde{D}_s (Log Z_t \Lambda_t) ds / \mathcal{Y}_t\right) \parallel_p \leq c_p$,

(ii) $\displaystyle\sup_{t \geq t_0} \frac{1}{t} \parallel E\left(\int_0^t ds \int_s^t \psi_1'(X_r, M_r)(\tilde{D}_s X_t)^{-1} \tilde{D}_s X_r dr / \mathcal{Y}_t\right) \parallel_p = 0(\varepsilon)$.

Démontrons (i).

$$\frac{1}{t} E\left(\int_0^t (\tilde{D}_s X_t)^{-1} \tilde{D}_s Log(Z_t \Lambda_t) ds / \mathcal{Y}_t\right) = \frac{1}{t} \frac{\tilde{E}\left(\int_0^t (\tilde{D}_s X_t)^{-1} \tilde{D}_s (Z_t \Lambda_t) ds / \mathcal{Y}_t\right)}{\tilde{E}(Z_t \Lambda_t / \mathcal{Y}_t)} .$$

Notons que

$$\begin{aligned} \tilde{D}_s X_t &= \zeta_{st} f(X_s) , \\ (\tilde{D}_s X_t)^{-1} &= \zeta_{ts} f^{-1}(X_s), \text{ avec } \zeta_{ts} = \zeta_{st}^{-1} . \end{aligned}$$

On a par ailleurs l'identité $\zeta_{ts} = \zeta_{t0} \zeta_{0s}$. On va utiliser ci-dessous deux formules d'intégration par parties analogues à celle de la section 5.3.2.

$$\begin{aligned} \tilde{E}\left[\int_0^t (\tilde{D}_s X_t)^{-1} \tilde{D}_s (Z_t \Lambda_t) ds / \mathcal{Y}_t\right] &= \\ &= \tilde{E}\left[\zeta_{t0} \int_0^t \zeta_{0s} f^{-1}(X_s) \tilde{D}_s (Z_t \Lambda_t) ds / \mathcal{Y}_t\right] \\ &= \tilde{E}\left[\int_0^t \zeta_{0s} f^{-1}(X_s) \tilde{D}_s (\zeta_{t0} Z_t \Lambda_t) ds / \mathcal{Y}_t\right] \\ &\quad - \tilde{E}\left[Z_t \Lambda_t \int_0^t \zeta_{0s} f^{-1}(X_s) \tilde{D}_s \zeta_{t0} ds / \mathcal{Y}_t\right] \\ &= \tilde{E}\left[Z_t \Lambda_t (\zeta_{t0} \int_0^t \zeta_{0s} f^{-1}(X_s) d\tilde{V}_s - \int_0^t \zeta_{0s} f^{-1}(X_s) \tilde{D}_s \zeta_{t0} ds) / \mathcal{Y}_t\right] . \end{aligned}$$

Donc :

$$\begin{aligned} \frac{1}{t} E\left(\int_0^t (\tilde{D}_s X_t)^{-1} \tilde{D}_s Log(Z_t \Lambda_t) ds / \mathcal{Y}_t\right) \\ = \frac{1}{t} E\left(\zeta_{t0} \int_0^t \zeta_{0s} f^{-1}(X_s) dV_s / \mathcal{Y}_t\right) - \frac{1}{t} E\left(\int_0^t \zeta_{0s} f^{-1}(X_s) \tilde{D}_s \zeta_{t0} ds / \mathcal{Y}_t\right) \\ - \frac{1}{\varepsilon t} E\left(\int_0^t [h(X_s) - h(M_s)] \zeta_{ts} f^{-1}(X_s) ds / \mathcal{Y}_t\right) \\ = \frac{1}{t} E\left(\int_0^t f^{-1}(X_s) \zeta_{0s} (D_s \zeta_{t0} - \tilde{D}_s \zeta_{t0}) ds / \mathcal{Y}_t\right) \\ - \frac{1}{\varepsilon t} E\left(\int_0^t [h(X_s) - h(M_s)] \zeta_{ts} f^{-1}(X_s) ds / \mathcal{Y}_t\right) . \end{aligned}$$

Il reste à montrer que les normes dans $L^p(\Omega, \mathcal{F}, P)$ des derniers termes sont bornées uniformément pour $t \geq t_0$, ce qui résulte des :

Lemme 6.5.4 *Pour tout $\varepsilon_0 > 0$, $p \geq 1$, il existe $a(p)$ et $b(p) > 0$ t.q.*

$$\parallel \zeta_{ts} \parallel_p \leq a(p) \exp\left[-\frac{b(p)}{\varepsilon}(t-s)\right], \; 0 \leq s \leq t .$$

Lemme 6.5.5

$$\zeta_{0s} D_s(\zeta_{t0}) = D_s(\zeta_{ts}) \,,$$
$$\zeta_{0s} \tilde{D}_s(\zeta_{t0}) = \tilde{D}_s(\zeta_{ts})$$

et pour tout $p \geq 1$, il existe $c(p)$ t.q.

$$\sup_{t \geq s} \| D_s \zeta_{ts} \|_p \leq c(p) \,,$$

$$\sup_{t \geq s} \| E[\tilde{D}_s(\zeta_{ts})/\mathcal{F}_s \vee \mathcal{Y}_t] \|_p \leq c(p) \,.$$

Notons que l'estimation des termes suivants résulte à nouveau des Lemmes, et du fait que les normes dans $L^p(\Omega, \mathcal{F}, P)$ des processus $\psi_i'(X_t, M_t)$ $(i = 2, 3)$ sont d'ordre $\sqrt{\varepsilon}$, et celle du processus $\psi_3(X_t, M_t)$ est d'ordre ε. Il nous reste donc à passer à la :

Preuve du lemme 6.5.4 : $\{\zeta_{ts}, \; t \geq s\}$ est l'unique solution de l'EDS :

$$\zeta_{ts} = 1 - \frac{1}{\varepsilon} \int_s^t f h'(X_r) \zeta_{rs} dr + \int_s^t (f'^2 - b')(X_r) \zeta_{rs} dr - \int_s^t f'(X_r) \zeta_{rs} dV_r$$

i.e. est donné par la formule :

$$(6.9) \quad \zeta_{ts} = \exp\left(-\frac{1}{\varepsilon} \int_s^t f h'(X_r) dr + \int_s^t \left(\frac{f'^2}{2}(X_r) - b'(X_r) \right) dr - \int_s^t f'(X_r) dV_r \right) \,.$$

Le Lemme résulte alors de ce que $f h'(x) \geq \alpha \delta > 0$ pour tout x, et les intégrands des autres intégrales dans l'exponentielle sont bornés.

Preuve du lemme 6.5.5. Les deux égalités se déduisent aisément de (6.9), dont on tire aussi :

$$\tilde{D}_s \zeta_{ts} = -\frac{1}{\varepsilon} \int_s^t \zeta_{tr} \varphi(X_r) f(X_s) ds + \int_s^t \zeta_{tr} \psi'(X_r) f(X_s) dr$$
$$- \zeta_{ts} f'(X_s) - \zeta_{ts} \int_s^t f''(X_r) f(X_s) \zeta_{sr} dV_r \,,$$
$$D_s \zeta_{ts} = \zeta_{ts} \left[-\frac{1}{\varepsilon} \int_s^t f h'(X_r) D_s X_r dr + \int_s^t \psi'(X_r) D_s X_r dr \right.$$
$$\left. - f'(X_s) - \int_s^t f''(X_r) D_s X_r dV_r \right] \,,$$

où $\varphi(x) = (f h')'(x) + f'(x) h'(x)$, $\psi(x) = \frac{1}{2} f'^2(x) - b'(x)$.

Dans l'expression de $D_s \zeta_{ts}$, le coefficient de ζ_{ts} ne dépend de ε que par le coefficient "explicite" $\frac{1}{\varepsilon}$. L'estimation de $D_s \zeta_{ts}$ est donc une conséquence facile du Lemme 6.5.4. En ce qui concerne $\tilde{D}_s \zeta_{ts}$, il résulte à nouveau du Lemme 6.5.4. que les normes dans $L^2(\Omega, \mathcal{F}, P)$

des trois premiers termes sont bornées, uniformément pour $t \geq s$. Pour le dernier terme, on a :

$$E\left(\zeta_{ts}\int_s^t f''(X_r)f(X_s)\zeta_{sr}dV_r/\mathcal{F}_s \vee \mathcal{Y}_t\right)$$
$$= E\left(\int_s^t f''(X_r)f(X_s)D_r(\zeta_{ts})\zeta_{sr}dV_r/\mathcal{F}_s \vee \mathcal{Y}_t\right)$$
$$= E\left(\int_s^t f''(X_r)f(X_s)D_r\zeta_{tr}dr/\mathcal{F}_s \vee \mathcal{Y}_t\right).$$

On a utilisé de nouveau un intégration par parties et (6.9). L'estimation du dernier terme résulte alors de l'estimation de $D_r\zeta_{tr}$, et du fait que pour $t - r$ grand, sa norme dans $L^p(\Omega, \mathcal{F}, P)$ décroît comme celle de ζ_{tr}. $\qquad\square$

Remarquons que l'opération de dérivation par rapport à $\tilde{D}$ (et le choix de $\tilde{P}$) ont joué un rôle crucial dans l'obtention des estimations.

Remarque 6.5.6 Pour le problème ci-dessus, le filtre de Kalman étendu donnerait également une approximation d'ordre ε du filtre optimal. Par contre, on peut trouver des filtres qui donnent une meilleure approximation du filtre optimal. Dans le cas où σ est constant et h est linéaire, le filtre du Théorème 6.5.3 donne une approximation d'ordre $\varepsilon^{3/2}$, et le filtre de Kalman étendu une approximation d'ordre ε^2 (voir Picard [73]).

Pour une étude du problème ci-dessus en temps discret, voir Milheiro [61].

Remarque 6.5.7 On s'est contenté d'étudier le cas où h est bijectif, i.e. le problème est tout à fait trivial pour $\varepsilon = 0$. Les cas plus intéressants sont ceux où h n'est pas bijectif. Un premier exemple est celui où $M = N$ (= 1 pour simplifier) et h est monotone par morceaux (voir l'Exemple 6.3). Ce cas est étudié dans Fleming, Ji, Pardoux [29], Fleming, Pardoux [28] et Roubaud [79]. Un deuxième exemple est celui où $M > N$, et h est une fonction bijective de N des coordonnées de x. Cette situation se subdivise en deux cas, suivant que la variance de la loi conditionnelle est petite (avec ε) ou pas. On trouvera des résultats sur ce problème dans Picard [77].

6.6 Un algorithme de filtrage non linéaire (dans un cas particulier)

Nous allons décrire un algorithme de type "méthode particulaire" de résolution de l'équation de Zakai, dans le cas particulièrement simple d'un problème de filtrage "sans bruit de dynamique" (qui inclut l'Exemple 1.1.2) :

$$(6.10) \qquad \begin{cases} X_t = X_0 + \int_0^t b(s, X_s)ds, \\ Y_t = \int_0^t h(s, X_s)ds + W_t. \end{cases}$$

La loi de X_0 est une probabilité quelconque Π_0 sur $\mathbb{R}^M$. On suppose que b est localement bornée et lipschitzienne en x, uniformément par rapport à t, et que h est bornée.

Notons $(\phi_t(x); t \geq 0, x \in \mathbb{R}^M)$ le flot associé à l'EDO satisfaite par $\{X_t\}$, i.e. :

$$\begin{cases} \dfrac{d\phi_t}{dt}(x) = b(t, \phi_t(x)), \ t \geq 0 \ , \\[2mm] \phi_0(x) = x \ . \end{cases}$$

Il est alors immédiat que si $\varphi \in C_b(\mathbb{R}^M)$,

$$\sigma_t(\varphi) = \int_{\mathbb{R}^M} \varphi(\phi_t(x)) \exp\left[\int_0^t (h(s, \phi_s(x)), dY_s) - \frac{1}{2} \int_0^t \mid h(s, \phi_s(x)) \mid^2 ds \right] \Pi_0(dx).$$

Si l'on approche Π_0 par une mesure de la forme :

$$\Pi_0^n = \sum_{i=1}^n a_i^n \delta_{x_i^n}$$

où $x_i^n \in \mathbb{R}^M$, $a_i^n \in [0,1]$, $1 \leq i \leq n$; $\sum_{i=1}^n a_i^n = 1$, alors σ_t est approchée par :

$$\sigma_t^n = \sum_{i=1}^n a_i^n(t) \delta_{x_i^n(t)}$$

avec $x_i^n(t) = \phi_t(x_i^n)$,

$$a_i^n(t) = a_i^n \exp\left[\int_0^t (h(s, x_i^n(s)), dY_s) - \frac{1}{2} \int_0^t \mid h(s, x_i^n(s)) \mid^2 ds \right] \ .$$

On a alors le :

Théorème 6.6.1 *Si $h \in C_b^1(\mathbb{R}_+ \times \mathbb{R}^M)$, et si b est bornée, alors $\Pi_0^n \Rightarrow \Pi_0$ quand $n \to \infty$ entraîne :*

$$\sigma_t^n \Rightarrow \sigma_t \ p.s. \ quand \ n \to \infty,$$

pour tout $t > 0$.

Preuve Le caractère p.s. de la convergence s'obtient en intégrant par parties

$$\int_0^t (h(s, \phi_s(x)), dY_s) \ .$$

Le reste est immédiat. □

Remarque 6.6.2 Cet algorithme est tout à fait parallélisable, puisque l'évolution de $\{(x_i^n(t), \ a_i^n(t)); \ t \geq 0\}$ se calcule pour chaque i indépendemment des autres.

L'évolution des points $x_i^n(t)$ est donnée par le modèle d'évolution de X_t, l'évolution des poids $a_i^n(t)$ est guidée par les observations. Notons que d'une part on a intérêt à normaliser périodiquement les $a_i^n(t)$ par $\sum_i a_i^n(t)$, et d'autre part que les valeurs relatives des $a_i^n(t)$ peuvent devenir très différentes les unes des autres. On peut avoir intérêt à rajouter des points dans la zone des $x_i^n(t)$ où les $a_i^n(t)$ sont "grands" et à en supprimer dans la zone des $x_i^n(t)$ où les $a_i^n(t)$ sont "très petits". En rendant ainsi le "maillage" "adaptatif", on peut espérer avoir une bonne approximation de la loi conditionnelle avec un petit nombre de points. □

Remarque 6.6.3 La méthode particulaire décrite ci-dessus peut s'adapter à une situation "avec bruit de dynamique", mais elle devient plus lourde à mettre en oeuvre. L'idée est en gros la suivante. On discrétise le temps. Sur chaque intervalle de temps élémentaire, chacune des n masses de Dirac à l'instant t_k est transformée en une mesure à densité à l'instant t_{k+1}. On approxime à nouveau la somme de ces n mesures à densité par une combinaison de n masses de Dirac, et on recommence sur l'intervalle de temps suivant. □

Bibliographie

[1] R.A. Adams : *Sobolev spaces*, Acad. Press 1975.

[2] F. Alinger, S.K. Mitter : New results on the innovations problem for non–linear filtering, *Stochastics* 4, 339-348, 1981.

[3] J. Baras, G. Blankenship, W. Hopkins : Existence, uniquenes and asymptotic behavior of solutions to a class of Zakai equations with unbounded coefficients, IEEE Trans. AC–28, 203–214, 1983.

[4] C. Bardos, L. Tartar : Sur l'unicité des équations paraboliques et quelques questions voisines, *Arkive for Rat. Mech. and Anal.* 50, 10-25, 1973.

[5] V. Beneš : Exact finite dimensional filters for certain diffusions with nonlinear drift, *Stochastics* 5, 65-92, 1981.

[6] A. Bensoussan : *Filtrage optimal des systèmes linéaires*, Dunod 1971.

[7] A. Bensoussan : On some approximation techniques in nonlinear filtering, in *Stochastic Differential Systems, Stochastic Control and Applications* (Minneapolis 1986), Springer 1988.

[8] A. Bensoussan : *Stochastic control of partially observable systems*, Cambridge Univ. Press, à paraître.

[9] J.M. Bismut : A generalized formula of Itô and some other properties of stochastic flows, *Z. Wahrsch.* 55, 331-350, 1981.

[10] J.M. Bismut : Martingales, the Malliavin calculus and hypoellipticity under general Hörmander's conditions, *Z. Wahrschein.* 56, 469-505, 1981.

[11] J.M. Bismut, D. Michel : Diffusions conditionnelles, *J. Funct. Anal.* 44, 174-211, 1981 et 45, 274-282, 1982.

[12] B.Z. Bobrovsky, M. Zakai : Asymptotic a priori estimates for the error in the nonlinear filtering problem, *IEEE Trans. Inform. Th.* 28, 371-376, 1982.

[13] N. Bouleau, F. Hirsch : Propriétés d'absolue continuité dans les espaces de Direchlet et applications aux équations différentielles stochastiques, in *Séminaire de Probabilités XX*, Lecture Notes in Math. 1204, 131-161, Springer 1986.

[14] R.W. Brockett : Remarks on finite dimensional estimation, in *Analyse des Systèmes*, Astéristique **75-76**, SMF 1980.

[15] F. Campillo, F. Le Gland : MLE for partially observed diffusions : direct maximization vs. the EM algorithm, *Stoch. Proc. and their Applic.*, à paraître.

[16] F. Campillo, F. Le Gland : Application du filtrage non linéaire en trajectographie passive, 12eme colloque GRETSI, 1989.

[17] M. Chaleyat-Maurel, D. Michel : Des résultats de non existence de filtre de dimension finie, *Stochastics* **13**, 83-102, 1984.

[18] M. Chaleyat-Maurel, D. Michel : Hypoellipticity theorems and conditional laws, *Zeit. Wahr. Verw. Geb.* **65**, 573-597, 1984.

[19] M. Chaleyat-Maurel, D. Michel : Une propriété de continuité en filtrage non linéaire, *Stochastics* **19**, 11-40, 1986.

[20] M. Chaleyat-Maurel, D. Michel, E. Pardoux : Un théorème d'unicité pour l'équation de Zakai, *Stochastics*, **29**, 1-13, 1990.

[21] M. Davis : Nonlinear filtering and stochastic flows, *Proc. Int. Cong. Math.* Berkeley 1986.

[22] M. Davis, M. Spathopoulos : Pathwise nonlinear filtering for nondegenerate diffusions with noise correlation, *Siam. J. Control* **25**, 260-278, 1987.

[23] H. Doss : Liens entre équations différentielles stochastiques et ordinaires, *Ann. Inst. H. Poincaré, Prob. et Stat.*, **13**, 99-125, 1977.

[24] S.D. Eidel'man : *Parabolic Systems*, North Holland 1969.

[25] N. El Karoui, D. Hu Nguyen, M. Jeanblanc–Picqué : Existence of an optimal Markovian filter for the control under partial observations, *Siam J. Control* **26**, 1025-1061, 1988.

[26] S. Ethier, T. Kurtz : *Markov Processes : Characterization and convergence*, J. Wiley 1986.

[27] W. Fleming, E. Pardoux : Optimal control for partially observed diffusions, *Siam J. Control* **20**, 261-285, 1982.

[28] W. Fleming, E. Pardoux : Piecewise monotone filtering with small observation noise *Siam J. Control* **27**, 1156-1181, 1989.

[29] W. Fleming, D. Ji, E. Pardoux : Piecewise linear filtering with small observation noise, in *Analysis and Optimization of Systems*, Lecture Notes in Control and Info.– Scie. **111**, Springer 1988.

[30] P. Florchinger : Filtrage non linéaire avec bruits corrélés et observation non bornée. Etude numérique de l'équation de Zakai, Thèse, Univ. de Metz, 1989.

[31] M. Fujisaki, G. Kallianpur, H. Kunita : Stochastic differential equations for the non-linear filtering problem, *Osaka J. Math.* **9**, 19-40, 1972.

[32] I. Gyöngy : Stochastic equations with respect to semimartingales III, *Stochastics* **7**, 231-254, 1982.

[33] U. Haussmann, E. Pardoux : A conditionnally almost linear filtering problem with non Gaussian initial condition, *Stochastics* **23**, 241-275, 1988.

[34] M. Hazewinkel, S.I. Marcus : On Lie algebras and finite dimensional filters *Stochastics* **7**, 29-62, 1982.

[35] M. Hazewinkel, S.I. Marcus, H.J. Sussmann : Non existence of exact finite dimensional filters for conditional statistics of the cubic sensor problem, *Systems and Control Letters* **5**, 331–340, 1983.

[36] M. Hazewinkel, J.C. Willems (eds.) : *Stochastic systems : The mathematics of filtering and identification and applications*, D. Reidel 1981.

[37] O. Hijab : Finite dimensional cansal functionals of Brownian motions, in *Non linear stochastic problems*, Proc. Nato. Asi Conf. on Nonlin. Stoch. Problems, D. Reidel 1982.

[38] L. Hörmander : Hypoelliptic second order differential equations, *Acta Math.* **119**, 147–171, 1967.

[39] N. Ikeda, S. Watanabe : *Stochastic Differential Equations and Diffusion Processes*, North-Holland/Kodansha 1981.

[40] A. Isidori : *Non linear Control Systems : an Introduction*, Lecture Notes in Control and Info. Scie. **72**, Springer 1985.

[41] G. Kallianpur : *Stochastic filtering Theory*, Springer 1980.

[42] G. Kallianpur, R.L. Karandikar : *White noise theory of prediction, filtering and smoothing*, Stochastics Monograph **3**, Gordon and Breach 1988.

[43] R. Katzur, B.Z. Bobrovsky, Z. Schuss : Asymptotic analysis of the optimal filtering problem for one-dimensional diffusions measured in a low noise channel, *Siam J. Appl. Math.* **44**, 591-604 and 1176-1191, 1984.

[44] N.V. Krylov, B.L. Rozovskii : On conditionnal distributions of diffusion processes, *Math. USSR Izvestija* **12**, 1978.

[45] N.V. Krylov, B.L. Rozovskii : Stochastic evolution equations, *J. of Soviet Math.* **16**, 1233-1277, 1981.

[46] H. Kunita : Asymptotic behaviour of the non linear filtering errors of Markov processes, *J. Multivariate Anal.* **1**, 365-393, 1971.

[47] H. Kunita : Cauchy problem for stochastic partial differential equation arising in non linear filtering theory, *Systems and Control Letters* **1**, 37-41, 1981.

[48] H. Kunita : Densities of a measure valued process governed by a stochastic partial differential equation, *Systems and Control Letters* **1**, 100-104, 1981.

[49] H. Kunita : Stochastic differential equations and stochastic flows of diffeomorphisms, in *Ecole d'été de Probabilités de St-Flour XII*, Lecture Notes in Math. **1097**, Springer 1984.

[50] T. Kurtz, D. Ocone : Unique characterization of conditional distributions in non linear filtering, *Annals of Prob.* **16**, 80-107, 1988.

[51] H. Kushner, H. Huang : Approximate and limit results for nonlinear filters with wide bandwidth observation noise, *Stochastics* **16**, 65-96, 1986.

[52] S. Kusuoka, D.W. Stroock : The partial Malliavin calculus and its application to non linear filtering, *Stochastics* **12**, 83-142, 1984.

[53] F. Le Gland : Estimation de paramètres dans les processus stochastiques en observation incomplète, Thèse, Univ. Paris-Dauphine, 1981.

[54] J. Lévine : Finite dimensional realizations of stochastic PDE's and application to filtering, à paraître.

[55] R.S. Liptser, A.N. Shiryayev : *Statistics of random processes* Vol. I, II, Springer 1977.

[56] C. Lobry : Bases mathémathiques de la théorie des systèmes asservis non linéaires, Notes polycopiées, Univ. de Bordeaux I, 1976.

[57] A. Makowski : Filtering formulae for partially observed linear systems with non Gaussian initial conditions, *Stochastics*, **16**, 1-24, 1986.

[58] S.I. Marcus : Algebraic and geometric methods in nonlinear filtering, *Siam J. Control* **22**, 817-844, 1984.

[59] D. Michel : Conditional laws and Hörmander's condition, in *Proc. Taniguchi Symposium on Stochastic Analysis*, K. Itô ed., 387-408, North–Holland 1983.

[60] D. Michel, E. Pardoux : An introduction to Malliavin's calculus and some of its applications, à paraître.

[61] P. Milheiro de Oliveira : Filtres approchés pour un problème de filtrage non linéaire avec petit bruit d'observation, Rapport INRIA, à paraître.

[62] S. K. Mitter : Filtering theory and quantum fields, in *Analyse des systèmes*, loc. cit.

[63] S.K. Mitter, A. Moro (eds.) : *Nonlinear filtering and stochastic control*, Lecture Notes in Math. **972**, Springer 1983.

[64] J. Norris : Sinplified Malliavin Calculus, in *Séminaire de Probabilités XX* loc. cit., 101-130.

[65] D. Ocone : Stochastic calculus of variations for stochastic partial differential equations *J. Funct. Anal.* **79**, 288-331, 1966.

[66] D. Ocone, E. Pardoux : A Lie-algebraic criterion for non existence of finite dimensionally computable filters, in *Stochastic Partial Differential Equations and Applications II*, G. Da Prato and L. Tubaro eds., Lecture Notes in Math. **1390**, Springer 1989.

[67] D. Ocone, E. Pardoux : Equations for the nonlinear smoothing problem in the general case, in Proc. 3^d Trento Conf. on SPDEs, G. Da Prato & L. Tubaro eds., Lecture Notes in Math. Springer, à paraître.

[68] E. Pardoux : Stochastic partial differential equations and filtering of diffusion processes, *Stochastics* **3**, 127-167, 1979.

[69] E. Pardoux : Equations du filtrage non linéaire, de la prédiction et du lissage, *Stochastics* **6**, 193-231, 1982.

[70] E. Pardoux : Equations of nonlinear filtering, and application to stochastic control with partial observation, in *Non linear filtering and stochastic control*, loc. cit., 208-248.

[71] E. Pardoux : Equations du lissage non linéaire, in *Filtering and control of random processes*, H. Korezioglu, G. Mazziotto & S. Szpirglas eds., Lecture Notes in Control and Info. Scie. **61**, 206–218, Springer 1984.

[72] E. Pardoux : Sur les équations aux dérivées partielles stochastiques de type parabolique, in *Colloque en l'honneur de L. Schwartz* Vol. 2, Astérisque **132**, 71-87, SMF 1985.

[73] J. Picard : Non linear filtering of one-dimensional diffusions in the case of a high signalto-noise ratio, *Siam J. Appl. Math.* **46**, 1098-1125, 1986.

[74] J. Picard : Filtrage de diffusions vectorielles faiblement bruitées, in *Analysis and Optimization of Systems*, Lecture Notes in Control and Info. Scie. **83**, Springer 1986.

[75] J. Picard : Nonlinear filtering and smoothing with high signal-to-noise ratio, in *Stochastic Processes in Physics and Engineering*, D. Reidel 1986.

[76] J. Picard : Asymptotic study of estimation problems with small observation noise, in *Stochastic Modelling and Filtering*, Lecture Notes in Control and Info. Scie. **91**, Springer 1987.

[77] J. Picard : Efficiency of the extended Kalman filter for nonlinear systems with small noise, à paraître.

[78] M. Pontier, C. Stricker, J. Szpirglas : Sur le théorème de représentation par rapport à l'innovation, in *Séminaire de Probabilité XX*, loc. cit., 34-39.

[79] M.C. Roubaud : Filtrage linéaire par morceaux avec petit bruit d'observation, à paraître.

[80] B. Rozovski : *Evolution stochastic systems*, D. Reidel 1990.

[81] D.W. Stroock : Some applications of stochastic calculus to partial differential equations, in *Ecole d'Eté de Probabilités de St Flour XI*, Lecture Notes in Math. **976**, 267–382, Springer 1983.

[82] H. Sussmann : On the gap between deterministic and stochastic ordinary differential equations, *Ann. of Prob.* **6**, 19-41, 1978.

[83] H. Sussmann : On the spatial differential equations of nonlinear filtering, in *Nonlinear partial differential equations and their applications*, Collège de France seminar **5**, 336-361, Research Notes in Math. **93**, Pitman 1983.

[84] J. Szpirglas : Sur l'équivalence d'équations différentielles stochastiques à valeurs mesures intervenant dans le filtrage markovien non linéaire, *Ann. Institut Henri Poincaré* **14**, 33-59, 1978.

[85] J. Walsh : An introduction to Stochastic Partial Differential Equations, in *Ecole d'été de Probabilité de St-Flour XIV*, Lecture Notes in Math. **1180**, 265–439, Springer 1986.

[86] M. Yor : Sur la théorie du filtrage, in *Ecole d'été de Probabilité de St-Flour IX*, Lecture Notes in Math. **876**, 239-280, Springer 1981.

[87] M. Zakai : On the optimal filtering of diffusion processes, *Zeit. Wahr. Verw. Geb.* **11**, 230-243, 1969.

[88] M. Zakai : The Malliavin Calculus, *Acta Applicandae Mat.* **3**, 175-207, 1985.

TOPICS IN PROPAGATION OF CHAOS

Alain-Sol SZNITMAN

The preparation of these notes was supported in part by NSF Grant No. DMS-8903858, and by an Alfred P. Sloan Foundation research fellowship.

TABLE DES MATIERES

A. S. SZNITMAN : "TOPICS IN PROPAGATION OF CHAOS"

0. INTRODUCTION

The terminology propagation of chaos comes from Kac. The initial motivation for the subject was to try to investigate the connection between a detailed and a reduced description of particles' evolution. For instance on the one hand one has Liouville's equations:

$$(0.1) \qquad \partial_t u + \sum_1^N v_i \partial_{x_i} u + \sum_{j \neq i} -\nabla V_N(x_i - x_j) \partial_{v_j} u = 0 \,,$$

where $u(t, x_1, v_1, \ldots, x_n, v_N)$ is the density of presence at time t, assumed to be symmetric of N particles, with position x_i and velocity v_i, and pairwise potential interaction $V_N(\cdot)$.

On the other hand one also has for instance the Boltzmann equation for a dilute gas of hard spheres

$$(0.2) \qquad \partial_t u + v.\nabla_x u = \int_{R^3 \times S_2} (u(x, \tilde{v}) u(x, \tilde{v}') - u(x, v) u(x, v')) |(v' - v) \cdot n| \, dv' dn$$

where $\tilde{v}, \tilde{v}'$ are obtained from v, v' by exchanging the respective components of v and v' in direction n, that is:

$$\tilde{v} = v + (v' - v) \cdot n \, n$$
$$\tilde{v}' = v' + (v - v') \cdot n \, n \,,$$

and $u(t, x, v)$ is the density of presence at time t of particles at location x with velocity v.

One question was to understand the nature of the connection between (0.1) and (0.2). There are now works directly attacking the problem (see Lanford [22], and more recently Uchiyama [54]). However at the time, Kac proposed to get insight into the problem, by studying simpler Markovian models of particles. The forward equations for the Markovian evolutions of N-particle systems come as a substitute for the Liouville equations. They are called master equations. Kac [15] traces the origin of these masters equations back to the forties with works of Norsdieck-Lamb-Uhlenbeck on a problem of cosmic rays, and of Siegert.

In the case of the Boltzmann equation, one can for instance forget about positions of particles and take as master equations:

$$(0.3)$$
$$\partial_t u_t^N(v_1, \ldots, v_N)$$
$$= \frac{1}{(N-1)} \sum_{1 \leq i < j \leq n} \int_{S_2} (u_t^N(v_1, .., \tilde{v}_i, .., \tilde{v}_j, .., v_N) - u_t^N(v_1, .., v_N)) |(v_i - v_j) \cdot n|) dn$$

One now tries to connect these master equations with the spatially homogeneous Boltzmann equations for hard spheres:

$$(0.4) \qquad \partial_t u = \int_{R^3 \times S_2} (u(\tilde{v}, t) u(\tilde{v}', t) - u(v, t) u(v', t)) |(v' - v) \cdot n| dv' dn .$$

The idea proposed by Kac, which motivates the terminology "propagation of chaos" is the following. If one picks a "chaotic" initial distribution of particles: $u_N(0, v_1, \ldots, v_N)$ $= u_0(v_1) \ldots u_0(v_N)$, for fixed N the evolution due to the master equations will in general destroy the independence property of $v_1, \ldots, v_N$ at time t. However if one focuses on the reduced distribution at time t of the first k components, it should approximately be given when N is large by $u_t(v_1) \ldots u_t(v_k)$, if $u_t(v)$ is the solution of equation (0.3) with initial condition $u_0(\cdot)$. So in this sense independence (or chaos) still propagates and eequation (0.4) emerges.

"Propagation of chaos" deals with symmetric evolution of particles, and this is not an innocent assumption. Among other things it tells us that the probability distribution of the first k particles $u^N(dv_1, \ldots, dv_k)$ is the normalized k-particle correlation measure, that is the intensity of the random measure $\frac{1}{N(N-1)\cdots(N-k-1)} \sum_{i_\ell \text{distinct}} \delta_{(v_{i_1}, \ldots, v_{i_k})}$. The consequence is that the study of one individual gives information on the behavior of the group. We have so far presented the N-particle system and the nonlinear equation. There is a third actor in the play the "nonlinear process". It describes the limit behavior of the trajectory of one individual. It is sometimes called the tagged particle process, however we refrained from using this terminology for it can have different meanings, (see for instance the end of Chapter I). The time marginals of the nonlinear process will evolve according to the nonlinear equation under study, and this motivates the name of "nonlinear process", although of course in some examples, where interactions vanish all can be very linear in fact.

We will present in these notes a selection of topics on "propagation of chaos" which by no means covers the huge literature on the subject.

Let us now close the introduction with an informal discussion, along the lines of the "BBGKY-hierarchical method" for a model of interacting diffusions due to McKean [27]. This will motivate why propagation of chaos should hold in this case.

One looks at N particles on R^d, with initial "chaotic" distribution $u_0^{\otimes N}$, satisfying the S.D.E.:

$$(0.5) \qquad dx_t^i = dB_t^i + \frac{1}{N} \sum_1^N b(x_t^i, x_t^j) \, dt , \quad 1 \leq i \leq N ,$$

where B^i are independent Brownian motions, and $b(\cdot, \cdot)$ is for instance smooth compactly supported. We are now going to explain in an informal way how the nonlinear equation

(0.6)
$$\partial_t u = \frac{1}{2}\Delta u - \text{div}\left(\int b(\cdot, y)\ u(t, y)\ dy\ u\right)$$
$$u_{t=0} = u_0\ ,$$

arises in the propagation of chaos effect.

Let us reinterpret equation (0.6). If P_t^0 denotes the Brownian transition density, we have the perturbation formula (obtained for instance by differentiating in s, $u_s P_{t-s}^0$):

$$u_t(x) - u_0\ P_t^0(x) = \int_0^t ds_1 \int dx_1 dx_2 u_{s_1}(x_1) u_{s_1}(x_2)\ b(x_1, x_2)\ \nabla_{x_1} P_{t-s_1}^0(x_1, x)\ .$$

Continuing the development of $u_{s_1}(x_1) u_{s_1}(x_2), \ldots$, by induction we find:
(0.7)
$$u_t = u_0 P_t^0 + \sum_{k=1}^m \int_{0 < s_k < \cdots < s_1 < t} ds_k .. ds_1\ u_0^{\otimes k+1} P_{s_k}^0\ B \cdot \nabla P_{s_{k-1}-s_k}^0 .. B \cdot \nabla P_{t-s_1}^0 + R_m\ ,$$

where $R_m = \int_{0 < s_{m+1} < s_m < \cdots < s_1 < t} ds_{m+1} .. ds_1\ u_{s_{m+1}}^{\otimes m+2}\ B \cdot \nabla P_{s_m - s_{m+1}}^0 .. \nabla P_{t-s_1}^0\ .$

Here we have adopted convenient notations, and P_t^0 acts naturally in a tensorial way on functions of an arbitrary number of variables, and $B \cdot \nabla$ maps functions of k variables into functions of $(k+1)$ variables, $k \geq 1$, by the formula:

(0.8)
$$[B \cdot \nabla] f(x_1, \ldots, x_{k+1}) = \sum_1^k b(x_i, x_{i+1})\ \nabla_i f(x_1, \ldots, x_k)\ .$$

If we use a similar perturbation method for the time marginals $u_{N,t}(x_1, \ldots, x_N)$ of the N-particle system, using the forward equation corresponding to (0.5), we now find:
(0.9)
$$u_{N,t} = u_0^{\otimes N} P_t^0 + \sum_{k=1}^m \int_{0 < s_k < \cdots < s_1 < t} u_0^{\otimes N}\ P_{s_k}^0\ B^N\ \nabla P_{s_k - s_{k-1}}^0 \cdots B^N \nabla P_{t-s_1}^0 + R_m^N\ ,$$

where $R_m^N = \int_{0 < s_{m+1} < \cdots < s_1 < t} u_{s_{m+1}}^N B^N \nabla P_{s_m - s_{m+1}}^0 \cdots B^N \nabla P_{t-s}^0$, and

(0.10)
$$[B^N \cdot \nabla] f(x_1, \ldots, x_N) = \frac{1}{N} \sum_{i,j=1}^N b(x_i, x_j)\ \nabla_i f(x_1, \ldots, x_N)\ .$$

Let us now see what happens as N goes to infinity when we consider $\langle u_{N,t}, f \rangle$ for a function $f(x_1)$ depending only on the first variable.

In (0.9), the first term

$$a_0^N = u_0^{\otimes N} P_t^0 f = u_0 P_t^0 f ,$$

coincides with the first term a_0 of (0.7).

The second term of (0.9) is

$$a_1^N = \int_0^t ds \; u_0^{\otimes N} \, P_{s_1}^0 B^N \cdot \nabla P_{t-s_1}^0 \, f \, ds_1 .$$

Since $P_{t-s_1}^0 f$ just depends on the x_1 variable

$$a_1^N = \int_0^t ds_1 \; u_0^{\otimes N} P_{s_1}^0 \left(\frac{1}{N} \sum b(x_1, x_j) \, \nabla_1 (P_{t-s_1}^0 f)(x_1) \right) ds_1$$

But $u_0^{\otimes N} P_{s_1}^0$ is symmetric. For $j \neq 1$, one can pick a permutation of $[1, N]$, leaving 1 invariant and mapping j on 2, so

$$a_1^N = \frac{N-1}{N} \int_0^t u_0^{\otimes N} P_{s_1}^0 \{ b(x_1, x_2)(\nabla_1 P_{t-s_1}^0 f)(x_1) \} \, ds_1 + o(N) ,$$

and in fact $u_0^{\otimes N}$ can be replaced by $u_0 \otimes u_0$ in the last expression so that a_1^N converges to the first term a_1 of (0.7).

The third term a_2^N of (0.9) is

$$a_2^N = \int_{0 < s_2 < s_1 < t} ds_2 \; ds_1 \; u_0^{\otimes N} P_{s_2}^0 B^N \, \nabla P_{s_1 - s_2}^0 \, B^N \nabla P_{t-s_1}^0 f \, ds_1 ds_2$$

Denote by ϕ the symmetric distribution $u_0^{\otimes N} P_{s_2}^0$ and by ψ the expression $u_0^{\otimes N} P_{s_2}^0 B^N \cdot \nabla P_{s_1-s_2}^0$. To obtain ψ one lets $P_{s_1-s_2}^0$ act on the right on:

$$- \sum_i \operatorname{div}_i(\phi(x_1, \dots, x_N) \frac{1}{N} \sum_{j=1}^N b(x_i, x_j)) ,$$

which is also symmetric. Applying the same trick as before, we find:

$$a_2^N = \frac{N-1}{N} \int_{0 < s_2 < s_1 < t} ds_2 ds_1 \langle \psi , \; b(x_1, x_2)(\nabla_1 P_{t-s_1}^0 f)(x_1) \rangle + o(N)$$

$$= \frac{(N-1)(N-2)}{N^2} \int_{0 < s_2 < s_1 < t} ds_2 ds_1 \langle u_0^{\otimes N} P_{s_2}^0, (b(x_1, x_3) \nabla_1 + b(x_2, x_3) \cdot \nabla_2)$$

$$\cdot P_{s_1-s_2}^0 (b(x_1, x_2) \nabla_1 (P_{t-s}^0 f)(x_1))$$

$$+ o(N) ,$$

and one sees that a_2^N converges to a_2 the third term in (0.7).

In the same way, one sees that there is a term by term convergence of a_k^N to a_k for each k. However, we cannot transform this into a bona fide propagation of chaos result, since we do not have a good control on our series.

Indeed, we have an estimate of the type

$$\|b(x_i, x_j)\nabla_i P_\tau^0\|_{L^\infty \to L^\infty} \leq \frac{c}{\sqrt{\tau}} ,$$

and the generic term a_k in (0.7) for instance is naturally estimated in terms of

$$k! c^k \int_{0 < s_2 < \cdots < s_1 < t} ds_k \ldots ds_1 [(t - s_1) \ldots (s_{k-1} - s_k)]^{-1/2} ,$$

but this quantity tends to infinity.

Of course this convergence problem comes from the fact that it is unreasonable to use a perturbation series of P_t^0 to take care of the drift term $b(\cdot, \cdot) \cdot \nabla$. Nevertheless this gives a flavor of the propagation of chaos result. In the next section, we will provide a proof by a probabilistic approach.

I. Generalities and first examples.

1) A laboratory example.

We now come back to McKean's example of interacting diffusions, and we are going to use a probabilistic method to attack the problem.

We suppose $b(\cdot, \cdot)$ bounded Lipschitz $R^d \times R^d \to R^d$, and construct on $(R^d \times C_0(R_+, R^d))^{N^*}$, with product measure $(u_0 \otimes W)^{\otimes N^*}$, ($u_0$ probability on R^d, W standard R^d-Wiener measure) the $X^{i,N}_\cdot$, $i = 1, \ldots, N$, satisfying

(1.1)
$$dX^{i,N}_t = dw^i_t + \frac{1}{N} \sum_{j=1}^N b(X^{i,N}_t, X^{j,N}_t)dt , \quad i = 1, \ldots, N$$
$$X^{i,N}_0 = x^i_0 ,$$

here x^i_0, $(w^i_\cdot)$, $i \geq 1$, are the canonical coordinates on the product space $(R^d \times C_0)^{N^*}$.

We are going to show that when N goes to infinity each $X^{i,N}_\cdot$, has a natural limit $\overline{X}^i_\cdot$. Each $\overline{X}^i_\cdot$ will be an independent copy of a new object: "the nonlinear process". Nonlinear process:

On a filtered probability space $(\Omega, F, F_t, (B_t)_{t \geq 0}, X_0, P)$, endowed with an R^d-valued Brownian motion $(B_t)_{t \geq 0}$, and an u_0-distributed, F_0 measurable R^d-valued random variable X_0, we look at the equation:

(1.2)
$$dX_t = dB_t + \int b(X_t, y) \, u_t(dy) \, dt$$
$$X_{t=0} = X_0 , \quad u_t(dy) \text{ is the law of } X_t$$

Theorem 1.1. *There is existence and uniqueness, trajectorial and in law for the solutions of* (1.2).

Remark 1.2. Let us notice that the nonlinear process has time marginals which satisfy in a weak sense the nonlinear equation

(1.3)
$$\partial_t u = \frac{1}{2}\Delta u - \operatorname{div}\left(\int b(\cdot, y) \, u_t(dy)u \right) .$$

Indeed, for $f \in C_b^2(R^d)$, applying Ito's formula:

$$f(X_t) = f(X_0) + \int_0^t f'(X_s)dB_s + \int_0^t \left(\frac{1}{2}\Delta f + \int_{R^d} b(X_s, y) \, u_s(dy) \, \nabla f(X_s) \right) ds ,$$

after integration this yields a weak version of (1.3).

$\square$

Proof: Let us now turn to the proof of Theorem 1.1. We introduce the Kantorovitch-Rubinstein or Vaserstein metric on the set $M(C)$ of probability measures on $C = C([0,T], R^d)$, defined by

$$(1.4) \quad D_T(m_1, m_2) = \inf\{ \int (\sup_{s \leq T} |X_s(\omega_1) - X_s(\omega_2)| \wedge 1) \, dm(\omega_1, \omega_2) ,$$

$$m \in M(C \times C), \quad p_1 \circ m = m_1, \quad p_2 \circ m = m_2\} .$$

Here X_s is simply the canonical process on C.

Formula (1.4) defines a complete metric on $M(C)$, which gives to $M(C)$ the topology of weak convergence. The proof of this fact can be found in Dobrushin [8].

Take now $T > 0$, and define Φ the map which associates to $m \in M(C([0,T], R^d))$ the law of the solution of

$$(1.5) \quad X_t = X_0 + B_t + \int_0^t (\int_C b(X_s, w_s) \, dm(w)) \, ds , \quad t \leq T .$$

Observe that this law does not depend on the specific choice of the space Ω, we use.

Next observe that if X_t, $t \leq T$, is a solution of (1.2), then its law on $C([0,T], R^d)$, is a fixed point of Φ, and conversely if m is such a fixed point of Φ, (1.5) defines a solution of (1.2), up to time T. So now our problem is translated into a fixed point problem for Φ, and one has the contraction lemma:

Lemma 1.3. *For $t \leq T$,*

$$D_t(\Phi(m_1), \Phi(m_2)) \leq c_T \int_0^t D_u(m_1, m_2) \, du , \quad m_1, m_2 \in M(C_T) ,$$

c_T *is a constant and,* $D_u(m_1, m_2)$ $(\leq D_T(m_1, m_2))$ *is the distance between the images of* m_1, *and* m_2 *on* $C([0,u], R^d)$.

Proof:

$$X_t^1 = X_0 + B_t + \int_0^t (\int b(X_s^1, w_s) dm_1(w)) ds , \quad t \leq T ,$$

$$X_t^2 = X_0 + B_t + \int_0^t (\int b(X_s^2, w_s) dm_2(w)) ds , \quad t \leq T .$$

So we find:

$$\sup_{s \leq t} |X_s^1 - X_s^2| \leq \int_0^t ds| \int b(X_s^1(\omega), w_s) dm_1(w) - \int b(X_s^2(\omega), w_s) dm_2(w)| .$$

But

$$| \int b(x, w_s) dm_1(w) - \int b(y, w_s) dm_2(w)| \leq K[|x - y| \wedge 1 + \int |w_s^1 - w_s^2| \wedge 1 \, dm(w^1, w^2)]$$

where m is any coupling of m_1, m_2 on $C([0,s], \mathbf{R}^d)$. From this

$$\sup_{s \leq t} |X_s^1 - X_s^2| \leq K \int_0^t ds |X_s^1(\omega) - X_s^2(\omega)| \wedge 1 + K \int_0^t D_s(m_1, m_2) ds .$$

Using Gronwall's lemma:

$$\sup_{s \leq t} |X_s^1 - X_s^2| \wedge 1 \leq K \, e^{KT} \int_0^t D_s(m_1, m_2) ds ,$$

from which the lemma follows.

□

From the lemma, we can immediately deduce weak and strong uniqueness for the solutions of (1.2). The existence part also follows now from a standard contraction argument.

Namely for $T > 0$, and $m \in M(C_T)$, iterating the lemma:

$$(1.6) \qquad D_T(\Phi^{k+1}(m), \Phi^k(m)) \leq c_T^k \frac{T^k}{k!} D_T(\Phi(m), m) .$$

So $\Phi^k(m)$ is a Cauchy sequence, and converges to a fixed point of $\Phi : P_T$. Now if $T' < T$, the image of P_T on $C([0, T'], \mathbf{R}^d)$ is still a fixed point, so the P_T are a consistent family, yielding a P on $C([0, \infty), \mathbf{R}^d)$. This provides the required solution.

□

Using Theorem 1.1, we now introduce on $(\mathbf{R}^d \times C_0)^{N^*}$, where we have constructed in (1.1) our interacting diffusions $X^{i,N}$, $i = 1, \ldots, N$, the processes $\overline{X}^i_{\cdot}$, $i \geq 1$, solution of:

$$(1.7) \qquad \overline{X}_t^i = x_0^i + w_t^i + \int_0^t \int b(\overline{X}_s^i, y) \, u_s(dy) \, ds ,$$

$$u_s(dy) = \mathrm{law}(\overline{X}_s^i) .$$

Theorem 1.4. *For any* $i \geq 1$, $T > 0$:

$$(1.8) \qquad \sup_N \sqrt{N} E[\sup_{t \leq T} |X_t^{i,N} - \overline{X}_t^i|] < \infty .$$

Proof: Dropping for notational simplicity the superscript N, we have:

$$X_t^i - \overline{X}_t^i = \int_0^t (\frac{1}{N} \sum_{j=1}^N b(X_s^i, X_s^j) - \int b(\overline{X}_s^i, y) u_s(dy)) \, ds$$

$$= \int_0^t ds \frac{1}{N} \sum_{j=1}^N \{(b(X_s^i, X_s^j) - b(\overline{X}_s^i, X_s^j)) + (b(\overline{X}_s^i, X_s^j) - b(\overline{X}_s^i, \overline{X}_s^j))$$

$$+ (b(\overline{X}_s^i, \overline{X}_s^j) - \int b(\overline{X}_s^i, y) \, u_s(dy))\} \, .$$

Writing $b_s(x, x') = b(x, x') - \int b(x, y) \, u_s(dy)$, we see that:

$$E[|X^i - \overline{X}^i|_T^*] \le K \int_0^T ds(E[|X_s^i - \overline{X}_s^i|] + \frac{1}{N} \sum_1^N E[|X_s^j - \overline{X}_s^j|] + E[|\frac{1}{N} \sum_{j=1}^N b_s(\overline{X}_s^i, \overline{X}_s^j)|])$$

Summing the previous inequality over i, and using symmetry, we find:

$$NE[|X^1 - \overline{X}^1|_T^*] = \sum_1^N E[|X^i - \overline{X}^i|_T^*]$$

$$\le K' \int_0^T \sum_{i=1}^N (E[|X_s^i - \overline{X}_s^i|] + E[|\frac{1}{N} \sum_{j=1}^N b_s(\overline{X}_s^i, \overline{X}_s^j)|]) \, ds \, .$$

Applying Gronwall's lemma, and symmetry, we find:

$$E[|X^i - \overline{X}^i|_T^*] \le K(T) \int_0^T ds E[|\frac{1}{N} \sum_{j=1}^N b_s(\overline{X}_s^i, \overline{X}_s^j)|] \, .$$

Our claim will follow provided we can show that:

$$E[|\frac{1}{N} \sum_{j=1}^N b_s(\overline{X}_s^i, \overline{X}_s^j)|] \le \frac{C(T)}{\sqrt{N}} \, .$$

But

$$E[(\frac{1}{N} \sum_{j=1}^N b_s(X_s^i, X_s^j))^2] = \frac{1}{N^2} E[\sum_{j,k} b_s(\overline{X}_s^i, \overline{X}_s^j) b_s(\overline{X}_s^i, \overline{X}_s^k)] \, ,$$

and because of the centering of $b_s(x, y)$ with respect to its second variable, when $j \ne k$

$$E[b_s(\overline{X}_s^i, \overline{X}_s^j) b_s(\overline{X}_s^i, \overline{X}_s^k)] = 0 \, .$$

The previous sum is then less than const./N. Our claim follows.

$\square$

Let us end this section with some comments.

— We have presented here a simple enough example. It is possible to let basically the same method work in the case of an additional interaction through the diffusion coefficient, see McKean [27], or [41]. The use in this context of the Vasershtein metric, can be found in Dobrushin [9]. Of course one can as well obtain higher moment estimates in (1.8). Theorem 1.4 suggests the possibility of a fluctuation theorem, which indeed exists, see for instance, Tanaka [51], Shiga-Tanaka [38], Kusuoka-Tamura [20], or [40], [41].

— As a result of the probabilistic proof we just described, we see that the introduction of a "nonlinear process", and not only of a nonlinear equation is fairly natural. For each t and k, the joint distribution of $(X_t^{1,N}, \ldots, X_t^{k,N})$ is converging to $u_t^{\otimes k}$, but in fact we have convergence at the level of processes.

— Let us also mention some interesting examples which arise in possibly singular cases, of the nonlinear equation

$$(1.9) \qquad \partial_t u = \frac{\sigma^2}{2}\Delta u - \mathrm{div}\Big(\int b(\cdot,y)\, u_t(dy)u\Big), \quad (\sigma = \text{constant})$$

a) When $\sigma = 0$, and $R^d = R^3 \times R^3 : (x,v)$, with $b((x,v),(x',v')) = (v, F(x-x'))$, one finds:

$$\partial_t u + v.\nabla_x \dot{u} + \int F(x-x')\, u(dx',dv') \cdot \nabla_v u = 0$$

This is Vlasov's equation (see Dobrushin [9]).

b) $R^d = R$, $\sigma = 1$, $b(x,y) = c\delta(x-y)$, we have

$$\partial_t u = \frac{1}{2}u'' - c(u^2)',$$

this is Burger's equation, we will come back to this (singular) example in Chapter II.

c) $R^d = R^2$, $b(x,y) = b(x-y)$ is the Biot-Savart kernel, with $b(z) = \frac{1}{2\pi|z|^2}(-z_2, z_1)$, one now finds:

$$\partial_t u + (b * u)\nabla u = \frac{\sigma^2}{2}\Delta u,$$

where we use $\mathrm{div}(b * u) = 0$. If one sets $v = (b * u)$, $\mathrm{curl}\, v = \partial_2 v_1 - \partial_1 v_2 = u$, and u satisfies the vorticity equation for the Navier-Stokes equation in dimension 2. For this example, see Goodman [12], Marchioro-Pulvirenti [29], Osada [35].

d) Scheutzow [37], gives an example of an equation like (1.3), with polynomial coefficients, in R^2, which admits some genuinely periodic solutions.

2) Some generalities

We will now give some definitions, that we will use for the study of propagation of chaos. $M(E)$ denotes here the set of probability measures on E.

Definition 2.1. E a separable metric space, u_N a sequence of symmetric probabilities on E^N. We say that u_N is u-chaotic, u probability on E, if for $\phi_1, \ldots, \phi_k \in C_b(E)$, $k \geq 1$,

$$(2.1) \qquad \lim_{N \to \infty} \langle u_N, \phi_1 \otimes \ldots \otimes \phi_k \otimes 1 \cdots \otimes 1 \rangle = \prod_1^k \langle u, \phi_i \rangle .$$

As mentioned in the introductory chapter, the symmetry assumption on the laws u_N is less innocent than one may think at first. As the next (easy) proposition shows, the notion of u-chaotic means that the empirical measures of the coordinate variables of E^N, under u_N tend to concentrate near u. This is a type of law of large numbers. Condition (2.1) can also be restated as the convergence of the projection of u_N as E^k to $u^{\otimes k}$ when N goes to infinity, see [2], p. 20. In the coming proposition, we suppose u_N symmetric.

Proposition 2.2.

i) u_N is u-chaotic is equivalent to $\overline{X}_N = \frac{1}{N} \sum_1^N \delta_{X_i}$ ($M(E)$-valued random variables on (E^N, v_N), X_i canonical coordinates on E^N) converge in law to the constant random variable u. It is also equivalent to condition (2.1), with $k = 2$.

ii) When E is a Polish space, the $M(E)$-valued variables $\overline{X}_N$ are tight if and only if the laws on E of X_1 under u_N are tight.

Proof:

i) First suppose u_N satisfies (2.1) with $k = 2$, and consequently with $k = 1$ as well. Take ϕ in $C_b(E)$,

$$E_N[(\overline{X}_N - u, \phi)^2] = \frac{1}{N^2} \sum_{i,j=1}^N E_N[\phi(X_i)\phi(X_j)] - \frac{2}{N} \sum_1^N E_N[\phi(X_i)]\langle u, \phi \rangle + \langle u, \phi \rangle^2 .$$

Using symmetry we find:

$$\frac{1}{N} E_N[\phi(X_1)^2] + \frac{(N-1)}{N} E_N[\phi(X_1)\phi(X_2)] - 2\langle u, \phi \rangle E_N[\phi(X_1)] + \langle u, \phi \rangle^2$$

which tends to zero by (2.1), with $k = 1, 2$. This implies that $\overline{X}_N$ converges in law to the constant random variable equal to u.

Conversely, suppose $\overline{X}_N$ converge in law to the constant u,

$$|\langle u_N, \phi_1 \otimes \ldots \otimes \phi_k \otimes 1 \ldots \otimes 1 \rangle - \prod_1^k \langle u, \phi_i \rangle|$$

(2.2)
$$\leq |\langle u_N, \phi_1 \otimes \ldots \otimes \phi_k \otimes 1 \cdots \otimes 1 \rangle - \langle u_N, \prod_1^k \langle \overline{X}_N, \phi_i \rangle \rangle|$$

$$+ |\langle u_N, \prod_1^k \langle \overline{X}_N, \phi_i \rangle \rangle - \prod_1^k \langle u_0, \phi_i \rangle| .$$

The second term in the right member of (2.2) goes to zero. The first term using symmetry can be written as:

$$|\langle u_N, \frac{1}{N!} \sum_{\sigma \in S_N} \phi_1(X_{\sigma(1)}) \cdots \phi_k(X_{\sigma(k)}) - \prod_1^k \langle \overline{X}_N, \phi_i \rangle \rangle| .$$

Observe now that if $M \geq \|\phi_i\|_\infty, 1 \leq i \leq k$.

(2.3)
$$\sup_{E^N} |\frac{1}{N!} \sum_{\sigma \in \rho_N} \phi_1(X_{\sigma(1)}) \cdots \phi_k(X_{\sigma(k)}) - \prod_1^k \langle \overline{X}_N, \phi_i \rangle|$$

$$\leq M^k \left[\left(\frac{(N-k)!}{N!} - \frac{1}{N^k} \right) \cdot \frac{N!}{(N-k)!} + \frac{1}{N^k} \left(N^k - \frac{N!}{(N-k)!} \right) \right]$$

$$= 2M^k \left(1 - \frac{N!}{N^k(N-k)!} \right) \to 0 .$$

Here we simply used that there are $N!/(N-k)!$ injections from $[1,k]$ into $[1,N]$, each of them has weight $(N-k)!/N!$ in the first sum of (2.3) and $1/N^k$ in the second sum, and in the second sum there are also $N^k - N!/(N-k)!$ terms where repetitions of coordinates occur. So we see that the first term of (2.2) goes to zero, and this proves i).
ii) For a probability $Q(dm)$ on $M(E)$, define the intensity $I(Q)$ as the probability measure

(2.4)
$$\langle I(Q), f \rangle = \int_{M(E)} \langle m, f \rangle \, dQ(m) ,$$

for $f \in bB(E)$.

We will in fact show the more general fact:

(2.5) tightness for a family of measure Q on $M(E)$ is equivalent to the tightness of their intensity measures $I(Q)$ on E.

Now in our situation of ii), by symmetry the intensity measure of the law of $\overline{X}_N$ is just the law of X_1, under E_N, so that our claim ii) follows from (2.5). Now the map

$Q \to I(Q)$ is clearly continuous for the respective weak convergence topologies. So, (2.5) will follow if we prove that whenever $I_n = I(Q_n)$ is tight, then Q_n is tight.

For each $\epsilon > 0$, denote by K_ϵ a compact subset of E, with $I_n(K_\epsilon^c) \leq \epsilon$, for every n. Now for $\epsilon, \eta > 0$, and any n, $Q_n(\{m, m(K_{\epsilon\eta}^c) \geq \eta\}) \leq \frac{1}{\eta} I_n(K_{\epsilon\eta}^c) \leq \epsilon$.

It follows that:

$$Q_n\Big(\bigcup_{k \geq 1} \{m, m(K_{\epsilon 2^{-k}/k}^c) \geq 1/k\} \Big) \leq \sum_{k \geq 1}^{\infty} \epsilon 2^{-k} = \epsilon \,,$$

this means that Q_n puts a mass greater or equal to $1 - \epsilon$ on the compact subset of $M(E)$: $\cap_{k \geq 1} \{m, m(K_{\epsilon 2^{-k}/k}^c) \leq 1/k\}$. This proves the Q_n are tight, and yields (2.5).

□

Remark 2.3.

1) It is clear from the proof of Proposition 2.2, that thanks to symmetry, the distribution under u_N of k particles chosen with the empirical distribution $\overline{X}_N$ is approximately the same as the law of $(X_1, \ldots, X_k)$ under u_N, when N is large. In fact this last distribution is the intensity of the empirical distribution of distinct k-uples: $\overline{X}_{N,k} = \frac{1}{N(N-1)\cdots(N-k+1)} \sum_{i_\ell \text{distinct}} \delta_{(X_{i_1}, \ldots, X_{i_k})}$.

2) Suppose the empirical measures $\overline{X}_N = \frac{1}{N} \sum_1^N \delta_{X_i}$ converge in law to a constant $u \in M(E)$, with an underlying distribution v_N on E^N non necessarily symmetric. Then the symmetrized distribution u_N on E^N, preserves $\overline{X}_N$, and u_N is then u-chaotic in the sense of Definition 2.1. This remark applies to the case of deterministic sequences X_i, with convergent empirical distributions for example.

□

We will now give a result which will be helpful when transporting results from a space E to a space F. E and F are separable metric spaces, and ϕ is a measurable map from E to F. One also has the natural diagonal or tensor map $\phi^{\otimes N}$ from E^N into F^N. If u_N is a (non necessarily symmetric) probability on E^N, we will write $v_N = \phi^{\otimes N} \circ u_N$.

Proposition 2.4.

2.6) If u_N is u-chaotic, and the continuity points C_ϕ of ϕ have full u-measure, v_N is v-chaotic if $v = \phi \circ u$.

2.7) If $\overline{u}_\infty$ is a limit point of the laws of the empirical distributions $\overline{X}_N$ as N tends to infinity, C_ϕ has full measure under the intensity measure $I(\overline{u}_\infty)$ of $\overline{u}_\infty$, and v_N is v-chaotic, then for $\overline{u}_\infty$-a.e. m in $M(E)$, $\phi \circ m = v$.

(2.8) If $F = R$, $Q \in M(M(E))$ and ϕ is bounded and C_ϕ has full $I(Q)$ measure, then for any Q_n converging weakly to Q and any continuous real function $h(\cdot)$,

$$E_Q[h(\langle m, \phi \rangle)] = \lim_n E_{Q_n}[h(\langle m, \phi, \rangle)].$$

Proof: It is not difficult to see that (2.6), (2.7), (2.8) follow from the remark: if $Q \in M(M(E))$ and $I(C_\phi) = 1$, then for any Q-convergent sequence Q_n, $\Psi \circ Q_n$ converges weakly to $\Psi \circ Q$, if $\Psi : M(E) \to M(F)$ is the map: $m \to \phi \circ m$. The proof of this last statement is that $I(C_\phi) = 1$ implies $Q[\{m, \langle, m, C_\phi \rangle = 1\}] = 1$, and when $\langle m, C_\phi \rangle = 1$, then m is a continuity point of Ψ (see Billingsley [2] p. 30). So the continuity points of Ψ have full Q-measure which yield the remark by a second application of the quoted result.

□

We will now give some comments on how the material of this section will be used in a propagation of chaos context.

In several cases one has a symmetric law P_N, on $C(R_+, R^d)^N$ or $D(R_+, R^d)^N$ for instance, describing the interacting particle system. The initial conditions are supposed to be u_0-chaotic, $u_0 \in M(R^d)$, and one tries to prove that the P_N are P-chaotic, for some suitable P on $C(R_+, R^d)$ or $D(R_+, R^d)$ with initial condition u_0, by (2.7).

One way to prove such a statement is to show that the laws $\overline{P}_N$ of the empirical distributions $\overline{X}_N$ converge weakly to δ_P. This can be performed by checking tightness of the $\overline{P}_N$, that is tightness of the laws of X^1 on C or D under P_N, and identifying all possible limit points $\overline{P}_\infty$ as being concentrated on P.

This last step will involve finding some denumerable collection of functionals on probabilities on C or D, $G(m)$ having some continuity property (see (2.8)), for which P is the only common zero. Then we will show that $\overline{P}_\infty$ is concentrated on the zeros of each functional.

The reason for using this method is that in many cases, $G(m)$ will be a quadratic function of m, and we will end up calculating $E_{\overline{P}_\infty}[G(m)^2]$, as $\lim_\ell E_{N_\ell}[G(\overline{X}_{N_\ell})^2]$, with the help of (2.8), for a suitable subsequence N_ℓ. But the study of this last limit when G is quadratic in m, by symmetry basically involves only four particles. In many examples, it yields a line of proof which is more pleasant than working directly with (2.1), even with $k = 2$.

3) Examples.

a) Our laboratory example:

$$dX_t^i = dB_t^i + \frac{1}{N} \sum_{1}^{N} b(X_t^i, X_t^j) \, dt \, , \quad i = 1, \ldots, N \, ,$$

$$X_0^i : \text{ independent } u\text{-distributed}$$

here the X^i have symmetric laws P_N on $C(R_+, R^d)^N$, which thanks to Theorem 1.4 are P-chaotic, where P is the law of the "nonlinear process" solution of

$$dX_t = dB_t + \int b(X_t, y) \, u_t(dy) \, dt$$

$$X_0 : u - \text{distributed}, \; u_t(dy) = \text{law of } X_t \, .$$

(A number of assumptions: independence of X_0^i, Lipschitz character of b can in fact be relaxed, see references at the end of section 1).)

b) Uniform measure on the sphere of radius $\sqrt{n}$ in R^n.

Proposition 3.1. *The uniform distribution $ds_n(x)$ on the sphere of radius $\sqrt{n}$ in R^n is u-chaotic, if $u = 1/\sqrt{2\pi} \exp\{-x^2/2\}dx$.*

Proof: s_n is clearly symmetric. We will show directly that the projection of s_n on the first k components of R^n converges weakly to $u^{\otimes k}$.

Call $\mu_n(dr)$ the law of the radius under $u^{\otimes n}$, and $s_{n,r}(dx)$ the uniform distribution on the sphere of radius r in R^n.

Take now f continuous with compact support on R^k. By the law of large numbers for $(x_1^2 + \cdots + x_n^2)/n$, under $u^{\otimes n}$, we know that for $0 < a < 1 < b$,

$$(3.1) \qquad \lim_{n \to \infty} \left| \int_{R^k} f(x) u^{\otimes k}(dx) - \int_{[a\sqrt{n}, b\sqrt{n}]} d\mu_n(r) \int ds_{n,r}(x) \, f(x_1, \ldots, x_k) \right| = 0 \, .$$

But one also has:

$$\int_{a\sqrt{n}}^{b\sqrt{n}} \mu_n(dr) \int ds_{n,r}(x) \, f(x_1, \ldots, x_k) = \int_{a\sqrt{n}}^{b\sqrt{n}} \mu_n(dr) \int ds_n(x) \, f(x_1 \frac{r}{\sqrt{n}}, \ldots, x_k \frac{r}{\sqrt{n}}) \, .$$

Now $r/\sqrt{n} \in [a, b]$ in the previous integral, and f is compactly supported and continuous, so that:

$$(3.2) \qquad \lim_{a \to 1, b \to 1} \sup_{n \geq k} \left| \int_{a\sqrt{n}}^{b\sqrt{n}} \mu_n(dr) \int ds_{n,r}(x) f - \mu_n([a\sqrt{n}, b\sqrt{n}]) \langle s_n, f \otimes 1 \cdots \otimes 1 \rangle \right| = 0 \, .$$

Now for each $a < 1 < b$, $\mu_n([a\sqrt{n}, b\sqrt{n}]) \to 1$ as n goes to infinity. By (3.1), (3.2), we see that when n goes to infinity $\langle s_n, f \otimes 1 \ldots \otimes 1 \rangle$ tends to $\langle u^{\otimes k}, f \rangle$. This finishes the proof of Proposition 3.1.

c) Variation on a theme.

We will now present an example closely related to the previous one, which explains why one had the previous result. In fact one could modify it to include the previous example, however as explained before we are not seeking the maximum generality here.

Take $x_1, \ldots, x_n$, which are iid, with law $\mu(dx) = f(x)\,dx$, on R^d, where $f > 0$, is C^1 and such that

(3.3) $$\int (f(x) + |\nabla f(x)|)\, e^{\lambda|x|} dx < \infty , \quad \text{for any } \lambda .$$

Then it is known that for any $a \in R^d$, there is a unique $\lambda = \lambda_a$ such that $\frac{1}{Z_\lambda} \cdot e^{\lambda x} \mu(dz)$ (Z_λ normalization factor) has mean a. In fact if $I(\cdot)$ is the convex conjugate of the logarithm of the Laplace transform of μ:

$$I(x) = \sup_\lambda (\lambda \cdot x - \log(\int e^{\lambda \cdot y} \mu(dy))) ,$$

we have: $\nabla I(a) = \lambda_a$.

Consider $s_n(dx) = \mu^{\otimes n}[(x_1, \ldots, x_n) \in dx / \frac{x_1 + \cdots + x_n}{n} = a]$, the conditional distribution of $(x_1, \ldots, x_n)$ given the mean $x_1 + \cdots + x_n/n = a$.

Proposition 3.2. $s_n(dx)$ *is ν-chaotic, where $\nu = \frac{1}{Z_\lambda} e^{\lambda \cdot x} \mu$ with $\lambda = \lambda_a$ determined by $\int x \cdot d\nu(x) = a$.*

Proof: We have, with obvious notations

(3.4)
$$s_n(dx) = \frac{1}{z_n} f(x_1) \ldots f(x_{n-1})\, f(an - x_1 - \cdots - x_{n-1}) dx_1 \ldots dx_{n-1}$$

$$= \frac{1}{e^{\lambda \cdot an} z_n} (e^{\lambda \cdot} f)(x_1) \ldots (e^{\lambda \cdot} f)(x_{n-1})(e^{\lambda \cdot} f)(an - x_1 \ldots - x_{n-1})\, dx_1 \ldots dx_{n-1}$$

$$= \nu^{\otimes n}[(x_1, \ldots, x_n) \in dx / x_1 + \cdots x_n = an] .$$

Now looking at $x_i - a = y_i$, we see that we can assume that μ is centered and $a = 0$, and (3.3) holds.

It is now enough to show that for $\phi_1, \ldots, \phi_k \in C_K(R^d)$,

(3.5) $$\langle s_n, \phi_1 \otimes \ldots \otimes \phi_k \otimes 1 \ldots \otimes 1 \rangle \longrightarrow \prod_1^k \langle \mu, \phi_i \rangle , \quad \text{as } N \text{ goes to infinity.}$$

Now from (3.4), the expression under study is

$$(\phi_1 f) * \cdots * (\phi_k f) * f^{(n-k)*}(0) \,/\, f^{n*}(0)$$

$$= \int \widehat{\phi_1 f}(\xi) \ldots \widehat{\phi_k f}(\xi)\, \hat{f}^{n-k}(\xi)\, d\xi \,/\, \int \hat{f}^n(\xi)\, d\xi \,,$$

using Fourier transforms.

Observe that $\int \hat{f}^n(\xi) d\xi = \frac{1}{n^{d/2}} \int \hat{f}^n(\frac{\xi}{\sqrt{n}})\, d\xi$, now $\hat{f}^n(\frac{\xi}{\sqrt{n}}) \overset{n\to\infty}{\longrightarrow} \exp\{-\frac{1}{2}\,{}^t\xi A\xi\}$, for each ξ. On the other hand (3.3) ensures that we have the domination:

$$|\hat{f}(\xi)| \le [1 + |\xi|^2]^{-\delta}\,, \quad \xi \in R^d\,,$$

for a suitable δ. So we have $|\hat{f}^n(\frac{\xi}{\sqrt{n}})| \le [(1+\frac{|\xi|^2}{n})^n]^{-\delta} \le (1+\frac{|\xi|^2}{n_0})^{-n_0\delta}$, when $n \ge n_0$, which is integrable if n_0 is large enough. So we see that as n tends to infinity

$$\int \hat{f}^n(\xi) d\xi \sim (\frac{n}{2\pi})^{-d/2}(\det A)^{-1/2} \text{ and } \int_{|\xi|\ge A} n^{d/2}|\hat{f}^{n-k}(\xi)| d\xi \to 0\,, \quad \text{for } A > 0\,.$$

So $\langle s_n, \phi_1 \otimes \cdots \otimes \phi_k \otimes 1 \ldots \otimes 1\rangle$ converges to $\widehat{\phi_1 f}(0)\ldots \widehat{\phi_k f}(0) = \langle \mu, \phi_1\rangle \ldots \langle \mu, \phi_k\rangle$, which yields our claim.

$\square$

4) Symmetric self exclusion process.

In this example there will be a microscopic scale at which particles interact (exclusion condition) and a macroscopic scale. The interaction will in fact disappear in the limit.

Our N particles move on $\frac{1}{N}Z \subset R$. We look at the simple exclusion process for these N particles from the point of view of N particles which evolve in a random medium given as follows: with the bonds $(i, i+\frac{1}{N})$, $i \in \frac{1}{N}Z$, we associate a collection of independent Poisson counting processes $N_t^{i,i+1/N}$, with intensity $\frac{N^2}{2}dt$. The motion of any particle in this random medium is given by the rule:

If a particle is at time t at the location i it remains there until the first of the two Poisson processes $N_{\cdot}^{i,i+1/N}$ or $N_{\cdot}^{i-1/N,i}$ has a jump, and then performs a jump across the corresponding bond.

As a consequence of this rule if we consider N particles with N distinct initial conditions, these particles will never collide (self exclusion). Here we pick an initial distribution u_N on $(\frac{1}{N}Z)^N \subset R^N$, which is u_0-chaotic, where $u_0 \in M(R)$. Using Remark 2.3, we can for instance take the deterministic locations $(\frac{1}{N}, \frac{2}{N}, \ldots, 1)$ and symmetrize them, in this case $u_0(dx) = 1_{[0,1]}dx$. Given the initial conditions and the random medium, this provides us with N trajectories $(X_{\cdot}^1, \ldots, X_{\cdot}^N)$ and, we denote by

P_N, the symmetric law on $D(R_+, R)^N$, one obtains, when the initial conditions are picked independent of the medium, and u_N distributed.

Theorem 3.3. P_N *is P-chaotic, if P is the law of Brownian motion with initial distribution* u_0.

Corollary 3.4. *For each* $s \geq 0$, *the law of* $(X_s^1, \ldots, X_s^N)$ *is* $u_0 * p_s$-*chaotic if* $p_s = (2\pi s)^{-1/2} \exp{-\frac{x^2}{2s}}$.

Proof: This is an immediate application of Proposition 3.4, where the map ϕ, is the coordinate at time s on $D(R_+, R)$.

$\square$

Before giving the proof of Theorem 3.3, let us give some words of explanation about the point of view we have adopted here. As mentioned before, for the propagation of chaos result one works with symmetric probabilities. This is why we have constructed the self-exclusion process, using the trajectories of particles interacting with a given random medium. If one instead is just interested in "density profiles", that is, the evolution of:

$$\eta_t^N = \frac{1}{N} \sum_1^N \delta_{X_t^i} \in M(R) \,,$$

one can directly build it as a Markov process evolving on the space of simple point measures, but one loses the notion of particle trajectories, especially when two neighboring particles perform a jump (see Liggett [24], DeMasi-Ianiro-Pellegrinotti-Presutti [7]). Notice however that Corollary 3.4 can be restated purely in terms of η_t^N. Indeed by Proposition 2.2 i), it means that when N tends to infinity η_t^N converges in law to the constant probability $u_0 * p_t$, that is belongs with vanishing probability to the complement of any neighborhood of $u_0 * p_t$ for the weak topology.

So we can derive a corollary of Theorem 3.3 purely in terms of "profile measures" η_t^N, stating that if η_0^N is a sum of N distinct atoms of mass $1/N$, and converges in law to the constant probability u_0 on R, then the random probabilities η_t^N converge in law to $u_0 * p_s$. Of course Theorem 3.3 has more in it. From a "profile point of view", we have introduced the symmetric variables through the trajectories $(X^1, \ldots, X^N)$.

Proof of Theorem 3.3: We consider the empirical measures $\overline{X}_N = \frac{1}{N} \sum_1^N \delta_{X^i} \in M(D)$, where $D = D(R_+, R)$. We will show the $\overline{X}_N$ concentrate their mass around P. We will first show that the laws of the $\overline{X}_N$ are tight, and then we will identify any possible limit point as being δ_P the Dirac mass at P. This will precisely mean the $\overline{X}_N$ converge in law to the constant P.

Tightness:

By Proposition 2.2 ii), it boils down to checking tightness for $X^1_\cdot$, under P_N. But here X^1 is distributed as a simple random walk on $\frac{1}{N}\mathbf{Z}$ in continuous time with jump intensity $= N^2 dt$, and initial distribution $x^1 \circ u_N$. Classically, the law of $X^1_\cdot$ under P_N converges weakly to P, and this yields tightness.

Identification of limit points:

Take $\overline{P}_\infty$ a limit point of the laws of the $\overline{X}_N$. We already know by the tightness step (see Proposition 2.2 ii) that $I(\overline{P}_\infty)$, the intensity of $\overline{P}_\infty$ is P. By Proposition 2.4, we know that $\overline{P}_\infty$ is concentrated on measures for which X_0 is u_0-distributed. If we introduce $F(m) = E_m[(e^{i\lambda(X_t - X_s) + (\lambda^2/2)(t-s)} - 1)\psi_s(X)]$, where $\lambda \in \mathbf{R}$, $t > s$, and $\psi_s(X) = \phi_1(X_{s_1})\ldots\phi_k(X_{s_k})$, with $0 \le s_1 < \ldots < s_k \le s$ and $\phi_1, \ldots, \phi_k \in C_b(\mathbf{R})$, it is enough to show that $F(m) = 0$, $\overline{P}_\infty$-a.s., for then varying over countable families of λ, $t > s$, s_i, ϕ_i, we will find that $X_t - X_s$ is independent of $\sigma(X_u, u \le s)$, and $N(0, t-s)$ distributed, for $\overline{P}_\infty$-a.e. m, which implies $\overline{P}_\infty = \delta_P$.

Using (2.8), since $(e^{i\lambda(X_t - X_s) + \lambda^2(t-s)} - 1)$ has continuity points of full measure under $P = I(\overline{P}_\infty)$, we find

$$E_{\overline{P}_\infty}[|F(m)|^2] = \lim_k E_{N_k}[|F(\overline{X}_{N_k})|^2]$$

$$= \lim E_{N_k}[|\frac{1}{N}\sum_1^N (e^{i\lambda(X^i_t - X^i_s) + (\lambda^2/2)(t-s)} - 1)\,\psi_s(X^i)|^2]\,,$$

using symmetry, the latter quantity equals:

(3.6)
$$\lim E_{N_k}[(e^{i\lambda(X^1_t - X^1_s) + (\lambda^2/2)(t-s)} - 1)(e^{-i\lambda(X^2_t - X^2_s) + (\lambda^2/2)(t-s)} - 1)\,\psi_s(X^1)\,\psi_s(X^2)]\,.$$

Observe that $E_{N_k}[e^{i\lambda(X^1_t - X^1_s) + (\lambda^2/2)(t-s)}\,\psi_s(X^1)\,\psi_s(X^2)] =$ $E_{N_k}[e^{i\lambda(X^1_t - X^1_s) + (\lambda^2/2)(t-s)}]\ E_{N_k}[\psi_s(X^1)\psi_s(X^2)]$, and the limit of the first term of the product is 1, by the weak convergence of the law of X^1 to P. In view of (3.6), to prove that $F(m) = 0$, $\overline{P}_\infty$ a.s., it is enough to check that:

$$A_N = E_N[(\exp\{i\lambda[(X^1_t - X^2_t) - (X^1_s - X^2_s)] + \lambda^2(t-s)\} - 1)\,\psi_s(X^1)\psi_s(X^2)]$$

tends to zero when N goes to infinity. Observe now that (X^1_t, X^2_t) is a pure jump process on $(\frac{1}{N}\mathbb{Z})^2$, with bounded generator:

$$Lf(x_1, x_2) = \frac{N^2}{2}(\Delta_N^1 + \Delta_N^2) \, f(x_1, x_2)$$

(3.7)
$$- \frac{N^2}{2} 1\{x_1 - x_2 = \frac{1}{N}\} \cdot D_1 D_2 f(x_1 - \frac{1}{N}, x_2)$$

$$- \frac{N^2}{2} 1\{x_2 - x_1 = \frac{1}{N}\} \cdot D_2 D_1 \, f(x_1, x_2 - \frac{1}{N}) \,,$$

here Δ^1, Δ^2, D_1, D_2, are the discrete Laplacian and the difference operators with respect to first and second variables. If we pick now $f(x_1, x_2) = \exp\{i\lambda(x_1 - x_2)\}$, we see that

(3.8) $\qquad |\frac{Lf(x_1, x_2)}{f} - N^2(e^{i\lambda/N} + e^{-i\lambda/N} - 2)| \leq \text{const } 1\{|x_1 - x_2| = \frac{1}{N}\} \,.$

Using now the fact that $f(X_t^1, X_t^2) \exp - \int_0^t \frac{Lf(X_u)}{f} du$ is a bounded martingale, we have

$A_N =$

$E_N[f(X_t^1, X_t^2)/f(X_s^1, X_s^2) \, e^{\lambda^2(t-s)} (1 - \exp - \int_s^t (\frac{Lf}{f} + \lambda^2)(X_u) du) \, \psi_s(X^1)\psi_s(X^2)] \,,$

using (3.8) we see that:

$$|A_N| \leq o(N) + \text{const } E_N[\int_0^t 1\{|X_s^1 - X_s^2| = \frac{1}{N}\} ds]$$

So Theorem 3.3 will follow from

Lemma 3.5.

$$\lim_{N \to \infty} E_N\Big[\int_0^\infty e^{-s} 1\{|X_s^1 - X_s^2| = \frac{1}{N}\} ds\Big] = 0 \,.$$

Proof: $Y_s = N|X_{s/N^2}^1 - X_{s/N^2}^2|$, as follows from (3.7) is a jump process on $\mathbb{N}^*$, with generator

$$L'f(k) = 1\{k \neq 1\}\Delta f(k) + 1\{k = 1\}[f(2) - f(1)] \,,$$

and the quantity under the limit sign in Lemma 3.5 equals:

(3.9) $\quad E_N[\int_0^\infty e^{-s/N^2} 1\{Y_s = 1\}\frac{ds}{N^2}] \leq \frac{C}{N^2} \sum_{k=0}^\infty Q_2[e^{-\tau/N^2}]^k = \frac{C}{N^2} \frac{1}{1 - Q_2[e^{-\tau/N^2}]} \,,$

where τ is the hitting time of 1, and Q_2 is the law of Y, starting from 2. If we now pick $a(N) < 1$, such that $a + a^{-1} - 2 = 1/N^2$, using the bounded martingale $a^{Y_{s \wedge \tau}} e^{-(s \wedge \tau)/N^2}$, we find

$$E_2[e^{-\tau/N^2}] = a(N) = 1 + \frac{1}{2N^2} - ((1 + \frac{1}{2N^2})^2 - 1)^{1/2} = 1 - \frac{1}{N} + o(\frac{1}{N}) .$$

It follows that $\frac{C}{N^2}(1-a)^{-1} \sim \frac{C}{N}$ which tends to zero, and in view of (3.9) proves the lemma.

$\square$

e) Reordering of Brownian motions.

We consider N independent Brownian motions $X_.^1, \ldots, X_.^N$ on R, with initial law u_0, which we suppose atomless. We then introduce the increasing reorderings $Y_t^1 \leq \ldots \leq Y_t^N$, of the $X_.^1, \ldots, X_.^N$, so that $Y_t^1 = \inf_i\{X_t^i\}$, $Y_t^2 = \sup_{|A|=N-1} \inf_{i \in A}\{X_t^i\}$, etc. Now the processes $(Y_.^1, \ldots, Y_.^N)$ are reflected Brownian motions on the convex $\{y_1 \leq y_2 \leq \ldots \leq y_N\}$, but they are not symmetric any more, so we consider the symmetrized processes $(Z_.^1, \ldots, Z_.^N)$ on the enlarged space $(\mathcal{S}_N \times C^N, d\nu_N \otimes P_{u_0}^{\otimes N})$, where $\mathcal{S}_N$ is the symmetric group on $[1, N]$, $d\nu_N$ the normalized counting measure, and $Z_t^i = Y_t^{\sigma(i)}$, σ being the $\mathcal{S}_N$ valued component on $\mathcal{S}^N \times C^N$.

The interest of this example comes from the fact that on the one hand the $(X_.^1, \ldots, X_.^N)$ and $(Z_.^1, \ldots, Z_.^N)$ have the same density profile:

(3.10) $$\frac{1}{N} \sum_1^N \delta_{X_t^i} = \frac{1}{N} \sum_1^N \delta_{Z_t^i} ,$$

but on the other hand we are going to prove that the Z^i are Q-chaotic, where Q is a different law from P_{u_0}. Of course the X^i are P_{u_0}-chaotic. As a result of (3.10), Q and P_{u_0} will share the same time marginals, namely $u_0 * (\frac{1}{\sqrt{2\pi s}} \exp\{-\frac{x^2}{2s}\})$. This emphasizes the fact that the limit behavior of the profile evolution is not enough to reconstruct the "nonlinear process".

Let us describe the law Q. The distribution function F_t of u_t, is strictly increasing for $t > 0$. If $C = \text{supp } u_0$, we can write the complement of C as a union of disjoint intervals (a_n, b_n). The points of $D = C \setminus \cup\{a_n, b_n\}$ are points of left and right increase of F_0, and D has full u_0 measure.

We define for $x \in D$,

(3.11) $$\psi_t(x) = F_t^{-1} \circ F_0(x) ,$$

one in fact has $\lim_{t \to 0} \psi_t(x) = x$, and $\psi : x \to (\psi_t(x))_{t \geq 0}$, defines a measurable map from D in $C(R_+, R)$.

Theorem 3.6. *The laws of $(Z_.^1, \ldots, Z_.^N)$ are Q-chaotic where $Q = \psi \circ u_0$.*

Proof: We first give a lemma, making precise the structure of $Y^1, \ldots, Y^N$ as reflected Brownian motion.

Lemma 3.7. *There are N independent $\sigma(X_u, u \leq t)$ Brownian motions on $C(R_+, R)^N$, $\beta^1, \ldots, \beta^N$, and $(N-1)$ continuous adpated increasing processes $\gamma_t^1, \ldots, \gamma_t^{N-1}$ such that*

(3.12)
$$\gamma_t^i = \int_0^t 1(Y_s^i = Y_s^{i-1}) \, d\gamma_s^i \,,$$
$$Y_t^1 = Y_0^1 + \beta_t^1 - \frac{1}{2}\gamma_t^1 \,,$$
$$Y_t^k = Y_0^k + \beta_t^k - \frac{1}{2}\gamma_t^k + \frac{1}{2}\gamma_t^{k-1}$$
$$Y_t^N = Y_0^N + \beta_t^N + \frac{1}{2}\gamma_t^{N-1} \,.$$

Proof: One uses induction, by stopping the processes at the successive times where distinct X^i, X^j meet, and applies Tanaka's formula. Since these stopping times tend to infinity, one then obtains the lemma. For details see [42].

Let us check tightness of the laws of Z^1.

Remark 3.8. Before proving this point, let us mention here that the symmetrized sampling of $Y^1, \ldots, Y^N$ by the random permutation σ is crucial for tightness. Suppose u_0 has support in $[0,1]$, if the X^i are constructed on the infinite product space, one knows that for $t > 0$, $\overline{\lim}_{N \to \infty} \frac{M_t^N}{\sqrt{2t \log N}} = 1$, a.s., with $M^N = \sup(X^1 - X_0^1, \ldots, X^N - X_0^N)$. But $Y_t^N = \sup(X_t^1, \ldots, X_t^N) \geq M_t^N$, so one cannot expect tightness for the law of Y^N.

□

By symmetry and (3.10), the law of Z_t^1 is u_t. So to prove our claim it is enough to prove

Lemma 3.9.
$$E[\sum_1^N |Y_t^i - Y_s^i|^4] \leq 3N(t-s)^2 \,.$$

Proof: We apply Ito-Tanaka's formula to

$$\|Y_t - Y_s\|^2 = \sum_1^N (Y_t^i - Y_s^i)^2$$

$$= 2\sum_i \int_s^t (Y_u^i - Y_s^i) d\beta_u^i$$

$$+ \sum_1^{N-1} \int_s^t [(Y_u^{i+1} - Y_s^{i+1}) - (Y_u^i - Y_s^i)] \, d\gamma_u^i + N(t-s) \,.$$

Now by (3.12), $Y_u^{i+1} = Y_u^i \ d\gamma_u^i$-a.s., and $Y_s^i \leq Y_s^{i+1}$, so after taking expectations

(3.13)
$$E[\sum_1^N (Y_t^i - Y_s^i)^2] \leq N(t-s) .$$

In the same way, we find:

$$\sum_1^N (Y_t^i - Y_s^i)^4 = 4 \sum_1^N \int_s^t (Y_u^i - Y_s^i)^3 \, d\beta_u^i$$

$$+ 2 \sum_1^{N-1} \int_s^t [(Y_u^{i+1} - Y_s^{i+1})^3 - (Y_u^i - Y_s^i)^3] \, d\gamma_u^i$$

$$+ 6 \sum_1^N \int_s^t (Y_u^i - Y_s^i)^2 \, du ,$$

and again $(Y_u^{i+1} - Y_s^{i+1})^3 - (Y_u^i - Y_s^i)^3 \leq 0$, $d\gamma_u^i$ a.s., so that taking expectation and (3.13),

$$E[\sum_1^N (Y_t^i - Y_s^i)^4] \leq 6N \int_s^t (u-s) du = 3N(t-s)^2 .$$

$\square$

Let us show now that for fixed k, the law of $(Z^1, \ldots, Z^k)$ converges to $Q^{\otimes k}$. Since for every t, $(Z_t^1, \ldots, Z_t^N)$ has the same reordered sequence $(Y_t^1, \ldots, Y_t^N)$ as $(X_t^1, \ldots, X_t^N)$, symmetry immediately implies that for each t, $(Z_t^1, \ldots, Z_t^N)$ has the same distribution as $(X_t^1, \ldots, X_t^N)$, that is $u_t^{\otimes N}$. Using this fact for $t = 0$, one sees easily that it is enough to prove that

14) for any limit point $\overline{P}$ of the laws of Z^1, and $s > 0$, $\overline{P}[Z_s = \psi_s(Z_0)] = 1$,

to deduce that any limit point of the laws of $(Z^1, \ldots, Z^k)$ is in fact $Q^{\otimes k}$. To check this statement observe that

$$E_{\overline{P}}[|F_0(Z_0) - F_s(Z_s)|] = \lim_{N_k} E_{N_k}[|F_0(Z_0^1) - F_s(Z_s^1)|]$$

$$\leq \overline{\lim}_k E_{N_k}[|F_0(Z_0^1) - \frac{1}{N} \sum_{i=1}^N 1(X_0^i \leq Z_0^1)|]$$

$$+ \overline{\lim}_k E_{N_k}[|F_s(Z_s^1) - \frac{1}{N} \sum_{i=1}^N 1(X_s^i \leq Z_s^1)|]$$

$$+ \overline{\lim}_k E_{N_k}[|\frac{1}{N} \sum_{i=1}^N 1(X_0^i \leq Z_0^1) - 1(X_s^i \leq Z_s^1)|] .$$

The first two terms go to zero. We have in fact

$$\lim E_N[\sup_x |F_0(x) - \frac{1}{N}\sum_i 1(X_0^i \le x)|] = 0 \ ,$$

since u_0 is atomless, and a similar result at time s. On the other hand we know that since $Z_\cdot^1 = Y_\cdot^{\sigma(1)}$,

$$|\sum_i 1(X_0^i \le Z_0^1) - 1(X_s^i \le Z_s^1)| \le 1 \ ,$$

so the last term goes to zero as well.

$\square$

If we set $x_t = \psi_t(x)$, x_t satisfies the differential equation for $t > 0$,

$$(3.15) \qquad \frac{dx_t}{dt} = -\frac{\partial_t F}{\partial_x F}(t, x_t) = -\frac{1}{2}(\log u_t)'(x_t) \ ,$$

as is seen from the implicit equation $F(t, x_t) = $ const. So Q is the law of a deterministic evolution with initial random distribution u_0, and u_t is a solution of the corresponding forward equation

$$\partial_t u - \partial_x(\frac{1}{2}(\log u_t)' u) = 0$$

$$v(t = 0, \cdot) = u_0 \ .$$

Let us mention finally that when x is a "bad point" in the convex hull of the support of C, that is $x \in [a_n, b_n]$, with $-\infty < a_n < b_n < \infty$, then one sees easily that $\lim_{t \to 0} \psi_t(x) = (a_n + b_n)/2$, on the equation:

$$0 = F_t(x_t) - F_0(x) = u_+ * p_t(-\infty, x_t] - u_- * p_t[x_t, \infty) \ , \quad t > 0 \ ,$$

with $u_- = 1_{(-\infty, a_n]} \cdot u_0$, $u_+ = 1_{[b_n, \infty)} \cdot u_0$.

f) Colored particles and nonlinear process.

Now we look at P_N on $C(R_+, R^d)^N$, which are P-chaotic. Let I_0 be a subset of R^d such that $\{X_0 \in I_0\}$ is a continuity set for P, for instance a product of intervals if $u_0 = X_0 \circ P$ has a density. We color in blue the particles which are in I_0 at time zero, and we are interested in the empirical measure at time t of the blue particles:

$$(3.16) \qquad \nu_t^N(dx) = \frac{1}{N}\sum_i 1(X_0^i \in I_0)\delta_{X_t^i} \ .$$

Then Proposition 2.4, easily implies that

$$\lim_{N \to \infty} E_N[\nu_t^N \in U^c] = 0 \ ,$$

for any neighborhood for the weak topology on $M_{+,b}(R^d)$ of

(3.17) $$\nu_t(dx) = P[X_0 \in I_0 , \quad X_t \in dx] .$$

The coloration of particles is one way to recover some trajectorial information, and gain some knowledge on P, if one uses profile measures.

For applications of coloring of particles, in a propagation of chaos context, to stochastic mechanics, we refer the reader to Nagasawa-Tanaka [31].

g) Loss of Markov property and local fluctuations.

The reader might be tempted to think that when the N-particle system follows a symmetric Markovian evolution, which is chaotic, the law P of the "nonlinear process" will inherit a Markov property. We will now give an example showing that this is not the case.

Heuristically, what happens, is that we have an N-particle system having local interactions, but there is no mechanism to "average out" the local fluctuations of the interaction. One interest of the example is that the presence of these local fluctuations does not prevent the chaotic behavior of the N-particle system (so there is propagation of chaos in this sense). The limit law P does not have the Markov property, and the limit of the density profile of particles which is governed by the time marginals of P, does not correspond now to a nonlinear forward equation.

We consider the N-particle system $(Z^1, \ldots, Z^N) \in C(R_+, E)^N$, where $E = R/Z \times R$, and $Z^i = (Y^i, X^i)$. It follows a symmetric Markovian evolution E^N, given as follows:

– The Y^i are constant in time, and the Y_0^i are i.i.d. dx-distributed, on R/Z.
– $X_t^i = \sigma_i B_t^i$, $t \geq 0$, $1 \leq i \leq N$, where B^i, $1 \leq i \leq N$, are i.i.d. standard Brownian motions independent of the (Y^i), $1 \leq i \leq N$, and $\sigma_i = (1 + \sum_{j \neq i} 1\{Y_j \in \Delta_N(Y_i)\})^{1/2}$,

$\Delta_N(x)$ denoting the only interval $[\frac{k}{N}, \frac{k+1}{N})$, $0 \leq k < N$, containing $x \in R/Z$.

Let us now describe the law P on $C(R_+, E)$, which appears in the propagation of chaos result. Under P:

– $Y_t \equiv Y_0$ is dy distributed on R/Z.
– $X.$ is independent of $Y.$ and distributed as a mixture of Brownian motions starting from zero, with trajectorial variance $\sigma^2 = 1 + m$, m being distributed as a Poisson, mean one, variable.

Let us right away observe that the law P is not Markovian. Indeed if (F_t) denotes the natural filtration on $C(R_+, E)$, $\sigma^2 = \overline{\lim}_{t \to 0} \frac{X_t^2}{2t \log \log 1/t}$ is F_1-measurable, and we have:

$$E^P[X_2^2 \mid F_1] = X_1^2 + \sigma^2 ,$$

whereas:

$$E^P[X_2^2 \mid X_1] = X_1^2 + E^P[\sigma^2 \mid X_1]$$
$$= X_1^2 + (\sum_{m \geq 0} \frac{(1+m)}{m!} p_{1+m}(X_1)) \Big/ (\sum_{m \geq 0} \frac{1}{m!} p_{1+m}(X_1)) .$$

Here $p_t(x)$ denotes $(2\pi t)^{-1/2} \exp\{-\frac{x^2}{2t}\}$.

We have

Theorem 3.10.

(3.18) The laws P_N of $(Z^1_\cdot, \ldots, Z^N_\cdot)$ are P-chaotic.

(3.19) For $t \geq 0$, the random measures $\frac{1}{N} \sum_1^N \delta_{(Y^i, X_t^i)} \in M(R/\mathbb{Z} \times R)$

("density profiles") converge in law to the constant

$dy \otimes (e^{-1} \sum_{m \geq 0} \frac{1}{m!} p_{t(1+m)}(x) \, dx)$.

Proof: (3.19) is an immediate consequence of (3.18). Let us prove (3.18). In view of Proposition 2.2, we will simply show that the law of $(Z^1_\cdot, Z^2_\cdot)$ converges weakly to $P^{\otimes 2}$. Anyway, the case of $(Z^1_\cdot, \ldots, Z^k_\cdot)$ is also obvious from our proof. It is enough to show for $f_i \in C(R/\mathbb{Z})$, $g_i \in C_b(C(R_+, R))$, $i = 1, 2$, that:

$$(3.20) \qquad \lim_N E_N[f_1(Y^1_\cdot)g_1(X^1_\cdot)f_2(Y_0^2)g_2(X^2_\cdot)] = \prod_{i=1}^2 E^P[f_i(Y_0)g_i(X_\cdot)] .$$

The left member of (3.20) equals

$$(3.21) \quad \begin{aligned} &\lim_N E_N[f_1(Y_0^1)g_1(\sigma_1 B^1_\cdot)f_2(Y_0^2)g_2(\sigma_2 B^2_\cdot)] \\ &= \lim_N E_N[f_1(Y_0^1)\phi_1(\sigma_1)f_2(Y_0^2)\phi_2(\sigma_2)] , \end{aligned}$$

where $\phi_i(a) = E[g_i(aB_\cdot)]$, $i = 1, 2$ (Wiener expectation), are continuous bounded functions. The expression inside the limit in (3.21) involves an expectation on the Y^i variables, $1 \leq i \leq N$, alone. To prove (3.20), it suffices to show that conditionally on $Y^1 = y_1$, $Y^2 = y_2$, $y_1 \neq y_2$, the law of $m_i = \sum_{j \neq i} 1\{Y_j \in \Delta_N(Y_i)\}$, $i = 1, 2$, converges weakly to the law of two independent mean one Poisson variables. To check this last point observe that for large N, $\Delta_N(y_1) \cap \Delta_N(y_2)$ is empty. So for $a_1, a_2 > 0$,

$$\lim_N E_N[e^{-a_1 m_1 - a_2 m_2}/Y_1 = y_1 \, , \, Y_2 = y_2]$$

$$= \lim \left(\int_0^1 \exp\{-a_1 1_{\Delta_N(y_1)}(x) - a_2 1_{\Delta_N(y_2)}(x)\} \, dx \right)^{N-2}$$

$$= \lim_N \left(1 + \frac{1}{N}(e^{-a_1} - 1 + e^{-a_2} - 1)\right)^{N-2} = \exp\{(e^{-a_1} - 1) + (e^{-a_2} - 1)\} \, ,$$

from which our claim follows. □

An example of a situation with a loss of the Markov property can also be found in Uchiyama [54].

1) Tagged particle: a counterexample.

In the introduction, we mentioned that we refrained from calling the "nonlinear process", the "tagged particle process", because this expression has a variety of meanings. We will give here an easy example where the tagged particle is the trajectory of the particle with initial starting point nearest to a certain point. It will turn out that the law of the tagged particle will converge to a limit distinct from the natural law of the nonlinear process conditioned to start from this point.

We consider the N-particle system $(Z^1, \ldots, Z^N) \in C(R_+, E)$, where $E = [0, 1) \times R$. The processes $Z^i = (Y^i, X^i)$, $1 \leq i \leq N$, will be independent Markov processes, satisfying:

- The Y^i are constant in time, and the Y_0^i are i.i.d. dy-distributed, on $[0,1)$.
- $X_t^i = \sigma_N(Y_0^i)B_t^i$, $t \geq 0$, $1 \leq i \leq N$, where B^i, $1 \leq i \leq N$, are i.i.d. standard Brownian motions independent of the (Y^i), $1 \leq i \leq N$, and

$$\sigma_N(y) = 1\{y \in \bigcup_{k \text{ even}} [\frac{k}{N}, \frac{k+1}{N})\} + \sqrt{2}\, 1\{y \in \bigcup_{k \text{ odd}} [\frac{k}{N}, \frac{k+1}{N})\} \, .$$

The tagged particle process $\overline{Z}^N$ will be defined by $\bar{Z}^N = (Y^i, X^i)$ on $\{Y_0^i = \min Y_0^j\}$, $1 \leq i \leq N$. This defines the tagged particle a.s., with no ambiguity, and since the starting point of Z^i is $(Y_0^i, 0)$, the tagged particle corresponds to the particle with initial starting point closest to $(0,0)$ in $[0,1) \times R$.

If we still denote by dy the measure on trajectories Y, constant in $[0,1)$, with initial point dy distributed, and by W^1, W^2 Wiener measure with respective variance 1 and 2, it is easy to see that the laws of the Z^i, which are independent, are P-chaotic, if $P = dx \otimes (\frac{1}{2}W^1 + \frac{1}{2}W^2)$ on $C(R_+, E)$.

Notice, by the way, that the nonlinear process has also lost its Markov property.

We are now going to show:

Proposition 3.11: $\overline{Z}^N$ *converges in law to* $Q = \delta_0 \otimes ((1 + e^{-1})^{-1} W^1 + e^{-1}(1 + e^{-1})^{-1} W^2)$ *(here* δ_0 *denotes the Dirac mass on the constant trajectory equal to* 0).

Proof: It is enough to show that for $f \in C_b([0,1))$, $g \in C_b(C(R_+, R))$,

$$(3.22) \quad \lim_N E[f(\overline{Y}_0^N)g(\overline{X}_\cdot^N)] = f(0) \times \left((1 + e^{-1})^{-1} E^{W^1}[g] + e^{-1}(1 + e^{-1}) E^{W^2}[g] \right).$$

Set $\tilde{Y}^N = N\overline{Y}^N$, we have

$$E[f(\overline{Y}_0^N)g(\overline{X}_\cdot^N)] = \sum_{0 \le k \text{ even } < N} E[f(\frac{\tilde{Y}^N}{N})\, 1\{k \le \tilde{Y}_N < k+1\}]\, E^{W^1}[g]$$

$$+ \sum_{0 \le k \text{ odd } < N} E[f(\frac{\tilde{Y}^N}{N})\, 1\{k \le \tilde{Y}_N < k+1\}]\, E^{W^2}[g].$$

Observe now that $\tilde{Y}^N$ converges in law to an exponential variable of parameter 1. Indeed, for $t > 0$, N large,

$$E[\tilde{Y}^N > t] = E[\bigcap_1^N (Y^i > \frac{t}{N}] = (1 - \frac{t}{N})^N,$$

which tends to e^{-t}.

From this it is easy to argue that the expression in (3.23) tends to:

$$\sum_{k \text{ even } \ge 0} (e^{-k} - e^{-(k+1)})\, f(0)\, E^{W^1}[g] + \sum_{1 \le k \text{ odd}} (e^{-k} - e^{-(k+1)})\, f(0) E^{W^2}[g],$$

which is equal to the right member of (3.22). This yields our claim. $\square$

So we see that the limit law for $\overline{Z}^N$ is given by Q which is distinct from the natural conditioning of the nonlinear process to be zero at time zero corresponding to $\delta_0 \otimes (\frac{1}{2} W^1 + \frac{1}{2} W^2)$. We also refer the reader to Guo-Papanicolaou [13], where the limit behavior of a tagged particle process for a system of interacting Brownian motions is studied.

II. A local interaction leading to Burgers' equation

The object of this chapter is to present one model of local interactions, proposed by McKean [27], which leads to Burgers' equation, as the forward equation of the nonlinear process.

The "laboratory example" we discussed in Chapter I, section 1),

$$dX_t^i = dw_t^i + \frac{1}{N} \sum_1^N b(X_t^i, X_t^j) \, dt \, , \quad i = 1, \ldots, N \, .$$

with $b(\cdot, \cdot)$ bounded Lipschitz, has an interaction term $\frac{1}{N} \sum_1^N b(X_t^i, X_t^j) dt$. Such a function b, independent of N and regular, corresponds to an interaction at a macroscopic distance. In this chapter we will be dealing with the one dimensional situation, when $b(\cdot, \cdot) = \text{const } \delta(x - y)$, and the interaction will be local in nature. We will start first in section 1) with a warm up calculation, of δ-like interaction terms, for independent particles.

1) A warm up calculation.

In our "laboratory example", as a result of the Lipschitz property of $b(\cdot, \cdot)$, and Theorem 1.4 of Chapter I, one sees easily that for each t,

$$\lim_{N \to \infty} E\Big[\frac{1}{N} \sum_{i=1}^N \Big(\frac{1}{N} \sum_{j=1}^N b(X_t^i, X_t^j) - \int b(X_t^i, y) u_t(dy) \Big)^2 \Big] = 0 \, .$$

In other words, the instantaneous drift term $\frac{1}{N} \sum_j b(X_t^i, X_t^j)$ seen by particle i, is getting close to the quantity $\int b(X_t^i, y) \, u_t(dy)$ which simply depends on X_t^i.

We are now going to analyze similar quantities when $b(\cdot, \cdot)$ is replaced by $\phi_{N,a}(x - y) = N^{ad}\phi(N^a(x-y))$, with $\phi(\cdot) \geq 0$, smooth, compactly supported, on R^d, $\int \phi(x)dx = 1$. The X_t^i will be independent d-dimensional Brownian motion, with initial distribution $u_0(dx) = u_0(x)dx$ having smooth compactly supported density. The quantities $Z_i = \frac{1}{(N-1)} \sum_{j \neq i} \phi_{N,a}(X_t^i - X_t^j)$ will play the role of an instantaneous "pseudo drift" seen by particle i. We will denote by $p_s(x,y)$ the Gaussian transition density. Now for $0 < a < \infty$, we will look at the $N \to \infty$, behavior of

(1.1) $$a_N = E\Big[\frac{1}{N} \sum_{i=1}^N \Big(\frac{1}{(N-1)} \sum_{j \neq i} \phi_{N,a}(X_t^i - X_t^j) - u_t(X_t^i) \Big)^2 \Big] \, .$$

The interpretation of the parameter a, is that now the interaction range between particles is of order N^{-a}. Thanks to symmetry,

$$a_N = E\left[\left(\frac{1}{N-1}\sum_{j=2}^{N}\phi_{N,a}(X_t^1 - X_t^j) - u_t(X_t^1)\right)^2\right]$$

$$= E\left[\left(\frac{1}{(N-1)}\sum_{j=2}^{N}\phi_{N,a}(X_t^1 - X_t^j) - \phi_{N,a} * u_t(X_t^1)\right)^2\right]$$

$$+ E\left[\left(\phi_{N,a} * u_t(X_t^1) - u_t(X_t^1)\right)^2\right].$$

The second term of the previous expression clearly tends to zero. Expanding the square in the first term we find

$$a_N = \frac{1}{(N-1)}E\left[\left(\phi_{N,a}(X_t^1 - X_t^2) - \phi_{N,a} * u_t(X_t^1)\right)^2\right] + o(N)$$

(1.2)
$$= \frac{1}{(N-1)}E\left[\phi_{N,a}^2(X_t^1 - X_t^2)\right] + o(N)$$

$$= \frac{1}{(N-1)}N^{ad}E\left[N^{ad}\phi^2(N^a(X_t^1 - X_t^2))\right] + o(N).$$

So we see that

$$— \text{ when } \quad 0 < a < 1/d, \quad \lim_N a_N = 0,$$

(1.3) $\qquad — \text{ when } \quad a = 1/d, \qquad \lim_N a_N = \int \phi^2 dx \times \|u_t\|_{L^2}^2 > 0,$

$$— \text{ when } \quad a > 1/d, \qquad \lim_N a_N = \infty.$$

The case $0 < a < 1/d$ corresponds to "moderate interaction" (see Oelschläger [34]). In fact when $a = 1/d$, we are in a "Poisson approximation" regime, and conditionally on $X_t^1 = x$, the sum

$$\frac{1}{(N-1)}\sum_{j=2}^{N}\phi_{N,a}(x - X_t^j) = (1 + o(N)) \times \sum_{j=2}^{N}\phi(N^{1/d}(x - X_t^j)),$$

converges in law to the distribution of $\int_{\mathbf{R}^d} M(dy)\phi(y)$, with $M(dy)$ Poisson point process of intensity $u_t(x)dy$ (there is no misprint here). So conditioned on X_t^i there is a true fluctuation of the quantity $\frac{1}{(N-1)}\sum_{j\neq i}\phi_{N,1/d}(X_t^i - X_t^j)$, for each i.

When $a > 1/d$, conditionally on $X_t^1 = x$, the quantity $\frac{1}{(N-1)}\sum_{j=2}^{N}\phi_{N,a}(x - X_t^j)$ is zero with a probability going to 1 uniformly in x, but has conditional expectation approximately $u_t(x)$. So now we really have huge fluctuations.

Let us by the way mention that even in the presence of fluctuations a propagation of chaos result may hold. One can in fact see that the symmetric variables $Z_i =$

$\frac{1}{N} \sum_{j \neq i} \phi_{N,a}(X_t^i - X_t^j)$, $s \leq i \leq N$, are v_a-chaotic. Here v_a stands for the law of $u_t(X_t)$, when $a < 1/d$, the law of $\int_{R^d} M(dy) \phi(y)$, where conditionally on X_t, M is a Poisson point process with parameter $u_t(X_t)$, when $a = 1/d$, and trivially the Dirac mass in 0 when $a > 1/d$. The case $a = 1/d$ is somewhat comparable to example g) in Chapter I, section 3).

Since we are interested in interactions going very fast to δ (in fact being δ), if we hope to see our "pseudodrift" seen by particle i close for N large to a quantity just depending on particle i, some helping effect has to come to rescue us. This helping effect will be integration over time.

Integration over time as a smoothing effect: We are now going to replace the quantity $\frac{1}{(N-1)} \sum_{j \neq i} \phi_{N,a}(X_t^i - X_t^j)$, for each i, by $\frac{1}{(N-1)} \sum_{j \neq i} \int_0^t \phi_{N,a}(X_s^i - X_s^j) ds$. Correspondingly we are now interested in the limit behavior of:

1.4)
$$b_N = E\Big[\frac{1}{N} \sum_{i=1}^N \Big(\frac{1}{(N-1)} \sum_{j \neq i} \int_0^t \phi_{N,a}(X_s^i - X_s^j) \, ds - \int_0^t u_s(X_s^i) ds\Big)^2\Big].$$

An analogous calculation as before yields:

$$b_N = \frac{1}{(N-1)} E\Big[\Big(\int_0^t ds \phi_{N,a}(X_s^1 - X_s^2) ds\Big)^2\Big] + o(N).$$

If we introduce $W_s = X_s^1 - X_s^2$, W_s is a Brownian motion with initial distribution $\check{u}_0 * \check{u}_0 = v_0$, and transition density $p_{2s}(x, y)$. We find that

1.5)
$$b_N = \frac{2}{(N-1)} \int v_0(dx) \, dy \, dz \int_0^t du \, p_{2u}(x, y) \, \phi_{N,a}(y) \int_0^{t-u} dv \, p_{2v}(y, z) \, \phi_{N,a}(z) + o(N).$$

It is clear that in dimension $d = 1$, for any value of a, (and formally in (1.5), even if $\phi_{N,a} = \delta$), $\lim_N b_N = 0$. In dimension $d \geq 2$, taking as new variables $N^a y$, and $N^a z$, we find

$$b_N = \frac{2}{(N-1)} \int v_0(dx) dy \, dz \int_0^t du$$
$$p_{2u}(x, \frac{y}{N^a}) \, \phi(y) \int_0^{(t-u)} dv \, p_{2v}(\frac{y}{N^a}, \frac{z}{N^a}) \, \phi(z) + o(N).$$

In dimension $d = 2$, we see again from the logarithmic Green's function singularity appearing in the term $\int_0^{(t-u)} dv \, p_{2v}(y/N^a, z/N^a)$, that for any $a < \infty$, $\lim_N b_N = 0$, and in fact it is clear that b_N will not be vanishing unless the interaction range is exponentially small. In dimension $d \geq 3$, using the transition density scaling

$$b_N = \frac{2}{(N-1)} N^{a(d-2)} \int v_0(dx)dy\ dz \int_0^t du$$

$$p_{2u}(x, \frac{y}{N^a})\ \phi(y) \int_0^{(t-u)N^{2a}} dv\ p_{2v}(y,z)\ \phi(z) + o(N)\ .$$

It is now clear that

(1.6)
$$\begin{array}{lll} \text{for} & a < \dfrac{1}{d-2}\ , & \lim_N b_N = 0\ , \\[2mm] \text{for} & a = \dfrac{1}{d-2}\ , & \lim b_N = \text{const} > 0\ . \\[2mm] \text{for} & a > \dfrac{1}{d-2}\ , & \lim b_N = \infty\ . \end{array}$$

So we see that the integration over time has removed the existence of a critical exponent $1/d$, in dimension 1. In dimension 2 the new critical regime corresponds to an exponentially small range of interaction. In dimension $d \geq 3$, the critical exponent $a = 1/d$ is raised to $1/(d-2)$.

In Chapter III, we will see that there is a "Poissonian picture" corresponding to these critical regimes in dimension $d \geq 2$.

As for dimension 1, we have seen that $\lim b_N = 0$, for any a. In fact it is the consequence of an even stronger result, namely:

$$\lim E\Big[\frac{1}{N(N-1)} \sum_{i \neq j} \big(\int_0^t \phi_{N,a}(X_s^i - X_s^j)ds - \frac{1}{2}L^0(X^i - X^j)_t\big)^2\Big] = 0\ .$$

We will use these type of ideas in the next sections, in our approach to the propagation of chaos result. For other approaches, we refer the reader to Gutkin [14], Kotani-Osada [21].

2. The N-particle system and the nonlinear process.

The N-particle system will be given by the solution

(2.1)
$$dX_t^i = dB_t^i + \frac{c}{N} \sum_{j \neq i} dL^0(X^i - X^j)_t\ ,$$

$$(X_0^i)_{1 \leq i \leq N} \in \Delta_N\ , \quad u_N - \text{distributed}, \quad c > 0\ .$$

Here $L^0(X^i - X^j)_t$ denotes the symmetric local term in 0 of $X^i - X^j$, B^i are independent 1-dimensional Brownian motions, Δ_N is the subset of R^N where no three coordinates are equal, and of course the initial conditions X_0^i are independent of the Brownian motions. Existence and uniqueness of solutions of (2.1) holds trajectorially and in law.

The solution is Δ_N valued, strong Markov. It has the Brownian like scaling property: $\lambda X^x_{\cdot/\lambda^2} \overset{\text{law}}{\sim} X^{\lambda x}_{\cdot}$, for an initial starting point $x \in \Delta_N$. The δ-interaction interpretation comes from the fact that the solution of

$$dX^{i,\epsilon}_t = dB^i_t + \frac{2c}{N}\sum_{j\neq i} \phi_\epsilon(X^{i,\epsilon}_t - X^{j,\epsilon}_t)\, dt\,, \quad \phi_\epsilon(\cdot) = \frac{1}{\epsilon}\phi(\frac{\cdot}{\epsilon})\,,$$

(2.2)
$$X^{i,\epsilon}_0 = X^i_0\,,$$

converge a.s. uniformly on compact intervals to the solution of (2.1), as ϵ goes to zero. Finally if u_N is symmetric the law of the (X^i) is symmetric as well. For these results we refer the reader to [42] and [46].

Let us now present the nonlinear process. One might be tempted to define the process as the solution of

(2.3)
$$dX_t = dB_t + 2c\, u(t, X_t)\, dt$$
$$X_0\,,\ u_0 - \text{distributed},$$

for $u(t,x)$ the density of X_t at time t, from which one would deduce that $u(t,x)$ is the solution of Burgers' equation:

$$\partial_t u = \frac{1}{2}\partial^2_x u - 2c\, \partial_x(u^2)$$
$$u_{t=0} = u_0\,,$$

However the density of the law at time t of a process is an ill behaved function for the weak topology on $M(C(\check{R}_+, R^d))$, and such a characterization of the nonlinear process is not best suited to apply the strategy explained in Chapter I 2). From the previous section, we know that quantities integrated over time are better behaved, and we are going to characterize the law of the nonlinear process as the unique law of continuous semimartingales $X_\cdot$, on some filtered probability space endowed with a Brownian motion $B_\cdot$, solution of:

.4) $X_t = X_0 + B_t + A_t$, X_0, u_0-distributed, A_t continuous adapted, of integrable variation with $A_t = cE_Y[L^0(X-Y)_t]$. $Y_t = Y_0 + \overline{B}_t + \overline{C}_t$ defined on an independent filtered space endowed with a Brownian motion $\overline{B}_\cdot$ has the same law as $(X_\cdot)$, and $\overline{C}$ is a continuous adapted integrable variation process. Here $L^0(X - Y)_\cdot$ denotes the symmetric local time in zero of $X_\cdot - Y_\cdot$, defined on the product space.

For the moment we will be concerned with the uniqueness statement. Indeed the more important role of the nonlinear process, in the proof of "propagation of chaos" comes in the identification of limit points of laws of empirical measures. We will see in the course of the proof that solutions of (2.4) have indeed the structure of (2.3). We will start with some a priori estimates:

Proposition 2.1.

i) *If*

$$(2.5) \qquad X_t = X_0 + B_t + A_t \text{ ,} (B. \text{ Brownian motion, } A. \text{ integrable variation)},$$

the image of $1_{[0,T]} ds\, dP$ under the map $(s,\omega) \to (s, X_s(\omega)) \in [0,T] \times R$, for $T < \infty$, has an L^2 density $u(s,x)$, with $\|u\|_{L^2([0,T] \times R)} \le C(E[\|A\|_T])$.

ii) *If Y. on some independent space has a decomposition as in (2,5), then*

$$(2.6) \qquad E_Y[L^0(X-Y)_t] = 2 \int_0^t u(s, X_s)\, ds \text{ ,}$$

and this defines a continuous increasing integrable process which only depends on the law of Y. Here u is the density associated to Y in i).

Proof:

i) Take $\phi(\cdot) \ge 0$, smooth, symmetric supported in $[-1/2, 1/2]$, $\int \phi = 1$. Define $\psi = \phi * \phi$ so that $\psi \ge 0$, is symmetric supported in $[-1,1]$ and $\int \psi = 1$. If we define $\psi_n(\cdot) = n\, \psi(n\cdot)$, $\phi_n(\cdot) = n\phi(n\cdot)$, we have $\psi_n(\cdot) = \phi_n * \phi_n$. We now use ψ_n to define our test function

$$(2.7) \qquad F_n(x) = \int_{-\infty}^x \int_{-\infty}^u (\psi_n(v) - \psi_1(v))\, dv\, du \text{ ,}$$

so that $F_n'(x) = 0$, for x outside $[-1, +1]$, $|F_n'| \le 1$, and $|F_n| \le 2$.

Take $X_t(\omega')$ an independent copy of $X_t(\omega)$, and set on the product space $Z_t = X_t(\omega) - X_t(\omega')$. Applying Ito's formula to $F_n(Z_t)$ we find after integration:

$$(2.8) \qquad \begin{aligned} E[F_n(Z_t)] =& E[F_n(Z_0)] + E\Big[\int_0^t F_n'(Z_s)\, d(A_s(\omega) - A_s(\omega'))\Big] \\ &+ E\Big[\int_0^t (\psi_n(Z_s) - \psi_1(Z_s))ds\Big] \text{ .} \end{aligned}$$

From this it follows that:

$$(2.9) \qquad E\Big[\int_0^T \psi_n(Z_s)ds\Big] \le 4 + \|\psi\|_\infty T + 2E[\|A\|_T] \text{ .}$$

But

$$\begin{aligned} E\Big[\int_0^T \psi_n(Z_s)ds\Big] &= \int_0^T \langle u_s \otimes u_s, \int \phi_n(y-z)\phi_n(y'-z)dz\rangle \\ &= \|u_n\|_{L^2([0,T] \times R)}^2 \text{ ,} \end{aligned}$$

where u_s denotes the law of X_s at time s, and $u_n(s,x) = \int u_s(dy)\,\phi_n(x-y)$.

i) easily follows from (2.9) now.

ii) Denote by u and v the densities corresponding to Y and X, using i). The quantity $\int_0^T ds\, u(s,X_s)$ does not depend on the version of u which is used and

$$E\Big[\int_0^T u(s,X_s)ds\Big] = \langle u,v\rangle_{L^2([0,T]\times R)} < \infty .$$

In our case the corresponding uniform estimate to (2.9), shows that

$$E[L^0(X-Y)_T] < \infty ,$$

and in fact working with the limit $F_\infty(\ )$ of the functions F_n, applying to it Tanaka's formula, one finds easily by studying $F_\infty(Z_t) - F_n(Z_t)$ that

(2.10) $$L^0(X-Y)_t = \lim_{n\to\infty} 2\int_0^t \psi_n(X_s - Y_s)\,ds \quad \text{in } L^1 .$$

The same with approximations of δ supported in R_+ or R_- gives the corresponding result for the right and left continuous local times $L^{0,r}(X-Y)$, and $L^{0,\ell}(X-Y)$.. From (2.10), after integration over Y we find, u_n denoting now the ψ_n regularization of u,

$$\lim_n E\Big[|E_Y[L^0(X-Y)_t] - 2\int_0^t u_n(s,X_s)ds|\Big] = 0 .$$

On the other hand

$$\lim_n E\Big[|\int_0^t u_n(s,X_s)ds - \int_0^t u(s,X_s)ds|\Big] \le \lim_n E\Big[\int_0^t |u_n - u|(s,X_s)ds\Big]$$
$$\le \lim_n \|u - u_n\|_2 \|v\|_2 = 0 .$$

ii) easily follows. From our proof it is also clear that similarly

(2.11) $$E_Y[L^{0,r}(X-Y)_t] = E_Y[L^{0,\ell}(X-Y)_t] = 2\int_0^t u(s,X_s)ds ,$$

the approximating density from the right or from the left just as well converging in L^2 to u.

$\square$

We now state our required uniqueness statement, in fact existence will be proved later.

Theorem 2.2. *Let $S(u_0)$ be the set of laws of solutions of (2.4). $S(u_0)$ has at most one element. If P is the law of a solution of (2.4), $u_t = X_t \circ P$, satisfies*

(2.12) $\exp\{-4cF_t(x)\} = \exp\{-4cF_0\} * p_t(x)$, F_t distribution function of u_t, (that is u_t satisfies Burgers' equation).

Proof: By Proposition 2.1, we know that $X_t = X_0 + B_t + 2c\int_0^t u(s, X_s)ds$, where $u \in L^2([0,T] \times R)$ is the density of the law of X_s for a.e. s. Applying Ito's formula to $f(T, X_T)$, with $f \in C_K^\infty((0,T) \times R)$, we see that

$$0 = E[\int_0^T (\partial_s f + \frac{1}{2}\partial_x^2 f + 2cu\partial_x f)(s, X_s)ds] ,$$

using the definition of u we deduce that:

(2.13) $-\partial_s u + \frac{1}{2}\partial_x^2 u - 2c\partial_x(u^2) = 0$, in the distribution sense on $(0,T) \times R$.

If we now set: $F(t,x) = \int_{-\infty}^x u_t(dy)$, $-\partial_t F + \frac{1}{2}\partial_x^2 F - 2cu^2$ is a distribution invariant under spatial translations. The value of this distribution tested against $f \in C_K^\infty((0,T) \times R)$ is equal to

$$\int_{(0,T)\times R} dt\, dx\, F(t, x-z)(\partial_t f + \frac{1}{2}\partial_x^2 f)(t,x) + 2cu^2(t, x-z)\partial_x f(t,x)$$

for any z in R. Letting z tend to $+\infty$, we see this last expression goes to zero, so that

$$-\partial_t F + \frac{1}{2}\partial_x^2 F - 2cu^2 = 0 \text{ in the distribution sense.}$$

Now $w'/w = -4cu$ is a change of unknown function which linearizes Burger's equation into $\partial_t w = \frac{1}{2}\partial_x^2 w$. So we intoduce a regularization by convolution in space time F_λ in $(\epsilon, T - \epsilon)$, and set $w_\lambda(t,x) = \exp\{-4cF_\lambda(t,x)\}$. One now has: $\partial_t w_\lambda - \frac{1}{2}\partial_x^2 w_\lambda = 8c^2 w_\lambda[(u^2)_\lambda - (u_\lambda)^2]$, but as $\lambda \to 0$ $u_\lambda^2 \to u^2$ in $L^1((\epsilon, T-\epsilon), R)$, $u_\lambda \to u$ in $L^2((\epsilon, T-\epsilon) \times R)$. From this letting λ go to zero we see that $\partial_t w - \frac{1}{2}\partial_x^2 w = 0$, in the distribution sense, so that by hypoellipticity, w is in fact smooth, and bounded, and from this for $0 < s < t < T$, $w_t = w_s * p_{t-s}$. Letting now s go to zero we find (2.12). So u is in fact the solution of Burgers' equation. Now $u(v,x)$, $v \geq s > 0$, is Lipschitz and bounded, and

$$X_t = X_s + (B_t - B_s) + \int_s^t 2cu(v, X_v)\, dv , \quad t \geq s ,$$

X_s is $u_s(dx)$ distributed.

It follows that any two solutions of (2.4) generate the same law on $C([s, +\infty), R)$. Since s is arbitrary our claim follows.

$\square$

Remark 2.3. When $u_0(dx) = u_0(x)\ dx$, with u_0 bounded measurable, one sees easily that the solution of Burgers' equation given by (2.12) is bounded measurable. Now one has trajectorial uniqueness for the equation

$$X_t = X_0 + B_t + 2c \int_0^t u(s, X_s)\ ds\ ,$$

see Zvonkin [58]. The proof of Theorem 2.2 now yields a trajectorial uniqueness for the solution of (2.4), when X_0, $(B.)$ are given.

$\square$

3) The Propagation of chaos result.

In this section we will prove the propagation of chaos result:

Theorem 3.1. *If u_N, supported on Δ_N, is u_0-chaotic, then $(X^1, \ldots, X^N)$ solutions of (2.1) are P_{u_0}-chaotic, where P_{u_0} is the unique element of $S(u_0)$.*

With no restriction of generality we will assume that $c > 0$.

We will use in this section tightness estimates, which will be proved in section 4), namely

Proposition 3.2. *There is a $K > 0$, such that for $N > 2c$, $1 \le i \neq j \le N$, $s \le t$,*

$$(3.1) \qquad E[|X_t^i - X_s^i|^4] \le K|t - s|^2$$
$$E[(L^0(X^i - X^j)_s^t)^4] \le K|t - s|^2\ .$$

Theorem 3.1 is in fact the corollary of a stronger statement, that will be presented now. Let us first introduce some notation. We denote by $\tilde{H}$ the closed subset of $C(R_+, R) \times C_0(R_+, R)$:

$$\tilde{H} = \{(X., B.) \in C \times C_0 : X. - X_0 - B. \in C_0^+\}\ .$$

C_0 and C_0^+ are respectively the space of continuous and continuous increasing functions from R_+ to R, with value zero at time zero. In the course of the proof we will show that in fact $S(u_0)$ has indeed one element. $\tilde{P}_{u_0}$ will stand for the joint distribution of $X., B.)$, if X is the nonlinar process and $B.$ its driving Brownian motion. Precisely $\tilde{P}_{u_0} \in M(H)$ will be the measure on $\tilde{H}$ image of P_{u_0} under the map: $X. \in C \to X., X. - X_0 - 2c \int_0^. u(s, X_s) ds) \in \tilde{H}$, where u is the solution of Burgers' equation with initial condition u_0. Let us mention that $u(s, x) \le \text{const } s^{-1/2}$, so that the map is in fact continuous. Finally we will consider the law Q image of $\tilde{P}_{u_0} \otimes \tilde{P}_{u_0}$ on the space $H = \tilde{H} \times \tilde{H} \times C_0^+$, under the map:

$$(X^1., B^1.)(X^2., B^2.) \to (X^1., B^1., X^2., B^2., L^0(X^1 - X^2).)\ ,$$

which is clearly $\tilde{P}_{u_0} \otimes \tilde{P}_{u_0}$ a.s. defined.

Using Proposition 2.4 of Chapter I, Theorem 3.1 is an easy consequence of

Theorem 3.3. *The empirical measures*

$$\overline{Y}_N = \frac{1}{N(N-1)} \sum_{i \neq j} \delta_{(X^i, B^i, X^j, B^j, L^0(X^i - X^j).)} \in M(H) \,,$$

converge in law to the constant Q.

Before embarking on the proof of Theorem 3.3, we are going to give some implications of a result such as Theorem 3.3, in terms of the quantities which were presented in section 1. In our present context the reader who is an afficionado of the "density profile" point of view is in fact interested merely in the behavior for large N of $\langle \eta_t^N, f \rangle = \frac{1}{N} \sum_i f(X_t^i)$, and using Ito's formula, only the bounded variation term $c \int_0^t \frac{1}{N^2} \sum_{i \neq j} f'(X_s^i) dL^0(X^i - X^j)_s$, is really problematic.

So we will give an implication of Theorem 3.3 in terms of this quantity.

Corollary 3.4.

$$(3.2) \qquad \lim_{N \to \infty} E\Big[\frac{1}{N} \sum_i \Big(\frac{1}{N} \sum_{j \neq i} L^0(X^i - X^j)_t - 2 \int_0^t u(s, X_s^i) ds\Big)^2\Big] = 0 \,, \quad t \geq 0 \,.$$

More generally for f continuous bounded $R \to R$

$$\lim_{N \to \infty} E\Big[\frac{1}{N} \sum_i \Big(\frac{1}{N} \sum_{j \neq i} \int_0^t f(X_s^i) \, dL^0(X^i - X^j)_s - 2 \int_0^t f(X_s^i) u(s, X_s^i) ds\Big)^2\Big] = 0 \,,$$

and

$$(3.4) \quad \lim_{N \to \infty} E\Big[\Big| \int_0^t \frac{1}{N^2} \sum_{i \neq j} f(X_s^i) dL^0(X^i - X^j)_s - 2 \int_0^t \int_R u^2(s, x) \, f(x) \, ds \, dx \Big|\Big] = 0 \,.$$

Proof: First (3.2) is a special case of (3.3), and (3.4) is an immediate consequence of (3.3), and the fact that thanks to Theorem 3.1

$$\lim_N E\Big[\Big|\frac{1}{N} \sum_i \int_0^t f(X_s^i) u(s, X_s^i) ds - \int_0^t \int_R u^2(s, x) \, f(x) \, ds \, dx \Big|\Big] = 0 \,.$$

Now to prove (3.3), notice that

$$(X., B.) \in \tilde{H} \to \int_0^t f(X_s) c^{-1} d(X. - X_0 - B.)_s \in R \,,$$

is a continuous map, and the expression under study is

$$(3.5) \qquad E[\frac{1}{N} \sum_i (\int_0^t f(X_s^i)c^{-1}d(X^i - X_0^i - B^i)_s - 2\int_0^t u(s, X_s^i)ds)^2] \ .$$

If we now replace the square in (3.5) by $(\ )^2 \wedge A$, using Theorem 3.3, the limit as N goes to infinity of the new quantity is

$$E_{\bar{P}_{u_0}}[(\int_0^t f(X_s)c^{-1}d(X. - X_0 - B.)_s - 2\int_0^t f(X_s)u(s, X_s)ds)^2 \wedge A] = 0 \ .$$

Using estimates (3.1), to remove the truncation, one easily proves (3.3).

$$\square$$

Proof of Theorem 3.3: The tightness of the laws of the $\overline{Y}_N$, comes from (3.1), together with the fact that u_N is u_0-chaotic. We now have to prove that any limit point $\overline{Q}_\infty$ of the laws of the $\overline{Y}_N$, is in fact δ_Q. The proof of the following lemma is easy and we refer to [42], for more details. G_t will denote the natural σ-field on H.

Lemma 3.5. For $\overline{Q}_\infty$ a.e. $m \in M(H)$, $(X^1_., B^1_.)$ and $(X^2_., B^2_.)$ are m independent identically distributed, (B^1, B^2) is a two dimensional G_t-Brownian motion, and the law of X_0^1 (or X_0^2) under m is u_0.

Let us first introduce some notation. We define the following functions on H:

$$A_t^i = X_t^i - X_0^i - B_t^i \ , \quad i = 1, 2 \ , \in C_0^+ \text{ (continuous increasing process)},$$

A_t is the C_0^+ valued component of H ,

$$H_t = |X_t^1 - X_t^2| - |X_0^1 - X_0^2| - \int_0^t \text{sign}^+(X_s^1 - X_s^2)\, dA_s^1$$

$$- \int_0^t \text{sign}^+(X_s^2 - X_s^1)dA_s^2 - A_t \ ,$$

$$D_t = |X_t^1 - X_t^2| - |X_0^1 - X_0^2| - \int_0^t \text{sign}^-(X_s^1 - X_s^2)dA_s^1$$

$$- \int_0^t \text{sign}^-(X_s^2 - X_s^1)dA_s^2 - A_t \ .$$

Lemma 3.6. For $\overline{Q}_\infty$ a.e. $m \in M(H)$,

$$(3.6) \qquad A_t^i = cE_m[A_t/\sigma(X_.^i, B_.^i)] \ , \quad i = 1, 2,$$

$$(3.7) \qquad H_t \text{ is a continuous } G_t - \text{supermartingale},$$

$$(3.8) \qquad D_t \text{ is a continuous } G_t - \text{submartingale}.$$

Proof: Let us first prove (3.6). Set $F(m) = \langle m, (cA_t - A_t^1)\, g(X_\cdot^1, B^1)\rangle$, where g is continuous bounded, and define $F_\alpha(m)$ by the same expression, replacing cA_t by $(cA_t) \wedge \alpha$ and A_t^1 by $A_t^1 \wedge \alpha$. It is enough to show that $F(m) = 0$, $\overline{Q}_\infty$-a.s. Observe now that by (3.1), for $k \leq \infty$,

$$E_{\overline{Q}_{N_k}}[|F(m) - F_\alpha(m)|] \leq \text{const } E_{Q_{N_k}}[\langle m, (cA_t - \alpha)_+ + (A_t^1 - \alpha)_+\rangle]$$

$$\leq \text{const } \alpha^{-1} .$$

It follows that

$$E_{Q_\infty}[|F(m)|] \leq \varliminf_k E_{\overline{Q}_{N_k}}[|F(m)|]$$

$$= c \varliminf_k E_{N_k}[|\frac{1}{N(N-1)} \sum_{i \neq j}\{(L^0(X^i - X^j)_t - \frac{1}{N}\sum_{k \neq i} L^0(X^i - X^k)_t)\, g(X^i, B^i)\}|]$$

$$= c \varliminf_k E_{N_k}[|\frac{1}{N}\sum_i (\frac{1}{N(N-1)}\sum_{j \neq i} L^0(X^i - X^j)_t)\, g(X^i, B^i)|] = 0 .$$

Let us now prove (3.7), (3.8) being proved in a similar fashion.

We now introduce for $t > s$

$$F(m) = \langle m, (H_t - H_s) \cdot g_s\rangle ,$$

where g_s is a nonnegative G_s-measurable continuous bounded function. Take now $K(m) \geq 0$, a continuous bounded function, it is enough to show that

$$E_{Q_\infty}[F(m)\, K(m)] \leq 0 ,$$

to be able to conclude that (3.7) holds. Observe now that the functional ($\alpha > 0$), on H,

$$-(\int_0^t \text{sign}^+(X_s^1 - X_s^2)\, d(A^1 \wedge \alpha)_s + \int_0^t \text{sign}^+(X_s^2 - X_s^1)\, d(A^2 \wedge \alpha)_s)$$

is bounded lower semicontinuous. Using very similar truncation arguments, we see that

$$E_{Q_\infty}[F(m)\, K(m)]$$

$$\leq \varliminf_k E_{N_k}[K(\overline{Y}_N)(\frac{1}{N(N-1)}\sum_{i \neq j}\{(|X_t^i - X_t^j| - |X_s^i - X_s^j|$$

$$- c\int_s^t \text{sign}^+(X_u^i - X_u^j) \times \frac{1}{N}\sum_{k \neq i} dL^0(X^i - X^k)_u$$

$$- c\int_s^t \text{sign}^+(X_u^j - X_u^i) \times \frac{1}{N}\sum_{h \neq j} dL^0(X^j - X^h)_u$$

$$- L^0(X^i - X^j)_t)\, g_s(X_\cdot^i, B_\cdot^i, X_\cdot^j, B_\cdot^j, L^0(X^i - X^j)).\}\}]$$

Since the process $(X_.^1, \ldots, X_.^N)$ is Δ_N valued, we can in fact replace sign$^+$ by sign in the previous expression, and find using Tanaka's formula:

$$\varliminf_k E_{N_k}\left[K(\overline{Y}_N) \times \frac{1}{N(N-1)} \sum_{i \neq j} \int_s^t \text{sign}(X_u^i - X_u^j)\, d(B_u^i - B_u^j) \cdot g_s^{ij}\right],$$

with obvious notations. This is less than:

$$\text{const} \cdot \varliminf_k E_{N_k}\left[\left\{\frac{1}{N(N-1)} \sum_{i \neq j} \int_s^t \text{sign}(X_u^i - X_u^j)\, d(B_u^i - B_u^j) \cdot g_s^{ij}\right\}^2\right]^{1/2},$$

which is easily seen to be zero after expanding the square and using the orthogonality of terms (i,j), (k,ℓ) with $\{i,j\} \cap \{k,\ell\} = \emptyset$.

□

Let us now continue the proof of Theorem 3.3. For an m satisfying the properties of Lemmas 3.5, 3.6, we know that

$$D_t = \int_0^t \text{sign}^+(X_s^1 - X_s^2)\, d(B^1 - B^2)_s + L^{0,\ell}(X^1 - X^2)_t - A_t + 2\int_0^t 1(X_s^1 = X_s^2)\, dA_s^1,$$

is a G_t-submartingale. From this we deduce that the bounded variation process

$$(3.9) \qquad K_t^+ = L^{0,\ell}(X^1 - X^2)_t - A_t + 2\int_0^t 1(X_s^1 = X_s^2)\, dA_s^1,$$

is continuous increasing. Similarly, we see that:

$$(3.10) \qquad K_t^- = L^{0,\ell}(X^1 - X^2)_t - A_t - 2\int_0^t 1(X_s^1 = X_s^2)\, dA_s^2,$$

is a continuous decreasing process. From section 2 (2.11), and the independence under n of $(X_.^1, B_.^1)$, $(X_.^2, B_.^2)$, we know that:

$$E_m[L^{0,\ell}(X^1 - X^2)_t / (X^1, B^1)] = 2\int_0^t u(s, X_s^1)\, ds.$$

Conditioning (3.9) with respect to (X^1, B^1), we see that

$$(3.11) \qquad \frac{1}{c}A_t^1 + C_t^1 = 2\int_0^t u(s, X_s^1)\, ds + 2\int_0^t p(s, X_s^1)\, dA_s^1,$$

where C_t^1 is a continuous increasing process depending on (X^1, B^1), and $p(s,x) = \int 1(X_s^i = x)dm$, for $i = 1, 2$. We can write

$$dA_t^1 = 1(p(t, X_t^1) < \frac{1}{4c})\, dA_t^1 + 1(p(t, X_t^1) \geq \frac{1}{4c})\, dA_t^1,$$

and from equation (3.11), we already know that

$$(\frac{1}{c} - 2p(t, X_t^1))\ 1(p(t, X_t^1) < \frac{1}{4c})\ dA_t^1$$

is absolutely continuous with respect to Lebesgue measure. Let us now study the measure $1(p(t, X_t^1) \geq \frac{1}{4c})dA_t^1$. It is supported by the closed set

$$F = \{t \geq 0\ ,\ \exists x \in R\ ,\ p(t, x) \geq \frac{1}{4c}\}\ ,$$

which has measure zero, since the law of X_t^1 (or X_t^2) has a density with respect to dx for almost every t.

Let us show that $F \subset \{0\}$. If not there is $I = (a, b) \subset F^c$, with $b < \infty$, $b \in F$. On I, $dA_t^1 = 1(p(t, X_t^1) < \frac{1}{4c})dA_t^1$, so that $1_I dA_t^1 << dt$. From this it immediately follows that $\int_I 1(X_s^2 = X_s^1)dA_s^1 = \int_I 1(X_s^2 = X_s^1)dA_s^2 = 0$, and by (3.9), (3.10) we get:

(3.12) $$1_I \cdot dA_t = 1_I \cdot dL^{0,\ell}(X^1 - X^2)_t = 1_I dL^0(X^1 - X^2)_t\ .$$

Now the process $\overline{X}_t^1 = X_{t+a^1}^1$, for $a^1 \in I$, and $0 \leq t < b - a^1$, satisfies with obvious notations:

$$\overline{X}_t^1 = \overline{X}_0^1 + \overline{B}_t^1 + c \int_{\overline{H}} L^0(\overline{X}^1 - \overline{X}^2)_t\ d\tilde{m}(\overline{X}^2, \overline{B}^2)\ .$$

From section 2), Theorem 2.2, this implies that

$$\exp\{-4c\overline{F}_t(x)\} = \exp\{-4c\overline{F}_0\}\ *\ p_t(x)\ ,\quad 0 \leq t < b - a^1\ .$$

So for t near $b - a^1$, $\exp\{-4c\overline{F}_t\}$ is uniformly continuous, bounded above and away from zero, so that $b \notin F$. This shows that $F \subset \{0\}$. Now the same reasoning we just made shows that $A_t = L^{\ell,0}(X^1 - X^2)_t = L^0(X^1 - X^2)_t$, and

(3.13) $$X_t^1 = X_0^1 + B_t^1 + cE_{\tilde{m}}^2[L^0(X^1 - X^2)_t]\ ,$$

where $\tilde{m}$ is the law of (X^1, B^1) or (X^2, B^2) under m, and a similar equation for X^2. So $S(u_0)$ is not empty and $m = Q$. This proves Theorem 3.3.

$\square$

Let us mention that for any $u_0 \in M(R)$ we can find a sequence u_N u_0-chaotic and concentrated in Δ_N, so that we have indeed $S(u_0) = \{P_{u_0}\}$, by any u_0.

From convergence in law to trajectorial convergence:

In the case where $u_0(dx)$ has a bounded density with respect to Lebesgue measure, we can in fact consider the trajectorial solutions

$$(3.14) \qquad \overline{X}_t^i = X_0^i + B_t^i + 2c \int_0^t u(s, \overline{X}_s^i) \, ds \, , \quad 1 \leq i \leq N \, ,$$

where u is the solution of Burgers' equation, with initial condition u_0. As already mentioned in Remark 2.3 we have pathwise uniqueness for the solution of (3.14). We can in fact obtain a trajectorial convergence in the fashion of Theorem 1.4 of Chapter I.

Theorem 3.7. *Suppose the (X_0^i) are independent u_0-distributed, for any i, T,*

$$(3.15) \qquad \lim_{N \to \infty} E[\sup_{t \leq T} |X_t^{i,N} - \overline{X}_t^i|] = 0$$

Proof: If one now defines

$$\overline{Z}_N = \frac{1}{N(N-1)} \sum_{i \neq j} \delta_{(X_.^i, B^i, \overline{X}_., X_.^j, B^j, \overline{X}_.^j, L^0(X^i - X^j).)} \in M(\tilde{H} \times C \times \tilde{H} \times C \times C_0^+) \, ,$$

by the same proof as before, one sees that $\overline{Z}_N$ converges in law to the Dirac mass on the law of $(\overline{X}_.^1, B^1, \overline{X}_., \overline{X}_.^2, B^2, \overline{X}_., L^0(\overline{X}^1 - \overline{X}^2).)$. Taking $F_\alpha(m) = \langle m, \sup_{s \leq T} |X_s^1 - \overline{X}_s^1| \wedge \alpha \rangle$, we see that $\lim_N E[\sup_{s \leq T} |X_s^1 - \overline{X}_s^1| \wedge \alpha] = 0$. Estimates (3.1) then allow us to prove (3.15).

□

4) Tightness estimates.

We are now going to explain how one derives the estimates (3.1) of Proposition 3.2. To obtain these estimates, we are going to use the increasing reordering $Y_t^1 \leq \ldots \leq Y_t^N$ of the processes $(X_t^1, \ldots, X_t^N)$, (see Chapter I section 3) example e)). On the one hand, we will be able to use techniques of reflected processes to derive estimates on $Y_.^1, \ldots, Y_.^N$ and on the other hand the identity $Y_t^1 + \cdots + Y_t^N = X_t^1 + \cdots + X_t^N$ will yield a piece of information on the bounded variation term of $X_t^1 + \cdots + X_t^N$.

In a first step, let us explain how an exponential control on $X_t^1 + \cdots + X_t^N$ gives an individual exponential control on the X_t^i, $1 \leq i \leq N$, and yields (3.1).

Proposition 4.1. *Suppose there exist $d_1, d_2 > 0$, such that for $N > 2c$, any $x \in \Delta_N$, and $t > 0$,*

$$(4.1) \qquad E_x[\exp\{\frac{d_1}{\sqrt{t}} \sum_{i=1}^N (X_t^i - x^i)\}] \leq d_2^N \, ,$$

then there exists $d, \overline{d} > 0$, such that for $N > 2c$, $i \in [1, N]$, $t > 0$,

$$(4.2) \qquad E_x[\exp\{\frac{d}{\sqrt{t}}(X_t^i - x^i)\}] \leq \overline{d} ,$$

and (3.1) holds.

Proof: Let us first prove (4.2). Using scaling we can assume $t = 1$. From the estimate for $i \in [1, N]$:

$$E[\exp\{d(B_1^i + \frac{c}{N}\sum_{j \neq i} L^0(X^i - X^j)_1)\}]$$

$$\leq \exp\{d^2\} \times E[\exp\{\frac{2dc}{N}\sum_{j \neq i} L^0(X^i - X^j)_1\}]^{1/2} ,$$

we see it is enough to focus on $E[\exp\{\frac{2d}{N}\sum_{j \neq i} L^0(X^i - X^j)_1\}]$. (4.2) will follow from a link between the individual terms $\frac{1}{N}\sum_{j \neq i} L^0(X^i - X^j)_1$ for $i \in [1, N]$ and the sums $\frac{1}{N}\sum_{1 \leq i \neq j \leq N} L^0(X^i - X^j)_1$. We introduce to this end: f such that f, f' are bounded, $f'' = \delta_0 +$ bounded function:

$$f(y) = [\arctan(y)]_+ ,$$

$$f'(y) = 0, \quad y < 0, 1/2 , \quad y = 0, (1 + y^2)^{-1}, \quad y > 0 .$$

$$g(y) = -\frac{2y_+}{(1 + y^2)^2} , \text{ the regular part of } f'' .$$

For simplicity, let us pick $i = 1$ as the individual term to be estimated. Set

$$(4.3)$$
$$S_t = \frac{-1}{N}\sum_{j \neq 1}\{f(X_t^1 - X_t^j) - f(X_0^1 - X_0^j)\} + \frac{c}{N^2}\sum_{\substack{j \neq 1 \\ k \neq 1}}\int_0^t f'(X_s^1 - X_s^j)\, dL^0(X^1 - X^k)_s$$

$$- \frac{c}{N^2}\sum_{\substack{j \neq 1 \\ k \neq j}}\int_0^t f'(X_s^1 - X_s^j)\, dL^0(X^j - X^k)_s + \frac{1}{N}\sum_{j \neq 1}\int_0^t g(X_s^1 - X_s^j)\, ds .$$

From Tanaka's formula, $S_t + \frac{1}{2N}\sum_{j \neq 1} L^0(X^1 - X^j)_t$ is a martingale with increasing process

$$(4.4) \quad U_t = \frac{1}{N^2}\sum_{j \neq 1}\int_0^t (f')^2(X_s^1 - X_s^j)\, 2\, ds + \frac{1}{N^2}\sum_{j \neq k}\int_0^t f'(X_s^1 - X_s^j)\, f'(X_s^1 - X_s^k)\, ds .$$

We now have for $\lambda > 0$,

(4.4)

$$E[\exp\{\frac{\lambda}{2N}\sum_{j\neq 1} L^0(X^1 - X^j)_1\}]$$

$$= E[\exp\{\frac{\lambda}{2N}\sum_{j\neq 1} L^0(X^1 - X^j)_1 + \lambda S_1 - \lambda^2 U_1 + \lambda^2 U_1 - \lambda S_1\}]$$

$$\leq E[\exp\{\frac{\lambda}{N}\sum_{j\neq 1} L^0(X^1 - X^j)_1 + 2\lambda S_1 - 2\lambda^2 U_1\}]^{1/2} E[\exp\{2\lambda^2 U_1 - 2\lambda S_1\}]^{1/2} .$$

Now by the exponential martingale property the first term of the last expression is smaller than 1. So we have:

(4.5) $$E[\exp\{\frac{\lambda}{2N}\sum_{j\neq 1} L^0(X^1 - X^j)_1\}] \leq E[\exp\{2\lambda^2 U_1 - 2\lambda S_1\}]^{1/2} .$$

Observe now that U_1 is bounded, and that the only two dangerous terms of S_1 are the second and third in (4.3). However the second term only comes with a negative sign in $-2\lambda S_1$, so that we only have to worry about the third:

$$\frac{2\lambda c}{N^2}\sum_{\substack{j\neq 1 \\ k\neq j}}\int_0^1 f'(X_s^1 - X_s^j)\, dL^0(X^j - X^k)_s \leq 2\lambda c\frac{\|f'\|_\infty}{N^2}\sum_{1\leq j\neq k\leq N} L^0(X^j - X^k)_1 .$$

But from the Cauchy-Schwarz inequality:

$$E[\exp\{2\lambda\|f'\|_\infty\frac{c}{N^2}\sum_{1\leq j\neq k\leq N} L^0(X^j - X^k)_1\}]$$

$$\leq E[\exp\{4\lambda\|f'\|_\infty\frac{1}{N}\sum_{i=1}^N (X_1^i - x^i)\}]^{1/2} E[\exp\{-4\lambda\|f'\|_\infty\frac{1}{N}\sum_i B_1^i\}]^{1/2}$$

$$\leq E[\exp\{4\lambda\|f'\|_\infty\sum_{i=1}^N (X_1^i - x^i)\}]^{1/2N}\exp\{4\lambda^2\|f'\|_\infty^2/N\} .$$

Picking λ small enough we see that (4.2) follows from (4.1).

Using scaling and the Cauchy Schwarz inequality, one easily deduces from (4.2), the first estimate $E[|X_t^i - X_s^i|^4] \leq K|t - s|^2$, in (3.1). On Tanaka's formula:

$$L^0(X^i - X^j)_s^t = |X_t^i - X_t^j| - |X_s^i - X_s^j| - \int_s^t \text{sign}(X_u^i - X_u^j)\, d(X_u^i - X_u^j) ,$$

it is easy to obtain the second estimate of (3.1)

$$E[\{L^0(X^i - X^j)_s^t\}^4] \leq K(t - s)^2 .$$

□

So we have reduced the proof of (3.1) to that of (4.1). Using scaling as before and the Cauchy-Schwarz inequality, it is enough to check that for some $d_1, d_2 > 0$:

$$(4.6) \qquad E[\exp\{\frac{d_1}{N} \sum_{i \neq j} L^0(X^i - X^j)_1\}] \leq d_2^N ,$$

for any initial point $x \in \Delta_N$ and $N > 2c$. As we mentioned before we are now going to use the increasing reordering $Y_t^1 \leq \ldots \leq Y_t^N$ of the processes $X_t^1, \ldots, X_t^N$. The semimartingale structure of the (Y^i) processes is given by the following lemma whose proof is very similar to that of Lemma 3.7, once one knows the $(X.)$ process is Δ_N valued (see [42] for details):

We suppose our process $X.$ is constructed on some filtered, probability space $(\Omega, F, F_t, (B_t^i), P)$, endowed with F_t-Brownian motions B^i, $1 \leq i \leq N$.

Lemma 4.2. *There are N independent F_t-Brownian motions $W_t^1, \ldots, W_t^N$, and $(N-1)$ continuous increasing processes $\gamma_t^1, \ldots, \gamma_t^{N-1}$ such that:*

$$(4.7) \qquad \begin{aligned} Y_t^1 &= Y_0^1 + W_t^1 - \frac{1}{2}a\gamma_t^1 , \\ Y_t^k &= Y_0^k + W_t^k - \frac{1}{2}a\gamma_t^k + \frac{1}{2}b\gamma^{k-1} , \quad 2 \leq k \leq N-1 , \\ Y_t^N &= Y_0^N + W_t^N \qquad\quad + \frac{1}{2}b\gamma_t^{N-1} , \end{aligned}$$

where $a = 1 - 2c/N$, $b = 1 + 2c/N$, and

$$(4.8) \qquad \gamma_t^i = \int_0^t 1(Y_s^i = Y_s^{i+1}) \, d\gamma_s^i .$$

□

The identity $Y_t^1 + \cdots + Y_t^N = X_t^1 + \cdots X_t^N$, now yields

$$(4.9) \qquad \frac{2}{N}(\gamma_t^1 + \cdots + \gamma_t^{N-1}) = \frac{1}{N} \sum_{i \neq j} L^0(X^i - X^j)_t .$$

So our estimate (4.6) can be rephrased in terms of the (γ^i) processes. Let us introduce some more convenient processes, namely:

$$(4.10) \qquad \begin{aligned} D_t^k &= b^{-(k-1)}(Y_t^{k+1} - Y_t^k) , \quad H_t^k = b^{-(k-1)}(W_t^{k+1} - W_t^k) , \\ C_t^k &= b^{-(k-1)}\gamma_t^k , \quad 1 \leq k \leq N-1 . \end{aligned}$$

One can see easily that:

$$D_t^1 = D_0^1 + H_t^1 + C_t^1 - \frac{\alpha}{2}C_t^2$$

$$D_t^k = D_0^k + H_t^k + C_t^k - \frac{\alpha}{2}C_t^{k+1} - \frac{1}{2}C_t^{k-1}, \quad 2 \le k \le N-2$$

$$D_t^{N-1} = D_0^{N-1} + H_t^{N-1} + C_t^{N-1} \qquad - \frac{1}{2}C_t^{N-2}, \quad \text{with } \alpha = ab = 1 - \frac{4c^2}{N^2}.$$

Moreover $C_t^i = \int_0^t 1(D_s^i = 0)dC_s^i$. It now follows from the solution to the Skorohod problem that: $(C^k)_{1 \le k \le N-1} = F((D_0^k + H^k), (C_.^k))$, where $F : (C \times C_0^+)^{N-1} \to (C_0^+)^{N-1}$, is the map:

$$F(v,c)_t^1 = \sup_{s \le t}(-v_.^1 + \frac{1}{2}\alpha c_.^2)_+,$$

$$F(v,c)_t^k = \sup_{s \le t}(-v_.^k + \frac{1}{2}\alpha c_.^{k+1} + \frac{1}{2}c_.^{k-1})_+$$

$$F(v,c)_t^{N-1} = \sup_{s \le t}(-v_.^k \qquad + \frac{1}{2}c_.^{N-2})_+$$

Set $|w|_t = \sum_{i=1}^{N-1} \sup_{s \le t} |w_s^i|$, for $w \in C^{N-1}$. It is not difficult to prove (see [42]) that:

Lemma 4.3.

(4.12) $$|F(v_., c_.) - F(v_., c_.')|_t \le \frac{1}{2}(1 + |\alpha|)|c_. - c_.'|_t.$$

If $c_. = F(v_., c_.)$, $\bar{c}_. = F(\bar{v}_., \bar{c}_.)$, then

(4.13) $$|c_. - \bar{c}_.|_t \le \frac{2}{1 - |\alpha|}|v_. - \bar{v}_.|_t, \quad t \ge 0,$$

and

(4.14) $$v_. \le \bar{v}_. \Rightarrow \bar{c}_. \le c_..$$

$\square$

Because of (4.12), when $N > 2c$, we can consider the fixed point solution $c_. = F(v_., c_.)$ for any $v_. \in C^{N-1}$, which is obtained by iteration.

Now because of (4.14), if one replaces in (4.11) D_0^k by 0, this increases the corresponding $c_.$ processes. If one then replaces D_0^k by $\frac{b}{N}^{-(k-1)}$ we see from (4.13) that the corresponding fixed point $\bar{c}_.$, satisfies

$$\frac{1}{N}(c_1^1 + \cdots + c_1^{N-1}) \le \frac{1}{N}(\bar{c}_1^1 + \cdots + \bar{c}_1^{N-1}) + \frac{1}{N}\frac{2}{1-\alpha} \times \sum_{k=1}^{N-1} \frac{b}{N}^{-(k-1)}$$

$$\le \frac{1}{N}(\bar{c}_1^1 + \cdots + \bar{c}_1^{N-1}) + \frac{N}{2c^2}.$$

The constant b^k, for $k \in [1, N]$ satisfy: $1 \le b^k \le e^{2c}$, and from the previous inequality we simply have to prove

$$(4.15) \qquad E[\exp\{\frac{d}{N}(\gamma_1^1 + \cdots + \gamma_1^{N-1})\}] \le \bar{d}^N , \qquad \text{for some } d, \bar{d} > 0 ,$$

when the initial point is now $x = (0, \frac{1}{N}, \ldots, \frac{N-1}{N})$. Set $\rho = (1 - 2c/N \: / \: 1 + 2c/N) = a/b$, one easily sees that for $i \in [1, N]$, $0 < \gamma \le \rho^i \le 1$, where γ is independent of N. By Ito's formula

$$\sum_1^N \rho^i (Y_t^i - Y_0^i)^2 = 2 \sum_{i=1}^N \int_0^t \rho^i (Y_u^i - Y_0^i) \, dW_u^i$$

$$+ \sum_{i=1}^{n-1} \int_0^t [\rho^{i+1} b(Y_u^{i+1} - Y_0^{i+1}) - \rho^i a(Y_u^i - Y_0^i)] \, d\gamma_u^i$$

$$+ t \sum_1^N \rho^i .$$

Using now the fact that $\rho^{i+1} b = \rho^i a$, $Y_u^{i+1} = Y_u^i \, d\gamma_u^i$-a.s., and $Y_0^{i+1} - Y_0^i = \frac{1}{N}$, we find:

$$(4.16) \qquad \sum_1^N \rho^i (Y_t^i - Y_0^i)^2 + \frac{1}{N} \sum_1^{N-1} \rho^i a \gamma_t^i \le 2 \sum_{i=1}^N \int_0^t \rho^i (Y_u^i - Y_0^i) \, dW_u^i + Nt .$$

So to obtain a control such as (4.15), it is enough to prove

$$E[\exp\{d \sum_1^N \int_0^1 \rho^i (Y_u^i - Y_0^i) dW_u^i\}] \le \bar{d}^N , \qquad \text{for some } d, \bar{d} > 0 .$$

Using an exponential martingale and Cauchy Schwarz inequality, as we did in (4.5), and then the convexity of the exponential, it is easily seen that it is enough to show that for some $d, \bar{d} > 0$,

$$(4.17) \qquad \int_0^1 E[\exp\{d \sum_1^N (Y_t^i - Y_0^i)^2\}] \, dt \le \bar{d}^N$$

Let us set $|Y_t - Y_0|^2 = \sum_1^N \rho^i (Y_t^i - Y_0^i)^2$, and $U_t = \exp\{\frac{\lambda}{t+1} |Y_t - Y_0|^2\}$. Applying Ito's formula, we find:

$$U_t = 1 + \sum_1^N \int_0^t 2\lambda\rho^i \frac{(Y_s^i - Y_0)}{s+1} U_s \, dW_s^i$$

$$+ \lambda \sum_{i=1}^{N-1} \int_0^t \{\rho^{i+1} b(Y_s^{i+1} - Y_s^i) - \rho^i a(Y_s^i - Y_0^i)\} (s+1)^{-1} U_s \, d\gamma_s^i$$

$$+ \frac{1}{2} \int_0^t U_s \left(\frac{4\lambda^2}{(s+1)^2} \sum_i \rho^{2i}(Y_s^i - Y_0^i)^2 - 2\lambda \frac{|Y_s - Y_0|^2}{(s+1)^2} \right) ds$$

$$+ \sum_1^N \lambda\rho^i \int_0^t (s+1)^{-1} U_s \, ds .$$

Now by the same reason as in (4.16), the third term of the last expression is nonpositive. For $\lambda \leq \frac{1}{2}$ since $\rho \leq 1$, the fourth term is as well nonpositive. It then follows that for $\lambda \leq 1/2$:

$$U_t \leq 1 + \lambda N \int_0^t U_s \, ds + \text{local martingale.}$$

It then follows using a familiar stopping time argument and Gronwall's inequality, that:

$$E[\exp\{\frac{1}{2(t+1)} \sum_1^N \rho^i(Y_t^i - Y_0^i)^2\}] \leq \exp\{\frac{N}{2}t\} .$$

From this (4.17) follow, and we get our claim (4.1).

□

5) Reordering of the interacting particle system.

We look now at the same problem in our present context as we did for independent Brownian motions, in section 3 example e) of Chapter I. As we will see a very similar result holds in our case as well. We again suppose u_0 atomless, and u_N u_0-chaotic, Δ_N supported. We introduce the symmetrized, reordered processes $(Z^1, \ldots, Z^N)$ defined by:

$$(5.1) \qquad\qquad Z_t^i = Y_t^{\sigma(i)} , \quad 1 \leq i \leq N ,$$

where σ is a uniformly distributed permutation of $[1, N]$, independent of the space where the $(X^1, \ldots, X^N)$ process is constructed.

We will see that the $(Z_\cdot^i)$ are R-chaotic where R is the law of a "deterministic" type evolution. Let us describe R. more precisely. We define for $x \in D = \text{supp } u_0 \setminus \cup_n \{a_n, b_n\}$, where $(\text{supp } u_0)^c = \cup_n (a_n, b_n)$, (a_n, b_n) disjoint intervals,

$$(5.2) \qquad\qquad \psi_t(x) = F_t^{-1} \circ F_0(x) \, ,$$

where F_t is the distribution function of the solution at time t of Burgers' equation, with initial u_0, that is;

$$\exp\{-4cF_t\} = \exp\{-4cF_0\} * p_t \, .$$

D has full measure under u_0, and for $x \in D$, $\lim_{t \to 0} \psi_t(x) = x$. The probability R is defined as $\psi \circ u_0$. We have

Theorem 5.1. *The laws* $(Z_\cdot^1, \ldots, Z_\cdot^N)$ *are R-chaotic.*

Proof: The proof is a repetition of the proof of Theorem 3.6 of Chapter I. The only point to explain is the tightness estimate. Recall that $p = \frac{a}{b} = \frac{1-2c/N}{1+2c/N}$, with the notations of section 4. Our required tightness estimates follow from

Lemma 5.2.

$$(5.3) \qquad\qquad E[\sum_1^N \rho^i (Y_t^i - Y_s^i)^4] \le 3N(t-s)^2 \, .$$

Proof: Basically by the same argument as in (4.16), we know that

$$(5.4) \qquad\qquad E[\sum_1^N \rho^i (Y_t^i - Y_s^i)^2] \le N(t-s) \, .$$

Applying then Ito's formula to $\sum_1^N \rho^i (Y_t^i - Y_s^i)^4$, we find;

$$\sum_1^N \rho^i (Y_t^i - Y_s^i)^4 = 4 \int_s^t \sum_i \rho^i (Y_u^i - Y_s^i)^3 \, dW_u^i$$

$$+ 2 \sum_1^{N-1} \int_s^t [\rho^{i+1} b(Y_u^{i+1} - Y_s^{i+1})^3 - \rho^i a(Y_u^i - Y_s^i)^3] \, d\gamma_u^i$$

$$+ 3 \int_s^t \sum_1^N \rho^i (Y_u^i - Y_s^i)^2 \, du \, .$$

By a now familiar argument the second term of the last expression is nonpositive, and since $\rho \le 1$, from (5.4) after integration we easily find our claim (5.3).

$\square$

The proof of Theorem 5.1 is then basically a repetition of that of Theorem 3.6, of Chapter I.

□

Before closing this section let us mention that now for $t > 0$, $x_t = \psi_t(x)$ satisfies the O.D.E.

$$(5.5) \qquad \dot{x}_t = -\frac{\partial_t F}{\partial_x F}(t, x_t) = -\frac{1}{u}(\frac{1}{2}\partial_x^2 F - 2cu^2) = (-\frac{1}{2}(\log u)' + 2cu)(t, x_t) \, .$$

So R is the law of the deterministic solution of O.D.E. (5.5) with initial random condition u_0 distributed.

III. The constant mean free travel time regime

In the warm up calculation of chapter II, section 1, we have seen that the interaction range $N^{-1/d-2}$, in dimension $d \geq 3$, and $\exp\{-\text{const}.N\}$ in dimension $d = 2$, is critical in the study of the quantity:

$$b_N = E\left[\frac{1}{N}\sum_{i=1}^{N}\left(\frac{1}{(N-1)}\sum_{j\neq i}\int_0^t \phi_{N,a}(X_s^i - X_s^j)\,ds - \int_0^t u_s(X_s^i)\,ds\right)^2\right],$$

for independent d-dimensional Brownian motions X^i, whereas there is no critical regime in dimension 1.

The object of this chapter is to study in more detail this critical regime, which we call "constant mean free travel time regime". The reason for this name is that in the limit regime, a "typical particle" does not feel any influence from the other particles before a positive time. We will see that there is a very algebraic Poissonian picture which governs the limiting regime.

In section 1 we will start with a study of annihilated Brownian spheres, motivated by the work of Lang-Nguyen [23]. However we will follow a different route from the hierarchy method they use (see introduction). Although the results of section 1 answer the "propagation of chaos" question, they somehow miss the deeper Poissonian limit picture which is then explained in sections 2 and 3.

1) Annihilating Brownian spheres

We consider $X^1, \ldots, X^N$, independent Brownian motions with initial distribution u_0, in R^d, $d \geq 2$, which are centers of "Brownian soap bubbles" of radius $\frac{1}{2}s_N$. This means that when the centers of two such spheres which are still intact come to a distance smaller than s_N, then both spheres are destroyed. We precisely pick

$$(1.1) \qquad \begin{aligned} s_N &= N^{-1/d-2}, \quad d \geq 3, \\ &= \exp\{-N\}, \quad d = 2, \end{aligned}$$

and we assume that u_0 has a bounded density V. We will denote by τ_i the death time of the ith sphere. We are going to study the chaotic behavior of the laws of the symmetric variables $(X^1, \tau_1), \ldots, (X^N, \tau_N)$ on $(C(R_+, R^d) \times [0, +\infty])^N$. The result we will obtain will already motivate the terminology of "constant mean free travel time regime".

Theorem 1.1. *The laws of the $((X^i, \tau_i))_{1 \leq i \leq N}$, are P-chaotic, where P on $C \times [0, +\infty]$ is defined by:*

– *X. is a Brownian motion with initial distribution u_0.*

$- P[\tau > t/X.] = \exp\{-c_d \int_0^t u(s, X_s)ds\}$, *where u is the unique bounded solution of*

(1.2)
$$\partial_t u = \frac{1}{2}\Delta u - c_d u^2$$
$$u_{t=0} = u_0$$

and

$$c_d = (d-2)\mathrm{vol}(S_d), \quad d \geq 3, \quad = 2\pi, \quad d = 2.$$

Remark 1.2. Let us first give some comments,

1) the nonlinear equation (1.2) is understood in the integral form:

$$u_t = u_0 P_t - c_d \int_0^t u_s^2 \, P_{t-s} \, ds,$$

where P_t is the Brownian semigroup. We let P_t act on the right in the previous integral formula to indicate that we are in fact dealing with a forward equation, although this is somehow obscured here by the self-adjointness of P_t.

2) An application of Ito's formula to $f(t, X_t) \exp\{-c_d \int_0^t u(s, X_s)ds\}$, when $f(s, x) = P_{t-s}\phi(x)$, $\phi \in C_k^\infty(R^d)$, easily shows that the subprobability distribution of the alive particle under P is u_t, that is:

$$E_P[\phi(X_t)\, 1(\tau > t)] = \int \phi(x)\, u_t(x)\, dx.$$

3) In fact one can get various reinforcements of the basic weak convergence result corresponding to Theorem 1.1. For instance if f_i, $i = 1, \ldots, k$ belong to $L^1(\mu)$ (μ: Wiener measure with initial distribution in u_0), and ϕ_i are bounded functions on $[0, +\infty]$, with a set of discontinuities included in $(0, \infty)$ having zero Lebesgue measure, then:

$$\lim_N E_N[f_1(X^1)\phi_1(\tau_1)\ldots f_k(X^k)\phi_k(\tau_k)] = \prod_1^k E_p[f_i(X)\phi_i(\tau)].$$

Also if one applies Theorem 2.2 of Baxter-Chacon-Jain [1], to $\tau_1 \wedge \cdots \wedge \tau_k$, one can also see that the density of presence of the first k particles, when they are all alive at time t, (that is $\tau_1 \wedge \ldots \wedge \tau_k > t$), converges in variation norm to $u(t, x_1)\ldots u(t, x_k)\, dx_1 \ldots dx_k$.

4) The tightness of the laws of $\overline{X}_N = \frac{1}{N}\sum_i \delta_{(X^i, r^i)}$ is clearly immediate, by Proposition 2.2 of Chapter I. This is an indication that this point will not be very helpful in the proof. One of the difficulties of the problem is that annihilation induces a fairly complicated and long range dependence structure. The proof has somehow to rely on the treatment of terms like $1\{\tau_i \leq t\}$.

One may be tempted to express an event like $\{\tau_1 \leq t\}$ in terms of the basic independent variables which are the processes $(X^1_\cdot, \ldots, X^N_\cdot)$, for instance see Lang-Nguyen [23], p. 244.

However we will refrain from doing this. Loosely speaking we will stop at the first step of the unraveling, where the killer of particle 1 is considered: $P_N[\tau_1 \leq t] \simeq (N-1)P[X^2$ had a collision with X^1 before time t, both were alive at that time]. Such an equality somehow bootstraps the law of (X^2, τ_2) in the law of (X^2, τ_1). The idea is that this should force limit points of empirical laws to be concentrated on measures m satisfying a self-consistent property which will be characteristic of the law P.

5) The collision region between two particles corresponds to the set $|x_1 - x_2| \leq s_N$, in R^{2d}.

It roughly corresponds to the region where the two-particle 1-potential generated by Lebesgue measure sitting on the diagonal of $R^d \times R^d$,

$$h(x_1, x_2) = \int_0^\infty e^{-s}ds \int_{R^d} p_s(x_1, z) \, p_s(x_2, z) \, dz$$

is larger than $c_d^{-1}N$. This notion of level set of potentials is appropriate to find the "right" collision sets in non Brownian situations, see [45]. One can see that if $T_{1,2}$ is the entrance time in the region $|x_1 - x_2| \leq s_N$, then as N goes to infinity for $x_1 \neq x_2$ $NE_{x_1,x_2}[T_{1,2} \in dt, \, (X^1_{T_{1,2}}, X^2_{T_{1,2}}) \in dx]$ converges vaguely to the measure $c_d p_s(x_1, z) \, p_s(x_2, z) \, dz \, ds$ on $(0, \infty) \times (R^d)^2$ (see [45]). Here dz stands for Lebesgue measure on the diagonal of $(R^d)^2$. This also makes plausible that we are dealing with a kind of Poisson limmit theorem.

□

We are going to prove Theorem 1.1, in a number of steps. The first step will be to give another characterization of the law P, given in Theorem 1.1, getting us closer to the actual form we will use to identify limit points of laws of empirical measures.

Lemma 1.2. *P is the only probability m on $C \times [0, \infty]$ such that*

– X is a Brownian motion with initial distribution u_0,

$$(1.3) \qquad - \text{for } t \geq 0, \quad E_m[1(\tau \leq t) - \int_0^t c_d \, 1(\tau > s) \, u(s, X_s) \, ds/X.] = 0 \,;$$

where $u(s, x) \in L^\infty(ds \, dx)$ is the density of the image of the measure $1(\tau \geq s) \, ds \, dm$ under (s, X_s).

Sketch of Proof: In view of Remark 1.2 2), which identifies the solution of non-linear equation (1.2), with the density of presence of the not yet destroyed particle, for P, one checks readily that P satisfies the required conditions.

Uniqueness:

Let $\nu(X., dt)$ be the conditional distribution on $[0, \infty]$ of τ given $X.$ which is μ-distributed, where μ is Wiener measure with initial condition u_0. Condition (1.3) implies that μ-a.s., for any t:

$$\nu_{X.}([0,t]) - \int_0^t c_d\, \nu_{X.}((s,\infty))\, u(s, X_s)\, ds = 0 ,$$

which implies that $\nu_{X.}((t,\infty)) = \exp\{-c_d \int_0^t u(s, X_s) ds\}$. It then follows that for $\phi \in C_K^\infty$, and a.e. t:

$$\langle u_t, \phi \rangle = E[\phi(X_t)\, 1(\tau \geq t)] = \langle u_0, P_t\phi \rangle - c_d \int_0^t (u_s^2, P_{t-s}\phi)\, ds ,$$

so that u is the unique bounded solution of the integral equation corresponding to (1.2). This yields that $m = P$.

$\square$

The present characterization of P has the advantage for us that u does not refer to the solution of (1.2) any more, but can be directly measured on m. We also got rid of the exponential term $\exp\{-c_d \int_0^t u(s, X_s) ds\}$, and deal instead with $c_d \int_0^t 1(\tau > s)\, u(s, X_s)ds$. This will reduce the complexity of computations. Even so $u(s, x)$ is a (very) ill behaved function of m under weak convergence. So later we are in fact going to reinterpret this quantity $u(s, x)$, using a priori knowledge on our possible limit points of empirical measures to obtain continuous enough functionals characterizing P.

Before that we introduce the idea of chain reactions, in order to restore some independence between particles. We set for $1 \leq i \neq j \leq N$

(1.4) $$T_{i,j} = \inf\{s \geq 0 , \ |X_s^i - X_s^j| \leq s_N\} .$$

The set $A_{N,t}^{i,j}$ will represent the occurrence between the Brownian trajectories $X^1, \ldots, X^N$, and forgetting about any destruction, of a chain reaction leading from i to j before time t. Precisely (for $i = 2, j = 1$):

(1.5)
$$A_{N,t}^{2,1} = \{T_{2,1} \leq t \text{ or } \exists k_1, \ldots, k_p \text{ distinct in } [3, N] \text{ such that}$$
$$S_1 = T_{2,k_1} \leq S_2 = T_{k_1,k_2} \leq \ldots \leq S_{p+1} = T_{k_p,1} \leq t\} .$$

One now introduces τ_j^i, the death time of particle j if one does not take into account the trajectory X^i for the determination of collisions. The interest for us of the chain reaction sets $A_{N,t}^{i,j}$ comes from

Lemma 1.3. *For x, y in R^d, $t > 0$:*

$$P_{x,y} - \text{ a.s. } \{t \wedge \tau_j \neq t \wedge \tau_j^i\} \subseteq A_{N,t}^{i,j} .$$

($P_{x,y}$ is the probability on C^N for which the X^ℓ are independent Brownian motions, with initial distribution u_0 for $\ell \neq i, j$, δ_x for $\ell = i$, and δ_y for $\ell = j$).

For the proof of Lemma 1.3, we refer to [43].

Since clearly τ_j^i is independent of X^i, we are now interested in showing that $P_N[A_{N,t}^{i,j}] \to 0$, when N goes to infinity. This will provide us with a tool to restore independence.

Proposition 1.4. *For $\eta > 0$, $\lim_N \sup_{|x-y|>\eta} P_{x,y}[A_{N,t}^{2,1}] = 0$.*

$$\lim_N P_N[A_{N,t}^{2,1}] = 0 . \tag{1.7}$$

Proof: The second statement is an immediate consequence of the first statement since u_0 has a bounded density. Let us prove the first statement. Using symmetry we have

$$P_{x,y}[A_{N,t}^{2,1}] \leq \sum_{p=0}^{N-2} N^p E_{x,y}[S_1 \leq \ldots \leq S_{p+1} \leq t] , \tag{1.8}$$

where now $k_1 = 3, \ldots, k_p = 2 + p \leq N$. For $\lambda > 0$, we set

$$h_N^\lambda(z) = E_{z,0}[\exp\{-\lambda T_{1,2}\}] = (g_\lambda(|z|)/g_\lambda(s_N)) \wedge 1 ,$$

where $g_\lambda(|x - y|)$ is the λ-Green's function for Brownian motion with covariance $2Id$. As $u \to 0$,

$$g_\lambda(u) \sim c_d^{-1} u^{2-d} , \quad d \geq 3 ,$$

$$\sim (2\pi)^{-1} \log(u^{-1}) , \quad d = 2 ,$$

It follows that for $N \geq N_0(\lambda)$:

$$N h_N^\lambda(z) \leq 2c_d \, g_\lambda(|z|) . \tag{1.9}$$

Set now $a_p^\lambda = N^p E_{x,y}[S_1 \leq \ldots \leq S_p \, e^{-\lambda S_p}]$. We have for $N \geq N_0(\lambda)$:

$$a_p^\lambda \leq N^p \, E_{x,y}[(S_1 \leq \ldots \leq S_{p-1}) \, e^{-\lambda S_{p-1}} \, h_N^\lambda(X_{S_{p-1}}^{p+2} - X_{S_{p-1}}^{p+1})]$$

$$\leq N^{p-1} \, E_{x,y}[(S_1 \leq \ldots \leq S_{p-1}) \, e^{-\lambda S_{p-1}} \, 2c_d \, g_\lambda(|X_{S_{p-1}}^{p+2} - X_{S_{p-1}}^{p+1}|)] .$$

If we integrate over X^{p+2} in the last expression, since $\int g_\lambda(|y|)dy = \lambda^{-1}$, we find:

$$a_p^\lambda \le N^{p-1} E_{x,y}[S_1 \le \cdots \le S_{p-1} \, e^{-\lambda S_{p-1}}] \frac{2c_d \|V\|_\infty}{\lambda} \, ,$$

and now by induction:

(1.10) $$\text{for } N \ge N_0(\lambda) \, , \quad a_p^\lambda \le \left(\frac{2c_d \|V\|_\infty}{\lambda} \right)^p .$$

Denote by a_p the pth term of the series in (1.8). Set $\lambda = 4c_d \|V\|_\infty$. For $N \ge N_0(\lambda)$:

$$a_0 \le e^{\lambda t} \, E_{x,y}[e^{-\lambda T_{2,1}}] \le e^{\lambda t} \frac{2c_d}{N} \, g_\lambda(|x-y|) \, .$$

Consider now $p \ge 1$. Pick $\epsilon \in (0,1)$.

For $N \ge N_0\left(\frac{2c_d \|V\|_\infty}{\epsilon} \right)$, by (1.10):

$$N^p \, E_{x,y}[S_1 \le \cdots \le S_p \le \frac{\epsilon}{2c_d \|V\|_\infty}] \le e\epsilon^p \, .$$

It follows that for $N \ge N_0(\lambda) \vee N_0(2c_d \|V\|_\infty/\epsilon)$:

$$N^p E_{x,y}[S_1 \le \ldots \le S_p \le S_{p+1} \le t]$$
$$\le N^p \, E_{x,y}[S_1 \le \ldots \le S_p \le \frac{\epsilon}{2c_d \|V\|_\infty}]$$
$$+ e^{\lambda t} N^p E_{x,y}[1(S_1 \le \ldots \le S_p \le S_{p+1}) \, 1(S_p \ge \frac{\epsilon}{2c_d \|V\|_\infty}) \, e^{-\lambda S_{p+1}}] \, .$$

The first term of the last expression is smaller than $e \, \epsilon^p$, and the second is smaller than:

(1.11) $$e^{\lambda t} N^{p-1} E_{x,y}[(S_1 \le \ldots \le S_p) \, 1(S_p \ge \frac{\epsilon}{2c_d \|V\|_\infty}) \, e^{-\lambda S_p} 2c_d g_\lambda(|X_{S_p}^1 - X_{S_p}^{2+p}|)]$$

The distribution of $X_{S_p}^1$ conditionally on $(X_\cdot^2, \ldots, X_\cdot^N)$, is Gaussian with covariance $S_p I_d$, and mean x. It has a density which is uniformly bounded when $S_p \ge \frac{\epsilon}{2c_d \|V\|_\infty}$ by $K = \left(\frac{c_d \|V\|_\infty}{\pi \epsilon} \right)^{d/2}$.

Integrating in (1.11) over $X_\cdot^1$, we find the upper bound $K \, e^{\lambda t} \, N^{p-1} \, E_{x,y}[(S_1 \le \ldots \le S_p) \, e^{-\lambda S_p}] \le \frac{K e^{\lambda t}}{N} (\frac{1}{2})^p$, by (1.10).

It follows that for $p \ge 1$, $\epsilon > 0$, and $N \ge N_0(\lambda) \vee N_0(2c_d \|V\|_\infty/\epsilon)$:

$$N^p E_{x,y}[S_1 \le \ldots \le S_p \le S_{p+1} \le t] \le e^{\lambda t} \frac{K}{N} (\frac{1}{2})^p + e \, \epsilon^p \, .$$

It now follows that:

$$P_{x,y}[A_{N,t}^{2,1}] \le e^{\lambda t} (\frac{2c_d}{N} g_\lambda(|x-y|) + \frac{K}{N}) + e \, \frac{\epsilon}{1-\epsilon} \, .$$

From this our claim follows.

□

As we mentioned already the tightness of the laws of the $\overline{X}_N = \frac{1}{N}\sum_i \delta_{(X^i,\tau_i)} \in M(C \times [0,\infty])$ is immediate. Let us now identify the possible limit points. In order to be able to identify such a limit point $\overline{P}_\infty$ as δ_P, we will use the following idea. The density $u(s,x)$ which appears in (1.3), is ill behaved for the weak convergence topology. However we know that for $\overline{P}_\infty$-a.e. m, the $X.$ component will be μ-distributed (that is Brownian motion with initial distribution u_0). So now for such an m:

$$\langle u_s, f \rangle = \langle m, f(X_s) \rangle - \langle m, f(X_s)\, 1(\tau < s) \rangle$$
$$= \langle V_s, f \rangle - \langle m, f(X_s)\, 1(\tau < s) \rangle \, ,$$

where $V(s,x) = u_0 * p_s(x)$. If now we anticipate the fact that for $\overline{P}_\infty$-a.e. m there should be a "Markov property" at time τ, then for such an m, the density $u(s,x)$ should be given by:

$$u(s,x) = V(s,x) - \langle m, p_{s-\tau}(X_\tau - x)\, 1(s > \tau) \rangle \, .$$

So we will interpret u, in (1.3), in terms of this last formula, and then we will check that for $\overline{P}_\infty$-a.e. m this expression defines the density of the image of $1(\tau \geq s)ds\, dm$ under (s, X_s). So the quantity $\int_0^t 1(\tau > s)\, u(s, X_s)ds$ is now replaced by $\int_0^{t\wedge\tau} V(s, X_s)ds - \tilde{E}_m[\int_0^{(t\wedge\tau-\tilde{\tau})+} p_s(X_{\tilde{\tau}+s} - \tilde{X}_{\tilde{\tau}})ds]$, which has nicer continuity properties. In view of these comments it is natural to try now to obtain

Proposition 1.5. *For $\overline{P}_\infty$-a.e. m, for $t \geq 0$:*

$$(1.12) \quad E_m[1(\tau \leq t) - c_d \int_0^{t\wedge\tau} V(s, X_s)ds + c_d \tilde{E}_m[\int_0^{(t\wedge\tau-\tilde{\tau})+} p_s(X_{\tilde{\tau}+s} - \tilde{X}_{\tilde{\tau}})ds/X.] = 0$$

Proof: We call D the at most denumerable set of t such that $I(\overline{P}_\infty)\,[\{\tau = t\}] \neq 0$. We then define for $h(\cdot) \in C_b(C)$, $t \notin D$:

$$G(m) = \langle m, \{1(\tau \leq t) - c_d \int_0^{t\wedge\tau} V(s, X_s)ds + \tilde{E}_m[c_d \int_0^{(t\wedge\tau-\tilde{\tau})+} p_s(X_{\tilde{\tau}+s} - \tilde{X}_{\tilde{\tau}})ds]\} h(X) \rangle$$

and we define the smoothed $G_\epsilon(m)$, $\epsilon > 0$, where the last integral in the previous expression is replaced by $\int_0^{(t\wedge\tau-\tilde{\tau}-\epsilon)+} p_{s+\epsilon}(X_{\tilde{\tau}+s+\epsilon} - \tilde{X}_{\tilde{\tau}})ds$. Since for $\overline{P}_\infty$-a.e. m, the $X.$ component is μ-distributed under m, one easily sees that the expression in the conditional expectation in (1.12) is integrable (and meaningful) and that

$$E_{\overline{P}_\infty}[|G(m)|] \leq \text{const.}\|V\|_\infty \epsilon + \lim_k E_{N_k}[|G_\epsilon(\overline{X}_{N_k})|] \, .$$

Now by considering for $1 \le i \ne j \le N$, separately X^i, X^j, and the remaining group of $(N-2)$ particles one sees that

$$1(\tau_i \le t) - \sum_{j \ne i} 1(T_{i,j} \le t)\, 1(\tau_j^i > T_{i,j})\, 1(\tau_i^j > T_{i,j})$$

equals zero except maybe when $\tau_i = 0$, but anyway the quantity remains bounded in absolute value by $\sum_{j \ne i} 1(T_{i,j} = 0)$. Since $N\, s_N^d$ converges to zero, the expectation of this quantity is easily seen to go to zero. From this remark, applying now the Cauchy Schwarz inequality and symmetry, we find:

$$(\lim_k E_{N_k}[|G_\epsilon(\overline{X}_{N_k})|])^2 \le$$

$$\overline{\lim}_N E_N[(N1(T_{1,2} \le t)\, 1(\tau_2^1 > T_{1,2})\, 1(\tau_1^2 > T_{1,2}) - c_d \int_0^{t \wedge \tau_1} V(s, X_s^1)\, ds$$

(1.13)
$$+ c_d \int_0^{(t \wedge \tau_1 - \tau_2 - \epsilon)^+} p_{s+\epsilon}(X_{\tau_2 + s + \epsilon}^1 - X_{\tau_2}^2)\, ds)\, h(X^1)$$

(same expression with particles 3 and 4) $h(X^3)]$

$$+ \overline{\lim}_N O(\frac{1}{N})\, E_N[N^2 1(T_{1,2} \le t)\, 1(T_{2,3} \le t) + N\, 1(T_{1,2} \le t) + 1]$$

$$+ \overline{\lim} O(\frac{1}{N^2})\, E_N[N^2\, 1(T_{1,2} \le t) + N1(T_{1,2} \le t) + 1]\,.$$

Now the last two terms are in fact zero. For instance integrating over X_0^1 and X_0^3, we have:

$$E_N[N^2\, 1(T_{1,2} \le t)\, 1(T_{2,3} \le t)] \le$$

$$\|V\|_\infty^2 E_N[N|W_{N,t}(X_0^2 + B^2 - B^1)| \times N|W_{N,t}(X_0^2 + B^2 - B^3)|]\,,$$

where $W_{N,t}$ denotes the Wiener sausage of radius s_N, in time t of the process inside the brackets, and $|\cdot|$ Lebesgue volume. But $N|W_{N,t}(X_0^2 + B^2 - B^1)| = N|W_{N,t}(B^2 - B^1)|$ is bounded in any L^p, $p < \infty$, uniformly in N, using usual estimates on the volume of Wiener sausage. It follows that $\frac{1}{N} E_N[N^2 1(T_{1,2} \le t)\, 1(T_{2,3} \le t)]$ converges to zero. The other terms are treated similarly.

Introduce now $\overline{\tau}_i$ the destruction time of particle i, $1 \le i \le 4$, if one replaces the set $[1, N]$, by $\{i\} \cup [5, N]$, when defining the collisions. Observe now that for instance

(1.14)
$$\{t \wedge \tau_1^2 \ne t \wedge \overline{\tau}_1\} \subseteq A_{N-2,t}^{3,1} \cup A_{N-2,t}^{4,1}\,,$$

where the subscript $(N-2)$ refers to the fact that the k_i in definition (1.5) are now supposed to belong to $[5, N]$. Indeed one simply uses the last occurrence of 3, 4, in any chain reaction $\{T_{3,k_1} < \ldots < T_{k_p,1} \leq t\}$, or $\{T_{4,k_1} < \ldots < T_{k_p,1} \leq t\}$, since on the set where $\{t \wedge \tau_1^2 \neq t \wedge \bar{\tau}_1\}$, one such chain reaction occurs. Similarly

$$\{t \wedge \tau_1 \neq t \wedge \bar{\tau}_1\} \subseteq A_{N-2,t}^{2,1} \cup A_{N-2,t}^{3,1} \cup A_{N-2,t}^{4,1} \,.$$

Let us now consider the quantity obtained by replacing in the first expression in the right member of inequality (1.13), the τ_i^j and τ_i by $\bar{\tau}_i$, i, j in $[1,4]$. The claim is that once "$\overline{\lim}_N$" is performed one does not change anything. There is a somewhat tricky point about this, that we will describe now. For the full details, however we refer to [43]. The point is that when getting bounds on the difference of the two expressions one obtains terms like:

$$E_N[N^2 1(T_{1,2} \leq t) \, 1(T_{3,4} \leq t) \, 1_{A_{N-2,t}^{3,1}}] \,, \quad E_N[N \, 1(T_{1,2} \leq t) \, 1_{A_{N-2,t}^{3,1}}] \,,$$
$$\text{or } E_N[N \, 1(T_{3,4} \leq t) \, 1_{A_{N-2,t}^{1,2}}] \,.$$

These terms go to zero with N because one can force volume of Wiener sausage in these terms. For instance, we can integrate over X_0^2 and X_0^4 in the first term and get an upper bound by $\|V\|_\infty^2 E_N[N|W_{N,t}(X_0^1 + B_. - B^2) \, N|W_{N,t}(X_0^3 + B^3 - B^4)| \, 1_{A_{N-2,t}^{1,2}}]$, which is easily seen to go to zero, thanks to the usual estimates on the volume of the Wiener sausage and the fact that $E_N[A_{N-2,t}^{1,2}] \to 0$. The point is that in the difference one does not need to generate terms like:

$$E[N^2 1(T_{1,2} \leq t) \, 1(T_{3,4} \leq t) \, 1_{A_{N-2,t}^{4,3}}] \,, \text{ or}$$
$$E[N 1(T_{1,2} \leq t) \, 1_{A_{N-2,t}^{2,1}}] \,,$$

for which we cannot use our previous reduction to an estimate on the volume of Wiener sausage. Indeed $\{T_{1,2} \leq t\} \subset A_{N-2,t}^{2,1}$, and the last term for instance does not go to zero.

Our last reduction step is to observe (see [43]) that $N \, 1(T_{1,2} \leq t) \, 1(\bar{\tau}_1 > T_{1,2}) \, 1(\bar{\tau}_2 > T_{1,2}) = N \, 1(T_{1,2} \leq t \wedge \bar{\tau}_1) - N \, 1(\bar{\tau}_2 \leq T_{1,2} \leq t \wedge \bar{\tau}_1)$ a.s. on the set where $\{T_{1,2} > 0\}$, and a similar equality holds for particles 3, 4, on $\{T_{3,4} > 0\}$. With this, one easily concludes that the first expression on the right of inequality (1.13) in fact equals:

$$\overline{\lim}_N E_N[(E_{\mu \otimes \mu}[\tilde{I}])^2] \,,$$

(1.15) where $\tilde{I} = (N 1(T_{1,2} \leq t \wedge \bar{\tau}_1) - c_d \int_0^{t \wedge \bar{\tau}_1} V(s, X_s^1) ds) \, h(X^1)$

$$- (N 1(\bar{\tau}_2 \leq T_{1,2} \leq t \wedge \bar{\tau}_1) - c_d \int_0^{(t \wedge \bar{\tau}_1 - \bar{\tau}_2 - \epsilon)+} p_{s+\epsilon}(X_{\bar{\tau}_2 + s + \epsilon}^1 - X_{\bar{\tau}_2}^2) ds) \, h(X^1)$$

In order to study expression (1.15), we now introduce the collision intensity for $w^1 \in C$ and $v \in M(R^d)$, $t \geq 0$:

$$(1.16) \qquad C_t^N(w^1, v) = E_v^2[N \, 1\{T_{1,2} \leq t\}] \ .$$

The expectation E_v^2 in (1.16) is performed with respect to the variable w^2, with Wiener measure having initial condition v.

Lemma 1.6. *Let $f(\cdot)$ belong to $C_0(R^d)$, for $T > 0$:*

$$\lim_N \sup_{u \in M(R^d)} E_u^1[\sup_{x \in R^d, t \leq T} |C_t^N(w^1, f(y-x)dy) - c_d \int_0^t f(s, X_s(w^1) - x)ds|] = 0 \ ,$$

where $f(s, x) = f * p_s(x)$.

Let us mention, although we will not really develop this point here, that one interest of the quantity $C_t^N(\cdot, v)$ is that it has nice limit properties for a wide range of processes, and does not really rely for its definition on the additive structure of R^d. For more details see [45].

Proof: Since C_t^N is nondecreasing in t and the limit $c_d \int_0^t f(s, X_s(w^1) - x)ds$ is, uniformly in x, continuous in t, by a well known technique (as in Dini's second theorem), it is enough to show that for fixed t

$$\limsup_N E_u^1[\sup_x |C_t^N(w^1, f(y-x)dy) - c_d \int_0^t f(s, X_s(w^1) - x)ds|] = 0 \ .$$

Now for each N and u the quantity we consider is smaller than:

$$E_u^1 \otimes E_{\delta_0}^2[\sup_x |N \int f(y-x)dy \, 1\{\inf_{0 \leq s \leq t} |X_0^1 + B_s^1 - B_s^2 - y| \leq s_N\}$$

$$- c_d \int_0^t f(X_0^1 + B_s^1 - B_s^2 - x)ds|]$$

where $B_\cdot^i = X_\cdot^i - X_0^i$, $i = 1, 2$. But the last expression equals

$$E_{\delta_0}^1 \otimes E_{\delta_0}^2[\sup_x |N \int f(y-x)dy \, 1\{\inf_{0 \leq s \leq t} |B_s^1 - B_s^2 - y| \leq s_N\}$$

$$- c_d \int_0^t f(B_s^1 - B_s^2 - x)ds|] \ .$$

This latter quantity converges to zero by a refinement on the usual limit result for the Wiener sausage of small radius, for details see [43], Lemma 3.4.

$\square$

Observe now that for any $(X^5, \ldots, X^N)$:

$$|E_{\mu \otimes \mu}[(C^N_{t \wedge \bar{\tau}_1}(X^1, u_0) - c_d \int_0^{t \wedge \bar{\tau}_1} V(s, X^1_s)ds)\, h(X^1)]|$$

$$= |E_\mu[(C^N_{t \wedge \bar{\tau}_1}(X^1, u_0) - c_d \int_0^{t \wedge \bar{\tau}_1} V(s, X^1_s)ds)\, h(X^1)]|$$

$$\le E_\mu[\sup_{s \le t} |C^N_s(X^1, u_0) - c_d \int_0^s V(u, X^1_u)du|]\, \|h\|_\infty$$

which converges to zero as N goes to infinity by a slight variant of Lemma 1.6 (here u_0 does not necessarily have a density in $C_0(R^d)$).

So we see that we can replace $E_{\mu \otimes \mu}[\tilde{I}]$ in (1.15) by

$$(1.17)\quad E_{\mu \otimes \mu}[(N1(\bar{\tau}_2 \le T_{1,2} \le t \wedge \bar{\tau}_1) - c_d \int_0^{(t \wedge \bar{\tau}_1 - \bar{\tau}_2 - \epsilon)_+} p_{s+\epsilon}(X^1_{\bar{\tau}_2 + s + \epsilon} - X^2_{\bar{\tau}_2})ds)\, h(X^1)]$$

without changing the limit result. The study of this last term is naturally more delicate. We cannot directly integrate over particle 2, and see a "collision intensity" term appear. The study of this last term will require some "surgery" on trajectories. Of course now $(X^5, \ldots, X^N)$ are held fixed and represent an outside random medium, and we are going to derive uniform estimates over this random medium. We start with the identity:

$$1\{\bar{\tau}_2 \le T_{1,2} \le t \wedge \bar{\tau}_1\} = 1\{\bar{\tau}_2 \le T_{1,2} \le t \wedge \bar{\tau}_1\} \times 1\{T_{1,2} < \bar{\tau}_2 + \epsilon\}$$
$$+ 1\{\bar{\tau}_2 + \epsilon \le T_{1,2} \le t \wedge \bar{\tau}_1\}$$

Moreoever we can write:

$$1\{\bar{\tau}_2 + \epsilon \le T_{1,2} \le t \wedge \bar{\tau}_1\} = (1 - 1\{\bar{\tau}_2 + \epsilon > t \wedge \bar{\tau}_1\})$$
$$\times (1 - 1\{t \wedge \bar{\tau}_2 < T_{1,2} < t \wedge \bar{\tau}_2 + \epsilon\} - 1\{T_{1,2} \le t \wedge \bar{\tau}_2\})$$
$$1\{T_{1,2} \circ \theta_\epsilon \circ \theta_{t \wedge \bar{\tau}_2} \le (t \wedge \bar{\tau}_1 - \bar{\tau}_2 - \epsilon)_+\}.$$

Using this decomposition it follows that the absolute value of the expression (1.17) is bounded by:

$$E_{\mu \otimes \mu}[(N1\{T_{1,2} \circ \theta_\epsilon \circ \theta_{t \wedge \bar{\tau}_2} \le (t \wedge \bar{\tau}_1 - \bar{\tau}_2 - \epsilon)_+\}$$

$$- c_d \int_0^{(t \wedge \bar{\tau}_1 - \bar{\tau}_2 - \epsilon)_+} p_{s+\epsilon}(X^1_{\bar{\tau}_2 + s + \epsilon} - X^2_{\bar{\tau}_2})ds)h(X^1)]|$$

$$(1.18)\qquad + \|h\|_\infty(2E_{\mu \otimes \mu}[N1\{\bar{\tau}_2 \wedge t \le T_{1,2} \le \bar{\tau}_2 \wedge t + \epsilon\}]$$

$$+ E_{\mu \otimes \mu}[N1\{T_{1,2} \circ \theta_\epsilon \circ \theta_{t \wedge \bar{\tau}_2} = 0\}]$$

$$E_{\mu \otimes \mu}[N1\{T_{1,2} \circ t \wedge \bar{\tau}_2\}1\{T_{1,2} \circ \theta_{t \wedge \bar{\tau}_2 + \epsilon} \le t\}])$$

It is now fairly standard to see that the last two terms converge to zero (uniformly over $X^5_., \ldots, X^N_.))$. As for the second term integrating out the initial condition X^1_0, it is bounded by

$$2\|h\|_\infty E_{\mu\otimes\mu}[N\int u_0(dy)\,1\{y \in W_{N,\epsilon}(X^2_{t\wedge\bar\tau_2} + \tilde{B}^2_. - \tilde{B}^1_. - B^1_{t\wedge\bar\tau_2})\}]\,,$$

where $\tilde{B}^i = X^i_{t\wedge\bar\tau_2+.} - X^i_{t\wedge\bar\tau_2}$, $i = 1, 2$, are standard independent Brownian motions. This last quantity is smaller than $2\|h\|_\infty\|V\|_\infty E_{\mu\otimes\mu}[N|W_{N,\epsilon}(B^2_. - B^1_.)|]$ which converges to $2\|h\|_\infty\|v\|_\infty c_d\epsilon$, as N converges to infinity.

Let us now look at the first term of (1.18). Conditioning over $X^1_.$, we have

$$E^2_\mu[N1\{T_{1,2} \circ \theta_\epsilon \circ \theta_{t\wedge\bar\tau_2} \le (t \wedge \bar\tau_1 - \bar\tau_2 - \epsilon)_+\}]$$
$$= E^2_\mu[N\tilde{E}^2_{X^2_{t\wedge\bar\tau_2}}[1\{T_{1,2} \circ \theta_\epsilon(w^1_{t\wedge\bar\tau_2+.}, \tilde{w}^2_.) \le (t \wedge \bar\tau_1 - \bar\tau_2 - \epsilon)_+\}]]$$
$$= E^2_\mu[C^N_{(t\wedge\bar\tau_1 - \bar\tau_2 - \epsilon)_+}(X^1_{t\wedge\bar\tau_2+\epsilon+.}, p_\epsilon(y - X^2_{t\wedge\bar\tau_2})dy)]\,.$$

After this surgery the first term of (1.18) is now

$$|E_{\mu\otimes\mu}[(C^N_{t\wedge\bar\tau_1 - \bar\tau_2 - \epsilon)_+}(X^1_{t\wedge\bar\tau_2+\epsilon+.}, p_\epsilon(y - X^2_{t\wedge\bar\tau_2})dy)$$
$$- c_d \int_0^{(t\wedge\bar\tau_1 - \bar\tau_2 - \epsilon)_+} p_{s+\epsilon}(X^1_{t\wedge\bar\tau_2+s+\epsilon} - X^2_{t\wedge\bar\tau_2})ds)\; h(X^1)]|$$
$$\le \|h\|_\infty E_{\mu\otimes\mu}[\sup_{v\le t,x}|C^N_v(X^1_{t\wedge\bar\tau_2+\epsilon+.}, p_\epsilon(y - x)dy) - c_d \int_0^v p_{s+\epsilon}(X^1_{t\wedge\bar\tau_2+s+\epsilon} - x)ds|]$$
$$\le \|h\|_\infty \sup_{u\in M(\mathbf{R}^d)} E^1_u[\sup_{v\le t,x}|C^N_v(X^1_., p_\epsilon(y - x)dy) - c_d \int_0^v p_{s+\epsilon}(X^1_s - x)ds|]$$

which is independent of $(X^5_., \ldots, X^N_.)$ and converges to zero as N goes to infinity thanks to Lemma 1.6.

So we have proved that

$$E_{P_\infty}[|G(m)|] \le \text{const. } \|V\|_\infty\epsilon + 2\|h\|_\infty\|V\|_\infty\epsilon\,.$$

Letting ϵ go to zero, we obtain our claim.

$\square$

As we mentioned previously, $\tilde{E}_m[\int_0^{(t\wedge\tau-\bar\tau)_+} p_s(X_{\bar\tau+s} - \hat{X}_{\bar\tau})ds]$
$= \int_0^t 1(s < \tau)\tilde{E}_m[1(\bar\tau < s)\, p_{s-\bar\tau}(X_s - \hat{X}_{\bar\tau})]ds$. So in view of the characterization of the law P given by Lemma 1.2, the fact that $\overline{P}_\infty = \delta_P$, will follow from

Proposition 1.6. For $\overline{P}_\infty$-a.e. m, for all $s \ge 0$, and $f \in bB(\mathbf{R}^d)$:

$$(1.19) \qquad \int f(x)\, E_m[p_{s-\tau}(x - X_\tau)\, 1(\tau < s)]\, dx = E_m[f(X_s)\, 1(s > \tau)]\,.$$

Proof: By the usual arguments we simply will check (1.19), for $s \in (0, \infty) \setminus D$, and $f \in C_b(R^d)$. We recall that $D = \{t : I(\overline{P}_\infty)[\{t = \tau\}] \neq 0\}$. Setting $H(m) = \langle m, (P_{(s-\tau)_+} f(X_\tau) - f(X_s) \, 1(\tau < s) \rangle$, we simply want to check that $H(m) = 0 \quad \overline{P}_\infty$-a.s. Now

$$E_{\overline{P}_\infty}[H(m)^2] =$$

$$\lim_k E_{N_k}[(P_{(s-\tau_1)_+} f(X_{\tau_1}^1) - f(X_s^1))(P_{(s-\tau_2)_+} f(X_{\tau_2}^2) - f(X_s^2)) \, 1(\tau_1 < s) \, 1(\tau_2 < s)] .$$

However on $(A_{N,s}^{1,2} \cup A_{N,s}^{2,1})^c$, $\tau_1 \wedge s = \tau_1^2 \wedge s$, and $\tau_2 \wedge s = \tau_2^1 \wedge s$. It follows that

$$E_{\overline{P}_\infty}[H(m)^2] = \lim_k E_{N_k}[E_\mu^1[(P_{(s-\tau_1^2)_+} f(X_{\tau_1^2}^1) - f(X_s^1)) 1(\tau_1^2 < s)]^2] = 0 ,$$

by an application of the strong Markov property at time τ_1^2, under the law E_μ^1.

$\square$

This now concludes the proof of Theorem 1.1. In the next section, we are going to give a different line of explanation on why Theorem 1.1 holds.

2. Limit picture for chain reactions in the constant mean free travel time regime.

The result which was presented in the last section is certainly satisfactory from a purely propagation of chaos frame of reference, however it misses the nice limit algebraic picture, corresponding to the "constant mean free travel time" regime, which underlies the result. We will somehow try to motivate and describe this limit picture.

The notion of chain reactions, leading to a specific particle, say particle one, corresponding to definition (1.5) played an important role in the derivation of Theorem 1.1. It simply involves the independent Brownian trajectories and completely forgets about destruction of particles. The idea we will follow here is that one should investigate the limit structure of the chain reactions leading to particle 1, in the constant mean free travel time regime. One then should construct the interaction (annihilation) on the limit object and somehow recapture the result of Theorem 1.1 in this scheme.

As a step in the study of the limit aspect of this somewhat loose notion of chain reaction between independent particles leading to particle 1, one may look at the limiting aspect of the first collision. For each $i \in [1, N]$, we set

$$T_i = \inf_{j \neq i} T_{i,j} , \quad \text{when} \quad \inf_{j \neq i} T_{i,j} \leq 1 , \quad +\infty \text{ otherwise,}$$

and Y^i "the first colliding trajectory", is defined on the set $0 < T_i \leq 1$, as X^j where j is the only index such that $T_i = T_{i,j}$, and otherwise, it is set to the constant trajectory 0.

Let us quote the following propagation of chaos result from [45].

Theorem 2.1. *The $(X^i_\cdot, T_i, Y^i_\cdot)_{1 \leq i \leq N}$ are Q-chaotic where Q is the law on $C \times ([0,1] \cup \{\infty\}) \times C$ such that under Q*

- X *has the distribution of a Brownian motion with initial law u_0.*
- $Q[T > t|X] = \exp\{-c_d \int_0^{t \wedge 1} V(s, X_s)ds\}$, $t \geq 0$.
- *Conditionally on $X, T, 0 < T \leq 1$, $Y_\cdot$ is distributed as a Brownian motion with initial law u_0 conditioned to be equal to X_T at time T, and otherwise it is the constant trajectory 0.*

This result motivates that the natural limit "chain reaction tree" should be constructed in the following way: One starts with one Brownian particle X_s, the ancestor, running until time t, with initial density u_0 (having the role of X^1). One then constructs the trajectories having a collision with the ancestor: conditionally on X_s, one picks a Poisson distribution of points on $[0,t]$ with intensity $c_d V(s, X_s)ds$. Now conditionally on the times $0 < t_1 < \ldots < t_n < t$, one considers independent Brownian bridges: $W^1, \ldots, W^n$ where W_ℓ for $\ell \leq n$ has the law of Brownian motion in time $[0, t_\ell]$, with initial distribution u_0 conditioned to be equal to X_{t_ℓ} at time t_ℓ. These $W^1, \ldots, W^n$ constitute the first generation in the chain reaction leading to $X_\cdot$. Note that we disregard the trajectory of W^ℓ, after time t_ℓ, since Theorem 1.1 indicates that there should not be any recollision in the limit picture.

Now one performs the same thing on each of the first generation trajectories W^ℓ, $1 \leq \ell \leq n$, as just described on $X_\cdot$. One obtains in this way the second generation of trajectories, and so on.

We are now going to give a precise description of this limit tree of chain reactions. As the explanation above indicates we will build a marked tree. Each "individual" being marked by a trajectory. We use Neveu's notion of tree [33], namely a tree is a subset of the set of finite sequences $U = \cup_{k \geq 0} (N^*)^k$, (the vertices) containing the sequence $\emptyset$ (only element of $(N^*)^0$), such that $u \in \pi$ when $uj \in \pi$, $j \in N^*$, and for $u \in \pi$ there is a $\nu_u \in N$, with $uj \in \pi$ if and only if $1 \leq j \leq \nu_u$. We will be interested in marked trees that is $\omega = (\pi, (\phi_u, u \in \pi))$, where $\phi_u \in D$, D the set of marks which is for us $\cup_{t>0} C([0,t], R^d) \sim (0, +\infty) \times C([0,1], R^d)$.

trajectorial picture: marked tree description:

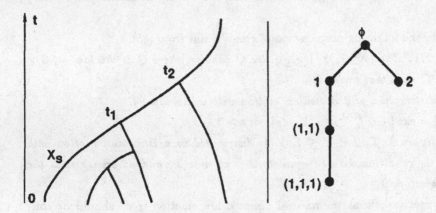

Following Neveu's notations, for the marked tree ω and $u \in \pi$, $T_u\omega$ is the tree translated at u, G_n is the nth generation of the tree: $\pi \cap (N^*)^n$, and F_n the σ-field generated by G_k, $0 \le k \le n$, and ϕ^u, $u \in G_k$, $0 \le k \le n$. For a trajectory $\psi. \in D$, we construct the probability R_ψ on the set of marked trees Ω which satisfies

- $R_\psi[\phi^\theta = \psi] = 1$ (the ancestor's mark is ψ).

- For f_u, $u \in U$ a collection of nonnegative measurable functions on Ω:

$$E_\psi[\prod_{u \in G_n} f_u \circ T_u \ / \ F_n] = \prod_{u \in G_n} E_{\phi^u}[f_u]$$

(Branching property).

The last requirement which describes the reproduction law is given in a somewhat more synthetic form than what was explained before:

- The random point measure on D: $\sum_{1 \le \ell \le \nu_\theta} \delta_{\phi^\ell}$ is a Poisson point process with intensity measure

$$(2.1) \qquad \int_0^t ds \, V(s, \psi_s) \, P^{s,\psi_s}(d\phi) \ ,$$

and $0 < \tau(\phi^1) < \cdots < \tau(\phi^\theta) < t = \tau(\psi)$, a.s. Here $\tau(\phi)$ denotes the time duration of a particle ($(0,\infty)$ valued component on D), and $P^{s,x}$ is the law of Brownian motion with initial u_0 on time $[0, s]$, conditioned to be equal to x at time s. We have normalized here c_d as being equal to 1.

Lemma 2.1.

- R_ψ-a.s. *the tree ω is finite.*

- $M(\omega) = \sum_{1 \le \ell \le \nu_\theta} \delta_{T_\ell(\omega)}$ *(random point measure on Ω)*

is a Poisson point measure with intensity

$$(2.2) \qquad Q_\psi = \int_0^t ds\, V(s, \psi_s)\, R^{s, \psi_s}\, ds\,,$$

where $R^{t, x} = \int R_\psi\, P^{t, x}(d\psi)$.

Proof: Let us check the first point: for $0 \le p \le n$, set

$$v_p = E_\psi \Big[\sum_{u \in G_{n-p}} \frac{1}{p!} (c\, \tau(\phi^u))^p \Big]\,, \qquad \text{with} \qquad c = \|V\|_\infty\,.$$

so that $v_0 = E_\psi[\#G_n]$ and $v_n = \frac{1}{n!}(ct)^n$. Then for $0 \le p < n$:

$$v_p = E_\psi\Big[E_\psi\Big[\sum_{\substack{u \in G_{n-p-1} \\ uj \in G_{n-p}}} \frac{1}{p!} (c\, \tau(\phi^{uj}))^p / F_{n-p-1} \Big]\Big]$$

$$= E_\psi\Big[\sum_{u \in G_{n-p-1}} E_{\phi^u}\Big[\sum_{1 \le j \le \nu_\theta} \frac{1}{p!} (c\, \tau(\phi^j))^p \Big]\Big]$$

$$= E_\psi\Big[\sum_{u \in G_{n-p-1}} \int_0^{\tau(\phi^u)} \frac{1}{p!} (ct)^p\, V(t, \phi_t^u) dt \Big] \le v_{p+1}\,.$$

From this $E_\psi[\#G_n] \le \frac{1}{n!}(ct)^n$, and summing over n, we find the result. As for the second statement, if F is a positive function on Ω:

$$E_\psi[\exp\{-\langle M, F\rangle\}] = E_\psi\Big[\prod_{1 \le \ell \le \nu_\theta} e^{-F(T_\ell(\omega))} \Big]$$

$$= E_\psi\Big[\prod_{1 \le \ell \le \nu_\theta} R_{\phi^\ell}[e^{-F}] \Big] \quad \text{by the branching property)}$$

$$= \exp\Big\{ \int_0^t V(s, \psi_s) ds \int dP^{s, \psi_s}(\phi)\, (R_\phi[e^{-F}] - 1)) \quad \text{by (2.1)),}$$

$$= \exp\Big\{ \int dQ(e^{-F} - 1)\Big\}$$

which proves our claim.

$\square$

Our tree ω is R_ψ a.s. finite. Now we can build the interaction on the limit tree corresponding to annihilation by adding a mark Z^u, $u \in \pi$, equal to zero or 1. $Z^u = 0$, will mean that the particle with trajectory ϕ^u is already destroyed by the time $\tau(\phi^u)$ at which it meets its direct ancestor, whereas $Z^u = 1$, will mean that it has not yet been destroyed.

So it is quite natural to impose the following recursive rule, to determine Z^u, $u \in \omega$: If $\nu_u = 0$ (no descendents), $Z^u = 1$. If $Z^{uj} = 1$, for some $1 \leq j \leq \nu_u$ (one of the direct descendents is alive), then $Z^u = 0$.

We now define $R^t = \int R^{t,x} V(t,x) \, dx$, for which the ancestor trajectory is distributed as a Brownian motion in time $[0,t]$, with initial condition u_0. Then we set

$$(2.3) \qquad u(t,x) = V(t,x) \, R^{t,x}[1(Z^\theta = 1)] \,, \quad t > 0 \,.$$

In fact u is the density of presence of ψ_t, under R^t, when the ancestor trajectory $\psi_\cdot$ is not already destroyed at time t.

Lemma 2.2. *For* $f \in bB(R^d)$:

$$\int u(t,x) \, f(x) \, dx = E_{R^t}[f(\psi_t) \, 1(Z^\theta = 1)] \,.$$

Proof:

$$\int u(t,x) \, f(x) dx = \int V(t,x) \, R^{t,x}[Z^\theta = 1] \, f(x) \, dx$$

$$= \int V(t,x) dx \int P^{t,x}(d\psi) \, R_\psi[Z = 1] \, f(\psi_t)$$

$$= E_{R^t}[f(\psi_t) \, 1(Z^\theta = 1)] \,.$$

$\square$

The result we show now tells us that one obtains the nonlinear equation (in integral form) (1.2), by constructing the interaction directly on the "limit chain reaction tree".

Theorem 2.3. $u(t,x)$, $0 \leq t \leq T$, *is the solution in* $L^\infty([0,T] \times R^d, ds \, dm)$ *of the integral equation:*

$$(2.4) \qquad w(t,x) = V(t,x) - \int_0^t \int w^2(s,y) \, p_{t-s}(y,x) \, ds \, dy \,.$$

Proof: The uniqueness is standard, see [44] for details. Let us check that u is a solution of (2.4). Observe that

$$(2.5) \qquad \begin{aligned} R_\psi[Z^\theta = 1] &= \exp\{-\int_0^t V(s,\psi_s) \, R^{s,\psi_s}[Z^\theta = 1] ds\} \\ &= \exp\{-\int_0^t u(s,\psi_s) ds\} \,. \end{aligned}$$

Then

$$u(t,x) = V(t,x)\, R^{t,x}[Z=1]$$

$$= V(t,x)\left(1 - \int_0^t ds\, P^{t,x}(d\psi)\, u(s,\psi_s)\, \exp\{-\int_0^s u(r,\psi_r)dr\}\right)$$

$$= V(t,x) - \int_0^t ds \int dP^t(\psi) p_{t-s}(\psi_s, x)\, u(s,\psi_s)\, \exp\{-\int_0^s u(r,\psi_r)dr\}$$

$$= V(t,x) - \int_0^t ds \int dy\, V(s,y)\, p_{t-s}(y,x)\, u(s,y) \int P^{s,y}(d\psi) \exp\{-\int_0^s u(r,\psi_r)dr\}$$

$$= V(t,x) - \int_0^t ds \int dy\, p_{t-s}(y,x)\, u^2(s,y)\,,$$

which shows that u satisfies (2.4).

$\square$

3) Some comments

Let us finally give some comments on the results presented in sections 1 and 2.

- The results presented in section 2, strongly suggest that in fact when one studies collisions (without any destructions) between independent particles $(X_.^1, \ldots, X_.^N)$, a better result than Theorem 2.1 should hold. Somehow one should have a propagation of chaos result at the level of the "trees of chain reactions" leading to each particle X^i, $1 \leq i \leq N$. The natural "limit" should in fact be precisely the noninteracting marked tree presented in section 2.

- The proof of Theorem 2.3 is very algebraic. The fact that equation (2.4) arises when one constructs the interaction directly on the marked tree remains true when one applies the same construction to a "general Markov process", (see [44]). One then obtains the integral form of the nonlinear equation:

$$(3.1) \qquad \partial_t u = L^* u - u^2 \, ,$$

where L is the formal generator of the Markov process. In fact by a slight variation on the construction of the marked tree, and of the interaction, one obtains the integral equation corresponding to

$$(3.2) \qquad \partial_t u = L^* u - u^{k+1} \, , \qquad (k \geq 1, \text{ integer}).$$

- Not only does the construction work for a "general Markov process" but for Brownian motion in one dimension as well! Consider for instance in this case, the random times $0 < t_1 < \cdots < t_n < t$, picked with a Poisson distribution of intensity $V(s, X_s)ds$, on $[0, t]$, when X_s, $s \in [0, t]$ is the ancestor trajectory. They are not at all the first "collision times" of $X_.$ with the first generation trajectories $W_.^1, \ldots, W_.^n$. Indeed the $W_.^\ell$ are distributed as independent bridges, being conditioned to be equal to X_{t_ℓ} at time t_ℓ. Of course in dimension $d \geq 2$, the trajectory $W_.^\ell$ meets the ancestor trajectory $X_.$ only at time t_ℓ, but in dimension $d = 1$, this need not be the case.

This point should be viewed as the fact that the limit structure constructed in section 2, exists very generally whether or not it comes as a limit picture from an approximating constant mean free travel time regime.

- One may wonder what the appropriate collision regime should be if one tries to handle $(k + 1)$-particle collisions with possible limit survival equation (3.2). As mentioned previously in Remark 1.2 5), the right guess should not be dictated by a notion of distance of interaction. Much more naturally it should be picked as a suitable level set in the $(k + 1)$ particle configuration space of some potential generated by a measure sitting on the "diagonal" (which is polar) (see [45]). A plausible guess would be to look at the set $h(x_1, \ldots, x_{k+1}) \geq N^k$, where

$$h(x_1, \ldots, x_{k+1}) = \int_0^\infty e^{-s} ds \int_{R^d} \prod_1^{k+1} p_s(x_i, z) \, dz = g_1^{(kd)}(D) \, .$$

where $g_1^{(kd)}(|x - x'|)$ is the 1-Green's function of Brownian motion in R^{kd} and D is the distance of $(x_1, \ldots, x_{k+1})$ to the diagonal $\{(z, \ldots, z)\}$ in $(R^d)^{k+1}$. If one wants to benefit from Brownian scaling, in dimension d, with $kd > 2$, one can use instead of h,

$$f(x_1, \ldots, x_{k+1}) = \int_0^\infty ds \int_{R^d} \prod_1^{k+1} p_s(x_i, z) \, dz = \frac{2}{c_{kd}} D^{2-kd} \, ,$$

$(c_{kd} = (kd - 2)vol(S_{kd}))$, which is continuous with values in $(0, \infty]$, finite except on the diagonal $x_1 = \cdots = x_{k+1}$, and homogeneous of degree $2 - kd$. The level set $\{f \geq N^k\}$ is then the homothetic of ratio $N^{-1/(d-2/k)}$ of the set $\{f \geq 1\}$.

IV. Uniqueness for the Boltzmann process

In this chapter, we are going to explain how ideas somewhat similar to those used in the "tree construction" of Chapter III, section 2, can be put at work to produce a uniqueness result of the nonlinear process associated to the spatially homogeneous Boltzmann equation for hard spheres.

The spatially homogeneous Boltzmann equation for hard spheres describes the time evolution of the density $u(t, v)$ of particles with velocity v, in a dilute gas, under an assumption of spectral homogeneity and hard spheres collisions as:

$$\partial_t u = \int_{R^n \times S_n} (u(t, \tilde{v})u(t, \tilde{v}') - u(t, v)\, u(t, v'))\, |(v' - v) \cdot n|\; dv'\, dn$$

with $\tilde{v} = v + (v' - v) \cdot nn$

$$\tilde{v}' = v' + (v - v') \cdot nn$$

If f is some "nice test function", disregarding integrability problems, a change of variable yields:

$$\partial_t \langle u,\ f \rangle = \langle u \otimes u,\ (f(v + (v' - v) \cdot nn) - f(v))\, |(v' - v) \cdot n| \rangle\ .$$

On this latter form, the equation naturally appears as the forward equation of a "nonlinear jump process", with Levy system:

$$\int M_t(v, d\tilde{v})\, h(\tilde{v}) = \int_{R^n \times S_n} h((v' - v) \cdot nn)\, |(v' - v) \cdot n|\, u(t, v')\, dv'\, dn$$

In section 2 we will introduce the nonlinear process as the solution to a certain (nonlinear) martingale problem. The unboundedness of the factor $|(v - v') \cdot n|$ will create a serious source of difficulty in seeing that this martingale problem is well posed.

1) Wild's formula

The theme of this section will be a formula of Wild [56], for the solution of the spatially homogeneous Boltzmann equation, which for instance covers the case of "cut-off hard spheres" (that is a collision intensity $|(v' - v) \cdot n| \wedge C$). We will also give some probabilistic interpretations of the formula for closely related to the formula.

The setting is the following: we suppose that we have a Markovian kernel Q_1 : $R^n \times R^n \times R^n$, and for μ_1, μ_2 two probabilities (or two bounded measures) on R^n, we define the probability (a bounded measure) $\mu_1 \circ \mu_2$ as:

(1.1) $$\langle \mu_1 \circ \mu_2 \rangle = \langle \mu_1 \otimes \mu_2, Q_1 f \rangle\ .$$

It is clear that for the variation norm we have the estimate $\|u_1 \circ u_2\| \le \|u_1\| \, \|u_2\|$. Wild's formula will give a series expansion for the solution of

(1.2)
$$\partial_t u = u \circ u - u$$
$$u_{t=0} = u_0 \in M(R^n) .$$

The spatially homogeneous Boltzmann equation for hard spheres with cutoff collision kernel $|(v' - v) \cdot n| \wedge 1$, for instance, will correspond to the choice of kernel:

$$Q_1 f(v, v') = \int_0^1 d\alpha \int_{S_n} dn [f(v + (v' - v) \cdot nn) \, 1(\alpha \le |(v' - v) \cdot n| \wedge 1)$$
$$+ f(v) \, 1(\alpha > |(v' - v) \cdot n| \wedge 1)] .$$

Proposition 1.1. *For any* $u_0 \in M(R^n)$, *here is a unique strongly continuous solution of*

(1.3)
$$u_t - u_0 = \int_0^t (u_s \circ u_s - u_s) \, ds ,$$

given by Wild's sum:

(1.4)
$$u_t = e^{-t} \sum_{k \ge 1} (1 - e^{-t})^{k-1} \, u^k ,$$

where $u^1 = u_0$ *and* $u^{n+1} = \frac{1}{n} \sum u^k \circ u^{n+1-k}$.

Proof: The uniqueness part follows from a classical O.D.E. result. Let us show that (1.4) does provide a solution to (1.3).

Define $u_t^1 \equiv u_0$, and for $n \ge 1$,

(1.5)
$$u_t^{n+1} = e^{-t} u_0 + \int_0^t e^{-(t-s)} \, u_s^n \circ u_s^n \, ds ,$$

then by induction one sees that for $n \ge 1$:

$$u_t^n \ge e^{-t} \sum_1^n (1 - e^{-t})^{k-1} \, u^k .$$

From this one easily concludes that u_t^n converges uniformly on bounded time intervals in variation norm to u_t given by (1.4). Because of (1.5), u_t is a solution of (1.3).

$\square$

Wild's formula expresses u_t as a barycenter of the sequence u^n, with weights $e^{-t}(1 - e^{-t})^{n-1}$. It can in several instances be used to derive asymptotic properties of u_t from those of the sequence u^n, see Ferland-Giroux [10], McKean [25], Tanaka [49], Murata-Tanaka [30].

Wild's formula has a nice interpretation in terms of continuous time binary branching trees, see also Ueno [55], [56]. Indeed $[e^{-t}(1 - e^{-t})^{n-1}]_{n \geq 1}$, is the distribution of the total number of particles of such a branching tree at time t, provided each particle branches with unit intensity. For each n, u^n is a convex combination of the various ways of inserting parentheses in a monomial $u_0 \circ \cdots \circ u_0$ of degree n. The ways of inserting parentheses in such a monomial of degree n are in natural correspondence with the tree subsets (in the sense given in section 2 of Chapter III) of the set of vertices $V = \cup_{k \geq 0}\{1, 2\}^k$, which possesses exactly n bottom vertices (with no descendants). For instance $u \circ (u \circ u)$ is associated to

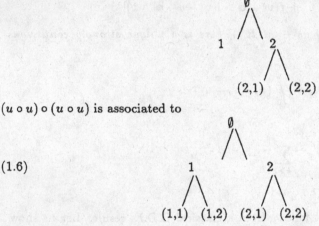

$(u \circ u) \circ (u \circ u)$ is associated to

(1.6)

The coefficients appearing in the convex combinations expressing u^n in terms of the various ways of inserting parentheses in a monomial of degree n of u yield precisely the conditional time binary tree, given that it has n bottom vertices at time t.

On the other hand one can also keep track of the time order at which branching occurs. For instance in (1.6) one distinguishes between two ordered trees depending on whether 1 or 2 branched first. Looking at the skeleton of successive jumps, it is easy to see that conditionally on the fact that there are n individuals at time t, there are $(n - 1)!$ equally likely such ordered trees.

If one uses a perturbation expansion of

$$u_t = u_0 e^{-t} + \int_0^t e^{-(t-s)} u_s \circ u_s \, ds \, ,$$

one finds

(1.7)

$$u_t = u_0 e^{-t} + e^{-t}(1 - e^{-t})u_0 \circ u_0 + \cdots + \frac{e^{-t}}{(n-2)!}(1 - e^{-t})^{n-2}\, u_0^{(n-1)}$$

$$+ e^{-t}\int_{0,s_{n-1}<\ldots<s_1<t} ds_{n-1}\ldots ds_1\, u_{s_{n-1}}^{(n)}\ .$$

Here for $v \in M(R^n)$, $v^{(n)}$ is defined as:

$$v^{(n)} = \sum_{\sigma \in S_{n-1}} v \,\overset{\circ}{\underset{\sigma(1)}{}}\, v \,\overset{\circ}{\underset{\sigma(2)}{}} \cdots \overset{\circ}{\underset{\sigma(n-1)}{}}\, v$$

and the permutation σ dictates the order in which the operation $\circ$ is performed. Observe for instance that $(u \circ u) \circ (u \circ u)$ corresponds both to $u \overset{\circ}{\underset{1}{}} u \overset{\circ}{\underset{3}{}} u \overset{\circ}{\underset{2}{}} u$ and to $u \overset{\circ}{\underset{2}{}} u \overset{\circ}{\underset{3}{}} u \overset{\circ}{\underset{1}{}} u$.

Of course by letting n go to infinity in formula (1.7), one finds (since the last term converges to zero):

(1.8)

$$u_t = \sum_{k \geq 1} \frac{e^{-t}}{(k-1)!}(1 - e^{-t})^{k-1}\, u_0^{(n-1)}\ ,$$

which only differs from Wild's formula because one uses the ordered trees as operation schemes. In Wild's formla one precisely forgets the ordering and simply keeps track of the skeleton subset of $\cup_{k \geq 0}\{1,2\}^k$ induced by the continuous time binary tree.

So we have somewhat informally presented the interpretation of Wild's formula in terms of a continuous time binary branching tree.

We are now going to explain a slightly modified point of view, which will be parallel to the construction of section 2 of Chapter III, and also close in spirit to the uniqueness proof for the Boltzmann process which will be provided in the next section.

First we suppose that we have a measurable map, $\psi(v, v', y)$, where y belongs to an auxiliary Polish space E, and a probability $\nu(dy)$, such that

(1.9)

$$Q_1 f(v, v') = \int_E f(\psi(v, v', y))\, d\nu(y)\ .$$

Here the variable y plays the role of a collision parameter.

We basically keep the notations of section 2 Chapter III, and as a first step we will construct a "noninteracting tree". As before we consider Ω the set of marked trees $\omega = (\pi, (\phi^u,\ u \in T))$, where the marks ϕ now take their values in $D = \{(\tau, v, y) \in [0, \infty) \times R^n \times E\}$. Here τ represents a trajectory duration, v an initial velocity and y a collision parameter.

Now for any $(\tau, v, y) \in D$, we consider the probability $R_{(\tau,v,y)}$ on Ω, such that:

- $R_{(\tau,v,y)}[\phi^\theta = (\tau, v, y)] = 1$ (the ancestor's mark is (τ, v, y)).
- For f_u, $u \in U$ a collection of nonnegative measurable functions on Ω,

$$E_{(\tau,v,y)}[\prod_{u \in G_n} f_u \circ T_u / F_n] = \prod_{u \in G_n} E_{\phi^u}[f_u]$$

(Branching property).

- The reproduction law is given by the fact that the point measure on $D : \sum_{1 \le \ell \le \nu_\phi} \delta_{\theta^\ell}$ is a Poisson point process with intensity measure on D:

$$1(s < \tau) \, ds \otimes u_0(dv) \otimes \nu(dy) \, ,$$

and $R_{(\tau,v,y)}$- a.s., $0 < \tau^1 < \ldots < \tau^{\alpha(1)} < \tau$.

Now very similarly to section 1 of Chapter III, we have:

Lemma 1.2.

- $R_{(\tau,v,y)}$-a.s., ω is a finite tree.
- Under $R_{(\tau,v,y)}$, the random point measure on $\Omega : M(\omega) = \sum_{1 \le \ell \le \nu_\phi} \delta_{T_\ell(\omega)}$, is a Poisson point measure with intensity

$$Q_{(\tau,v,y)} = \int_{[0,\tau] \times R^n \times E} d\tau' \, du_0(v') \, d\nu(y') \, R_{\tau',v',y'} \, .$$

We can now construct the interaction on the tree $\omega = (\pi, (\tau^u, v^u, y^u), \ u \in \pi)$ as a supplementary mark x^u, which will represent the velocity of the particle $u \in \pi$, at time τ^u. The mark v^u corresponds to the initial velocity of this particle and the collision parameter y^u, will be used in calculating the effect of paricle u, on its direct ancestor.

More precisely, using the fact that the tree is a.s. finite, we set: $z^u = v^u$ if u has no descendants, $z^u = z_{\nu_u}$, otherwise, where the sequence z_j, $0 \le j \le \nu_u$ is defined by:

(1.10)
$$z_j = \psi(z_j, \ z^{uj}, v^{uj}) \, , \quad 1 \le j \le \nu_u$$
$$z_0 = v^u \, .$$

We now denote by R^t for $t > 0$, the measure $\int du_0(v) \, d\nu(y) \, R_{t,v,y}$. With this notation, Lemma 1.2 says that $M(\omega)$ under R^t is a Poisson point process with intensity measure $\int_0^t ds \, R^s$.

The corresponding result to Theorem 2.3 of section 2, now tells us that under R^t, the law of the supplementary mark z^θ, solves the equation (1.3).

Proposition 1.3. *The law u_t of z^θ under R^t, satisfies*

$$u_t - u_0 = \int_0^t (u_s \circ u_s - u_s) \, ds \, .$$

Proof: With notation (1.10), working under R^t:

$$z^\theta = v^\theta \, 1(\nu_\theta = 0) + \psi(z_{\nu_\theta - 1}, z^{\nu_\theta}, y^{\nu_\theta}) \, 1(\nu_\phi \geq 1) \, .$$

Now, with the help of Lemma 1.2, conditinally on $\nu_\theta \geq 1$, and $\tau^{\nu_\theta} = s$, $z_{\nu_\theta - 1}, z^{\nu_\theta}, y^{\nu_\theta}$ are independent, with, $z_{\nu_\theta - 1}$, z^{ν_θ}, u_s distributed. So we find that $u_t = u_0 e^{-t} + \int_0^t e^{-(t-s)} u_s \circ u_s \, ds$.

Our claim follows from this.

$\square$

Proposition 1.3 provides a representation formula for the solution of spatially homogneous Boltzmann equations with cutoff, which is also giving an intuition for the proof off uniqueness of the hard sphere Boltzmann process given in the next section.

2) Uniqueness for the Boltzmann process.

In this section we will look at the Boltzmann process (spatially homogeneous for hard spheres), as the solution of a certain martingale problem. The unboundedness of the collision intensity $|(v' - v) \cdot n|$, is a source of difficulty for the uniqueness of the solution, especially if one does not want to assume too many moment integrability conditions on the solution.

The existence of the solution, in the theorem we are now going to state, comes from a tightness result on the solutions corresponding to the cutoff problems (with collision function $|(v' - v) \cdot n| \wedge N$). For details on the existence part, we refer the reader to [39].

Theorem 2.1. *Let* $u_0 \in M(R^n)$, *be such that* $\int |v|^3 \, u_0(dv) < \infty$. *There is a unique probability* P *on* $D(R_+, R^n)$, *such that*

i) *for* $T < \infty$, $\sup_{t \leq T} \int |X_t|^3 \, dP < \infty$

ii) $X_0 \circ P = u_0$

iii) *for* $f \in bB(R^n)$,

$$f(X_t) - f(X_0) - \int_0^t \int_{D \times S_n} [f(X_s + (X_s(\omega') - X_s) \cdot nn)$$
$$- f(X_s)] \, |X_s(\omega') - X_s) \cdot n| \, dP(\omega') \, dn \, ds$$

is a P-*martingale.*

We are now going to explain in a number of steps the proof of uniqueness.

Let us first recall that for a solution P of such a martingale problem, one can give a trajectorial representation as follows. Denote by $M_p(R_+ \times R_+ \times S_n \times D)$, the set of simple pure point measure on $R_+ \times R_+ \times S_n \times D$, finite on any compact restriction of the first two coordinates (with at mot one atom on each $\{t\} \times R_+ \times S_n \times D$, and none if $t = 0$).

Now on the product space $R^n \times M_p$ one can put the product of the probabilitiy u_0 and of Poisson measure with intensity $dt \otimes d\theta \otimes dn \otimes dP(\omega')$. P is now the law of the unique solution Z_s of the equation

(2.1)
$$Z_t = Z_0 + \int_0^t \int_{R_+ \times S_n \times D} (X_s(\omega') - Z_{s-}) \cdot n \; n \; 1\{\theta < |(Z_{s-} - X_s(\omega')) \cdot n|\}$$
$$N(ds \; d\theta \; dn \; d\omega')$$

with $\int_0^t \int_{R_+ \times S_n \times D} 1\{\theta < |(Z_{s-} - X_s(\omega')) \cdot n|\} \; N(ds \; d\theta \; dn \; d\omega') < \infty$,

for all $t < \infty$,

Here Z_0 is the R^n-valued coordinate on the space $R^n \times M_p$, and N is the canonical Poisson measure induced by the second coordinate.

Consider then P_1 and P_2 two solutions of the martingale problem. We are first going to construct a coupling measure P_0 on $D(R_+, R^n)^2$, of P_1 and P_2, whch will also satisfy a nonlinear martingale problem.

Lemma 2.2. *There exists a probability P_0 on $D(R_+, R^n)^2$ such that*

1) $X^1 \circ P_0 = P_1$, $X^2 \circ P_0 = P_2$.

2) *for $f \in bB(R^n \times R^n)$:*

$$f(X_t^1, X_t^2) - f(X_0^1, X_0^2) - \int_0^t \int_{D^2 \times S_n}$$
$$[f(\tilde{X}_s^1, \tilde{X}_s^2) \, | \, (X_s^1 - X_s^1(\omega')) \cdot n| \wedge |(X_s^2 - X_s^2(\omega')) \cdot n|$$
$$+ f(\tilde{X}_s^1, X_s^2) \, (|(X_s^1 - X_s^1(\omega')) \cdot n| - |(X_s^2 - X_s^2(\omega')) \cdot n|)_+$$
$$+ f(X_s^1, \tilde{X}_s^2) \, (|(X_s^2 - X_s^2(\omega')) \cdot n| - |(X_s^1 - X_s^1(\omega')) \cdot n|)_+$$
$$- f(X_s^1, X_s^2) \, (|(X_s^2 - X_s^1(\omega')) \cdot n| \vee |(X_s^2 - X_s^2(\omega')) \cdot n|)] \; dn \; dP_0(\omega') \; ds$$

is a P_0-martingale, with the notation for $i = 1, 2$:

$$\tilde{X}_s^i = X_s^i + (X_s^i(\omega') - X_s^i(\omega)) \cdot n \; n$$

3) $(X_0^2, X_0^2) \circ P_0 = \text{diag } u_0$, *the diagonal image of u_0 on $R^n \times R^n$.*

Proof: The set $\mathcal{C}$ of probabilities on $D(R_+, R^n)^2$ with respective projections on the first and second coordinates givenby P_1 and P_2, is a weakly compact and convex set.

Denote by F the map which associates to $Q \in \mathcal{C}$, the law on $D(R_+, R^n)^2$ of (Z^1, Z^2) solutions on $R^n \times R^n \times M_p(R_+ \times R_+ \times S_n \times D(R_+, R^n)^2)$ with measure diag $u_0 \otimes$ Poisson $(dt \otimes d\theta \otimes dn \otimes dQ)$ of the equations for $i = 1, 2$.

$$Z_t^i = Z_0^i + \int_0^t \int_{R_+ \times S_n \times D(R_+, R^n)^2} (X_s^i(\omega') - Z_{s-}^i) \cdot nn \, 1\{\theta < |(Z_s^i - X_s^i(\omega')) \cdot n|\}$$
$$N(ds \, d\theta \, dn \, d\omega'),$$

with

$$\int_0^t \int 1\{\theta < |(Z_{s-}^i - X_s^i(\omega')) \cdot n|\} \, N(ds \, d\theta \, dn \, d\omega') < \infty.$$

In view of representation (2.1) F maps $\mathcal{C}$ into $\mathcal{C}$. Moreover $F(Q)$ is characterized by 1), 2), 3), with the replacement of $dP_0(\omega')$ integration by $dQ(\omega')$ integration. From this one sees that F is weakly continuous, and the existence of a fixed point to the map F now follows from Tychonov's theorem. This yields our claim.

$\square$

Let us now introduce some notations convenient for what follows. We set for $x = (x_1, x_2)$, $x' = (x_1', x_2')$:

$$k(x, x') = (1 + |x_i| + |x_i'|) \vee (1 + |x_2| + |x_2'|),$$
$$\phi(x) = 9(1 + |x_1|^2 + |x_2|^2),$$
$$Qh(x, x') = \int h(x + \Phi(x, x', y), x' + \Phi(x', x, y)) \, d\nu(y)$$

where $y = (\alpha, n) \in [0, 1] \times S_n = E$, $d\nu = d\alpha \otimes dn$, and Φ is the map from $R^{2n} \times R^{2n} \times E$ into R^{2n}, given by:

$$\Phi(x, x', y) = ((x_1' - x_1) \cdot nn, (x_2' - x_2) \cdot nn), \text{ if } \alpha \le (a_1 \wedge a_2)/k,$$
$$= ((x_1' - x_1) \cdot nn) \text{ if } a_1/k \ge \alpha > (a_1 \wedge a_2)/k,$$
$$= (0, (x_2' - x_2) \cdot nn) \text{ if } a_2/k \ge \alpha > (a_1 \wedge a_2)/k,$$
$$= (0, 0) \quad \text{if } \alpha \ge (a_1 \vee a_2)/k,$$

with $a_i = |(x_i - x_i') \cdot n|$, $k = k(x, x')$. We will also write for $f \in bB(R^n)$: $Q_1 f = Q(f \otimes 1)$.

If we put the probability diag $u_0 \otimes$ Poisson$(ds \otimes d\theta \otimes d\nu \otimes dP_0(\omega'))$, on $R^n \times R^n \times M_p(R_+ \times R_+ \times E \times D)$ (with $D = D(R_+, R^n)^2$), P_0 can also be represented as the law of the solution of

$$(2.3) \qquad Z_t(\omega) = Z_0(\omega) + \int_0^t \int_{R_+ \times E \times D} \Phi(Z_{s-}(\omega), \omega'(s), y) \, 1(\theta < k(Z_{s-}(\omega), \omega'(s)))$$
$$N(\omega, ds \, d\theta \, dy \, d\omega'),$$

with

$$\int_0^t \int_{R_+ \times E \times D} 1(\theta < k(Z_{s-}(\omega), \omega'(s))) \, N(\omega, ds \, d\theta \, dy \, d\omega') \, .$$

Moreover we have the estimates:

(2.4)
$$k(x, x') \leq \phi(x)^{1/2} \, \phi^{1/2}(x'),$$
$$Q(\phi \oplus \phi) = \phi \oplus \phi \quad (\text{where } \phi \oplus \phi(x, x') = \phi(x) + \phi(x')).$$

Another nice feature of our coupling measure P_0, as seen from (2.2), is that if x and x' are "diagonal", that is $x_1 = x_2$ and $x'_1 = x'_2$, then $\Phi(x, x', y)$ is diagonal as well.

The representation formula (2.3) for the coupling measure P_0, should be viewed as a way to keep track thanks to the marks "ω'" of the first generation of trajectories, contributing to the determination of the bi-particle trajectory Z. on $[0, t]$. The scheme is now to construct the successive generations, using first a similar formula to (2.3) for each mark ω' (this yields the second generation), and iterating the procedure. On this "projective object", the problem will now be to see that the ancestor trajectory is in fact the result of a calculation on a finite tree whicch preserves the diagonal property of the initial conditions. This will prove that P_0 is in fact supported by the diagonal, and will yield uniqueness.

The first step (compare with Lemma 1.2, and Lemma 2.1 of Chapter III).

Lemma 2.3. *There is an auxiliary space* $(\tilde{\Omega}, P)$, *endowed with a Poisson point measure* $N(\tilde{\omega}, ds \, d\theta \, dy \, d\tilde{\omega}')$ *of intensity* $ds \otimes d\theta \otimes d\nu(y) \otimes dP(\tilde{\omega}')$, *and with a* P_0-*distributed process* $(Z.(\omega))$ *such that* Z_0 *is independent of the point measure* N *and* diag u_0 *distributed and*

$$\text{for } t > 0, \int_0^t \int_{R_+ \times E \times \tilde{\Omega}} 1\{\theta < k(Z_{s-}(\tilde{\omega}), \, Z_s(\tilde{\omega}'))\} \, N(\tilde{\omega}, ds \, d\theta \, dy \, d\tilde{\omega}') < \infty$$

$$Z_t(\tilde{\omega}) = Z_0(\tilde{\omega}) + \int_0^t \int_{R_+ \times E \times \tilde{\Omega}} \Phi(Z_{s-}(\tilde{\omega}), \, Z_s(\tilde{\omega}'), y)$$
$$1\{\theta < k(Z_{s-}(\tilde{\omega}), Z_s(\tilde{\omega}'))\} \, N(\omega, ds \, d\theta \, dy \, d\omega') \, ,$$

for the proof of this we refer the reader to [39]. The idea of the construction as alluded to before is that (2.3) gives a natural map from $R^{2n} \times M_p(R_+ \times R_+ \times E \times D)$ into D. Iterating this one has a natural map from $R^{2n} \times M_p(R_+ \times R_+ \times E \times R^{2n} \times M_p(R_+ \times R_+ \times E \times D))$ in the space $R^{2n} \times M_p(R_+ \times R_+ \times E \times D)$. The space $\tilde{\Omega}$ essentially arises as a projective limit of this scheme.

Now on our extended space $\widetilde{\Omega}$, we can define the number $K_t(\widetilde{\omega})$ of generations which contribute to the state of the "ancestor trajectory" $Z.(\widetilde{\omega})$ on time $[0,t]$, by stating:

$$\{K \geq 0\} = R_+ \times \widetilde{\Omega}$$

$$\{K \geq 1\} = \{(t,\widetilde{\omega})/\exists s \in (0,t], \ N(\widetilde{\omega}, \{s\} \times R_+ \times E \times \widetilde{\Omega}) = 1,$$

$$\text{and } {}^s\theta < k(Z_{s-}(\widetilde{\omega}), Z_s({}^s\widetilde{\omega}'))\}$$

here ${}^s\theta$ and ${}^s\widetilde{\omega}'$ are the marks of $N(\widetilde{\omega}, \cdot)$ at time s.

$$\{K \geq n+1\} = \{(t,\omega)/\exists s \in (0,t], \ N(\widetilde{\omega}, \{s\} \times R \times E \times \widetilde{\Omega}) = 1,$$

$${}^s\theta < k(Z_{s-}(\widetilde{\omega}), Z_s({}^s\widetilde{\omega}')) \text{ and } K_s({}^2\widetilde{\omega}') \geq n\},$$

and

$$\{K \geq \infty\} = \bigcap_n \{K \geq n\}.$$

Our next step is that only a finite number of generations play a role in the determination of $Z.(\widetilde{\omega})$.

Lemma 2.4. *For $t > 0$, $K_t(\omega) < \infty$, P-a.s.*

Proof: Set $\bar{Z}_t(\widetilde{\omega}) = Z_t(\widetilde{\omega})$ when $K_t < \infty$, δ otherwise. So $\bar{Z}.$ is $R^n \cup \{\delta\}$ valued. Now for $f \in bB(R^n)$ (equal to zero on δ):

$$f(\overline{Z}_t) - f(\bar{Z}_0) = \int_0^t \int_{R_+ \times E \times \widetilde{\Omega}} [f(Z_s(\widetilde{\omega})) \, 1\{K_{s-}(\widetilde{\omega}) \vee K_s(\widetilde{\omega}') < \infty\}$$

$$- f(\bar{Z}_{s-}(\widetilde{\omega}))] \, 1\{\theta < k(Z_{s-}(\widetilde{\omega}), Z_s(\widetilde{\omega}'))\} \, N(\omega, \ ds \, d\theta \, d\nu(y) \, dP(\widetilde{\omega}'))$$

From this after integration, we find:

$$(2.5) \quad E[f(\bar{Z}_t)] - E[f(Z_0)] = \int_0^t \int_{\Omega \times \Omega} Q_1 f(\bar{Z}_s(\widetilde{\omega}), \bar{Z}_s(\widetilde{\omega}')) \, k(\bar{Z}_s(\widetilde{\omega}), \bar{Z}_s(\widetilde{\omega}'))$$

$$- f(\bar{Z}_s(\widetilde{\omega})) \, k(\bar{Z}_s(\widetilde{\omega}), Z_s(\widetilde{\omega}')) \, dP(\widetilde{\omega}) \, dP(\widetilde{\omega}') \, ds$$

In view of our integrability assumptions on P_1, and P_2, we can apply (2.5) with ϕ instead of f. And by a very similar argument using $Q(\phi \oplus \phi) = \phi \oplus \phi$ as stated in (2.2), we get:

$$E[\phi(Z_t)] = E[\phi(Z_0)]$$

It then follows that

$$E[\phi(Z_t) - \phi(\bar{Z}_t)] = \int_0^t \int_{\Omega \times \Omega} -\frac{1}{2}[Q(\phi \oplus \phi) - \phi \oplus \phi] \times k(\bar{Z}_s(\widetilde{\omega}), \bar{Z}_s(\widetilde{\omega}'))$$

$$+ \phi(\bar{Z}_s(\widetilde{\omega}))[k(\bar{Z}_s(\widetilde{\omega}), Z_s(\widetilde{\omega}')) - k(\bar{Z}_s(\widetilde{\omega}), \bar{Z}_s(\widetilde{\omega}'))]$$

$$dP(\widetilde{\omega}) \, dP(\widetilde{\omega}') \, ds$$

Now the first term in the integral is zero, as for the second, using $k(x, x') \leq \phi^{1/2}(n) \, \phi^{1/2}(x')$, we find:

$$E[\phi(Z_t) \, 1\{K_t = \infty\}] \leq \int_0^t E[\phi^{3/2}(Z_s)] \, E[\phi^{1/2}(Z_s) \, 1\{K_s = \infty\}] \, ds$$

From Gronwall's lemma, we now find $E[\phi(Z_t) \, 1\{K_t = \infty\}] = 0$, from which our claim follows.

$\square$

We now have the required ingredients to see that our coupling probability P_0 is diagonally supported.

Indeed when $K_t(\tilde{\omega}) = 0$, then $Z_t^1(\tilde{\omega}) = Z_0^1(\tilde{\omega}) = Z_0^2(\tilde{\omega}) = Z_t^2(\tilde{\omega})$, since no jump occurs.

Suppose now that we know that $K_T(\tilde{\omega}) \leq n$ implies $Z_t^1(\tilde{\omega}) = Z_t^2(\tilde{\omega})$, for $t \leq T$. Take now $\tilde{\omega} \in \tilde{\Omega}$ with $K_T(\tilde{\omega}) = n+1$, then for $t \leq T$:

$$Z_t(\tilde{\omega}) = Z_0(\tilde{\omega}) + \int_0^t \int_{R^n \times E \times \tilde{\Omega}} \Phi(Z_{s-}(\tilde{\omega}), Z_s(\tilde{\omega}'), y) \, 1\{\theta < k(Z_{s-}(\tilde{\omega}), Z_s(\tilde{\omega}'))\}$$
$$1\{K_s(\tilde{\omega}') \leq n\} \, N(\tilde{\omega}, ds \, d\theta \, dy \, d\tilde{\omega}') \, ,$$

since only mark ${}^s\tilde{\omega}'$ for which $K_s({}^s\tilde{\omega}') \leq n$ come in the determination of $Z_t(\tilde{\omega})$, $t \leq T$. Now as observed already if x and x' are diagonal so is $\Phi(x, x', y)$. It then follows that $Z_t(\tilde{\omega})$, $t \leq T$ is diagonal. Now our claim follows by induction, thanks to Lemma 2.4.

So we obtain that the coupling probability P_0 is diagonally supported, and this proves that $P_1 = P_2$.

$\square$

References

1. Baxter, J. R. - Chacon, R. V. - Jain, N. C.: Weak limits of stopped diffusions, Trans. Amer. Math. Soc., 293, 2, 767-792, (1986).
2. Billingsley, P.: Convergence of Probability Measures, Wiley, New York (1968).
3. Calderoni, P. - Pulvirenti, M.: Propagation chaos for Burger's equation, Ann. Inst. H. Poincaré, série A, N.S. 39, 85-97 (1983).
4. Cercignani, C.: The grad limit for a system of soft spheres, Comm. Pure Appl. Math 26, 4 (1983).
5. Dawson, D. A.: Critical dynamics and fluctuations for a mean field model of cooperative behavior, J. Stat. Phys. 31, 29-85 (1978).
6. Dawson, D. A. - Gärtner, J.: Large deviations from the McKean-Vlasov limit for weakly interacting diffusions, Stochastics 20, 247-308 (1987).
7. De Masi, A. - Ianiro, N. - Pellegrinotti, A. - Pressutti, E.: A survey of the hydrodynamical behavior of many particle systems, in Nonequilibrium phenomena II, Ed.: J. L. Lebowitz, E. W. Montroll, Elsevier (1984).
8. Dobrushin, R. L.: Prescribing a system of random variables by conditional distributions, Th. Probab. and its Applic. 3, 469 (1970).
9. ————————— Vlasov equations, Funct. Anal. and Appl. 13, 115 (1979).
10. Ferland, R., Giroux, G.: Cutoff Boltzmann equation: convergence of the solution, Adv. Appl. Math. 8, 98-107 (1987).
11. Funaki, T.: The diffusion approximation of the spatially homogeneous Boltzmann equation, Duke Math. J. 52, 1-23, (1985).
12. Goodman, J., Convergence of the random value method, IMA, vol. 9, G. Papanciolaou ed., Hydrodynamic Behavior and Interacting Particles, 99-106, Springer, Berlin (1987).
13. Guo, M. - Papanicolaou, G. C.: Self diffusion of interacting Brownian particles, Taniguchi Symp., Katata 1985, 113-151.
14. Gutkin, E.: Propagation of chaos and the Hopf-Cole transformation, Adv. Appl. Math. 6, 413-421, 1985.
15. Kac, M.: Foundation of kinetic theory, Proc. Third Berkeley Symp. on Math. Stat. and Probab. 3, 171-197, Univ. of Calif. Press (1956).
16. ————————— Some probabilistic aspects of the Boltzmann equation, Acta Physica Austraiaca, suppl. X, Springer, 379-400 (1979).
17. Karandikar, R. L.,Horowitz, J.: Martingale problems associated with the Boltzmann equation, preprint, 1989.
18. Kipnis, C. - Olla, S. - Varadhan, S.R.S.: Hydrodynamics and large deviation for simple exclusion processes Comm. Pure Appl. Math., 42, 115-137, (1989).
19. Kurtz, T.: Approximation of population processes CMBS-NSF Reg. Conf. Sci. Appl. Math. Vol. 36, Society for Industrial and Applied Mathematics, Philadelphia (1981).
20. Kusuoka, S. - Tamura, Y.: Gibbs measures with mean field potentials, J. Fac. Sci. Tokyo Univ., sect. 1A, 31, 1, 223-245 (1984).
21. Kotani, S. - Osada, H.: Propagation of chaos for Burgers' equation, J. Math. Soc. Japan, 37, 275-294 (1985).
22. Lanford, O. E., Time evolution of large classical systems, Lecture Notes in Physics 38, 1-111, Springer, Berlin, (1975).
23. Lang, R. - Nguyen, X.X.: Smoluchowski's theory of coagulation in colloids holds rigorously in the Boltzmann-Grad limit, Z. Wahrscheinlichkeitstheor. Verw. Gebiete 54, 227-280 (1980).

24. Liggett, T. M.: Interacting particle systems, Springer, Berlin (1985).
25. McKean, H. P.: Speed of approach to equilibrium for Kac's caricature of a Maxwellian gas, Arch. Rational Mech. Anal. 21, 347-367 (1966).
26. ―――――――― A class of Markov processes associated with nonlinear parabolic equations, Proc. Nat. Acad. Sci. 56, 1907-1911 (1966).
27. ―――――――― Propagation of chaos for a class of nonlinear parabolic equations, Lecture series in differential equations 7, 41-57, Catholic University, Washington, D. C. (1967).
28. ―――――――― Fluctuations in the kinetic theory of gases, Comm. Pure Appl. Math. 28, 435-455 (1975).
29. Marchioro, C. – Pulvirenti, M.: Hydrodynamics in two dimensions and vortex theory, Comm. Math. Phys. 84, 483-504 (19820.
30. Murata, H. - Tanaka, H.: An inequality for certain functionals of multidimensional probability distributions, Hiroshima Math. J., 4, 75-81 (1974).
31. Nagasawa, M. – Tanaka, H.: Diffusion with interaction and collisions between colored particles and the propagation of chaos, Probab. Th. Rel. Fields 74, 161-198 (1987).
32. ―――――――― On the propagation of chaos for diffusion processes with coefficients not of average form, Tokyo Jour. Math. 10 (2), 403-418 (1987).
33. Neveu,J.: Arbres et processus de Galton-Watson, Ann. Inst. Henri Poincaré Nouv. Ser. B, 22, 2,199-208 (1986).
34. Oelschläger K.: A law of large numbers for moderately interacting diffusion processes, Z. Wahrscheinlichkeitstheor. Verw. Gebeite 69, 279-322 (1985).
35. Osada, H.: Limit points of empirical distributions of vortices with small viscosity, IMA, vol. 9, G. Papanicolaou ed., Hydrodynamic behavior and interacting particles, 117-126, Springer, Berlin (1987).
36. ―――――――― Propagation of chaos for the two dimensional Navier-Stokes equation.
37. Scheutzow, M.: Periodic behavior of the stochastic Brusselator in the mean field limit, Prob. Th. Re. Fields 72, 425-462, (1986).
38. Shiga, T. – Tanaka, H.: Central limit theorem for a system of Markovian particles with mean field interactions, Z. Wahrscheinlichkeitstheor. Verw. Gebiete 69, 439-445 (1985).
39. Sznitman, A. S.: Equations de type Boltzmann spatialement homogènes, Z. Wahrscheinlichkeitstheor. Verw. Gebiete 66, 559-592 (1984).
40. ―――――――― Nonlinear reflecting diffusion process and the propagation of chaos and fluctuations associated, J. Funct. Anal. 56 (3), 311-336 (1984).
41. ―――――――― A fluctuation result for nonlinear diffusions, infinite dimensional analysis, S. Albeverio, ed., 145-160, Pitman, Boston (1985).
42. ―――――――― A propagation of chaos result for Burgers' equation, Z. Wahrscheinlichkeitstheor. Verw. Gebiete 71, 581-613 (1986).
43. ―――――――― Propagation of chaos for a system of annihilating Brownian spheres, Comm. Pure Appl. Math. 60, 663-690 (1987).
44. ―――――――― A trajectorial representation for certain nonlinear equations, Astérisque, 157-158, 363-370 (1988).
45. ―――――――― A limiting result for the structure of collisions between many independent diffusions, Probab. Th. Rel Fields 81, 353-381 (1989).
46. Sznitman, A. S. – Varadhan, S.R.S.: A multidimensional process involving local time, Z. Wahrscheinlichkeitstheor. Verw. Gebiete 71, 553-579 (1986).

47. Spohn, H.: The dynamics of systems with many particles, statistical mechanics of local equilibrium states (preprint).

48. Tamura, Y.: On asymptotic behaviors of the solution of a nonlinear diffusion equation, J. Fac. Sci. Tokyo Univ., sect. IA, 31, 1, 195-221 (1984).

49. Tanaka, H.: Probabilistic treatment of the Boltzmann equation of Maxwellian molecules, Z. Wahrscheinlichkeitstheor. Verw. Gebiete 46, 67-105 (1978).

50. ——————— Some probabilistic problems in the spatially homogeneous Boltzmann equation, Proc. IFIP–ISI conf. on appl. of random fields, Bangalore, Jan. 82.

51. ——————— Limit theorems for certain diffusion processes with interaction, Taniguchi Symp. S. A. Katata, 469-488 (1982).

52. Tanaka, H. – Hitsuda, M.: Central limit theorem for a simple model of interacting particles, Hiroshima Math. J. 11, 415-423 (1981).

53. Uchiyama, K.: On the Boltzmann Grad limit for the Broadwell model of the Boltzmann equation, J. Stat. Phys. 52, 331-355 (1988).

54. ——————— Derivation of the Botlzmann equation from particle dynamics, Hiroshima Math. J. 18, 2 (1988).

55. Ueno, T.: A class of Markov processes with bounded nonlinear generators, Japanese J. Math. 38, 19-38 (1968).

56. ——————— A path space and the propagation of chaos for Boltzmann's gas model, Proc. Japan Acad. 6 (47) 529-533 (1971).

57. Wild, E.: On the Boltzmann equation in the kinetic theory of gases, Proc. Cambridge Phil. Soc., 47, 602-609 (1951).

58. Zvonkin, A. K.: A transformation of the phase space of a diffusion process that removes the drift, Math. USSR Sbornik, 22, 1, 129-149, (1974).

EXPOSES 1989

A. BENASSI — Calcul différentiel stochastique anticipatif
Martingales hiérarchiques d'ordre quatre

J. BROSSARD — Densité de l'intégrale d'aire

E. BUSVELLE — Filtres de dimension finie et immersions

L. CHEVALIER — Caractérisation de la classe LLogL en probabilité
et en analyse

F. COMETS — Estimation dans les champs de Markov

D. DE BRUCQ — Ordre d'un modèle et complexité

N. EL KAROUI — Contrôle partiellement observable avec contrôle
dans la fonction d'observation

A. ESTRADE — Exponentielle stochastique et semi-martingales à
valeurs dans un groupe de LIE

B. FERNANDEZ — Fluctuations of a particle system

G. GIROUX — Vitesse de convergence vers l'équilibre pour des
processus de Boltzmann

O. HIJAB — Control of degenerate diffusions in R^d

C. LANDIM — Equations hydrodynamiques pour des systèmes
de particules attractifs sur Z^d

J.F. LE GALL — Brownian excursions, trees and measure-valued
branching processes

C. LEONARD — Grandes déviations d'une équation différentielle
perturbée

M. MAJSNEROWSKA — Remarks on some Poisson approximation

U. MANSMANN — The Dirac Polaron

M. PONTIER — Existence d'un équilibre stochastique

L. PRATELLI — Invariability of the notions of semimartingale and
stochastic integral with respct to a change of space

E. RIO — Approximations fortes pour des processus de
sommes partielles

LISTE DES AUDITEURS

Mr.	ALABERT A.	Université de Barcelone (Espagne)
Mr.	AZEMA J.	Université de Paris VI
Mr.	BADRIKIAN A.	Université Blaise Pascal (Clermont II)
Mr.	BENASSI A.	Université Blaise Pascal (Clermont II)
Mr.	BERNARD P.	Université Blaise Pascal (Clermont II)
Mr.	BEZANDRY P.	Université de Strasbourg
Mr.	BOUGEROL P.	Université de Nancy I
Mr.	BRAY G.	Université de Ouagadougou (Burkina Faso)
Mr.	BROSSARD J.	Institut Fourier, Grenoble I
Mr.	BRUNAUD M.	Université de Paris-Sud (Orsay)
Mr.	BUSVELLE E.	Université de Rouen
Mr.	CAMPILLO F.	I.N.R.I.A., Centre de Sophia Antipolis
Mme	CHALEYAT-MAUREL M.	Université de Paris VI
Mr.	CHASSAING P.	Université de Nancy I
Mr.	CHEVALIER L.	Institut Fourier, Grenoble I
Mle	CHEVET S.	Université Blaise Pascal (Clermont II)
Mr.	COMETS F.	Université de Paris-Sud (Orsay)
Mr.	DALAUD L.	Université d'Abidjan (Côte d'Ivoire)
Mr.	DE BRUCQ D.	Université de Rouen
Mr.	DECAUWERT J.M.	Institut Fourier, Grenoble I
Mr.	DERMOUNE Azzouz	Université Blaise Pascal (Clermont II)
Mr.	DJEHICHE B.	Université de Stockholm (Suède)
Mr.	DEZA F.	I.N.S.A. de Rouen
Mme	DONATI-MARTIN C.	Université de Provence, Aix-Marseille
Mr.	EL-HOSSEINY H.	Institut Fourier, Grenoble I
Mme	ELIE L.	Université de Paris VII
Mme	EL KAROUI N.	Université de Paris VI
Mr.	EL KHARROUBI A.	Institut Fourier, Grenoble I
Mle	ESTRADE A.	Université d'Orléans
Mme	FERNANDEZ B.	Université de Mexico (Mexique)
Mr.	FLORCHINGER P.	Université de Metz
Mr.	FRANCOIS O.	Université de Grenoble I
Mr.	GALLARDO L.	Université de Brest
Mme	GERARDI A.	Université "La Sapienza" à Rome (Italie)
Mr.	GIROUX G.	Université de Sherbrooke (Canada)
Mr.	GOLDBERG J.	I.N.S.A. de Lyon
Mr.	GRORUD A.	Université de Provence, Aix-Marseille
Mr.	HENNEQUIN P.L.	Université Blaise Pascal (Clermont II)
Mr.	HIJAB O.	Temple University à Philadelphia (U.S.A.)
Mr.	KERKYACHARIAN G.	Université de Nancy I
Mle	KOUKIOU F.	Université de Lausanne (Suisse)
Mr.	LANDIM C.	Ecole Polytechnique de Palaiseau
Mr.	LE GALL J.F.	Université de Paris VI
Mr.	LE GLAND F.	I.N.R.I.A., Centre de Sophia Antipolis
Mr.	LENGLART E.	I.N.S.A. de Rouen
Mr.	LEONARD C.	Université de Paris-Sud (Orsay)
Mme	MAJSNEROWSKA M.	Université de Wroclaw (Pologne)
Mr.	MANSMANN U.	Université de Berlin (R.F.A.)
Mr.	MEKETE T.	I.N.S.A. de Rouen

Mme MELEARD S.	Université du Maine, Le Mans
Mme MICHEL D.	Université de Toulouse III
Mr. MICLO L.	Ecole Normale Supérieure à Paris
Mme MILHEIRO DE OLIVEIRA P.	I.N.R.I.A., Centre de Sophia Antipolis
Mr. NOBLE J.	Université de Warwick (Grande-Bretagne)
Mr. OLIVAR G.	Université de Barcelone (Espagne
Mr. PESCE H.	Université de Grenoble
Mr. PETRITIS D.	Université de Lausanne (Suisse)
Mme PICARD D.	Université de Paris VII
Mme PIETRUSKA-PALUBA K.	Courant Institute, Université de New-York (U.S.A.)
Mme PONTIER M.	Université d'Orléans
Mr. PRATELLI L.	Université de Pise (Italie)
Mr. PRATELLI M.	Université de Pise (Italie)
Mr. RAKOTOPARA D.	Centre de Recherche SHELL de Grand-Couronne
Mr. REVUZ D.	Université de Paris VII
Mr. RIO E.	Université de Paris VI
Mme ROELLY-COPPOLETTA S.	Université de Paris VI
Mme ROUBAUD M.C.	I.N.R.I.A., Centre de Sophia-Antipolis
Mr. ROYNETTE B.	Université de Nancy II
Mr. RUIZ DE CHAVEZ J.	Université de Paris VI
Mr. RUSSO F.	E.N.S.T. à Paris
Mr. SCHAUMLOFFEL K.	Université Bremen (R.F.A.)
Mr. SHEU S.	National Tsing Hua University, Hsinchu (Chine)
Mr. SZAJOWSKI K.	Université de Wroclaw (Pologne)
Mr. TAYLOR J.	Université de Montréal (Canada)
Mme THIEULLEN M.	Université de Provence, Aix-Marseille
Mr. VAN BIESEN J.	Université d'Anvers à Wilrijk (Belgique)
Mr. VAN CASTEREN J.	Université d'Anvers à Wilrijk (Belgique)

LIST OF PREVIOUS VOLUMES OF THE "Ecole d'Eté de Probabilités"

| 1977 | D. DACUNHA-CASTELLE | (LNM 678) |

"Vitesse de convergence pour certains problèmes
statistiques"
H. HEYER
"Semi-groupes de convolution sur un groupe localement
compact et applications à la théorie des probabilités"
B. ROYNETTE
"Marches aléatoires sur les groupes de Lie"

| 1978 | R. AZENCOTT | (LNM 774) |

"Grandes déviations et applications"
Y. GUIVARC'H
"Quelques propriétés asymptotiques des produits de
matrices aléatoires"
R.F. GUNDY
"Inégalités pour martingales à un et deux indices :
l'espace H^p"

| 1979 | J.P. BICKEL | (LNM 876) |

"Quelques aspects de la statistique robuste"
N. EL KAROUI
"Les aspects probabilistes du contrôle stochastique"
M. YOR
"Sur la théorie du filtrage"

| 1980 | J.M. BISMUT | (LNM 929) |

"Mécanique aléatoire"
L. GROSS
"Thermodynamics, statistical mechanics and
random fields"
K. KRICKEBERG
"Processus ponctuels en statistique"

| 1981 | X. FERNIQUE | (LNM 976) |

"Régularité de fonctions aléatoires non gaussiennes"
P.W. MILLAR
"The minimax principle in asymptotic statistical theory"
D.W. STROOCK
"Some application of stochastic calculus to partial
differential equations"
M. WEBER
"Analyse infinitésimale de fonctions aléatoires"

| 1982 | R.M. DUDLEY | (LNM 1097) |

"A course on empirical processes"
H. KUNITA
"Stochastic differential equations and stochastic
flow of diffeomorphisms"
F. LEDRAPPIER
"Quelques propriétés des exposants caractéristiques"

83	D.J. ALDOUS "Exchangeability and related topics" I.A. IBRAGIMOV "Théorèmes limites pour les marches aléatoires" J. JACOD "Théorèmes limite pour les processus"	(LNM 1117)

83 D.J. ALDOUS
"Exchangeability and related topics"
I.A. IBRAGIMOV
"Théorèmes limites pour les marches aléatoires"
J. JACOD
"Théorèmes limite pour les processus"   (LNM 1117)

84 R. CARMONA
"Random Schrödinger operators"
H. KESTEN
"Aspects of first passage percolation"
J.B. WALSH
"An introduction to stochastic partial differential
equations"   (LNM 1180)

85-87 S.R.S. VARADHAN
"Large deviations"
P. DIACONIS
"Applications of non-commutative Fourier
analysis to probability theorems
H. FOLLMER
"Random fields and diffusion processes"
G.C. PAPANICOLAOU
"Waves in one-dimensional random media"
D. ELWORTHY
Geometric aspects of diffusions on manifolds"
E. NELSON
"Stochastic mechanics and random fields"   (LNM 1362)

86 O.E. BARNDORFF-NIELSEN
"Parametric statistical models and likelihood"   (LNS   50)

88 A. ANCONA
"Théorie du potentiel sur les graphes et les variétés"
D. GEMAN
"Random fields and inverse problems in imaging"
N. IKEDA
"Probabilistic methods in the study of asymptotics"   (LNM 1427)

89 D.L. BURKHOLDER
"Explorations in martingale theory and its applications"
E. PARDOUX
"Filtrage non linéaire et équations aux dérivées partielles
stochastiques associées"
A.S. SZNITMAN
"Topics in propagation of chaos"   (LNM 1464)

1290: G. Wüstholz (Ed.), Diophantine Approximation and Tran-
scendence Theory. Seminar, 1985. V, 243 pages. 1987.

1291: C. Mœglin, M.-F. Vignéras, J.-L. Waldspurger, Correspon-
ces de Howe sur un Corps p-adique. VII, 163 pages. 1987

1292: J.T. Baldwin (Ed.), Classification Theory. Proceedings,
5. VI, 500 pages. 1987.

1293: W. Ebeling, The Monodromy Groups of Isolated Singulari-
of Complete Intersections. XIV, 153 pages. 1987.

1294: M. Queffélec, Substitution Dynamical Systems – Spectral
lysis. XIII, 240 pages. 1987.

1295: P. Lelong, P. Dolbeault, H. Skoda (Réd.), Séminaire
nalyse P. Lelong – P. Dolbeault – H. Skoda. Seminar, 1985/1986.
283 pages. 1987.

1296: M.-P. Malliavin (Ed.), Séminaire d'Algèbre Paul Dubreil et
ie-Paule Malliavin. Proceedings, 1986. IV, 324 pages. 1987.

1297: Zhu Y.-l., Guo B.-y. (Eds.), Numerical Methods for Partial
rential Equations. Proceedings. XI, 244 pages. 1987.

1298: J. Aguadé, R. Kane (Eds.), Algebraic Topology, Barcelona
6. Proceedings. X, 255 pages. 1987.

1299: S. Watanabe, Yu.V. Prokhorov (Eds.), Probability Theory
Mathematical Statistics. Proceedings. 1986. VIII, 589 pages.
8.

1300: G.B. Seligman, Constructions of Lie Algebras and their
ules. VI, 190 pages. 1988.

1301: N. Schappacher, Periods of Hecke Characters. XV, 160
es. 1988.

1302: M. Cwikel, J. Peetre, Y. Sagher, H. Wallin (Eds.), Function
ces and Applications. Proceedings, 1986. VI, 445 pages. 1988.

1303: L. Accardi, W. von Waldenfels (Eds.), Quantum Probability
Applications III. Proceedings, 1987. VI, 373 pages. 1988.

1304: F.Q. Gouvêa, Arithmetic of p-adic Modular Forms. VIII, 121
es. 1988.

1305: D.S. Lubinsky, E.B. Saff, Strong Asymptotics for Extremal
nomials Associated with Weights on $\mathbb{R}$. VII, 153 pages. 1988.

1306: S.S. Chern (Ed.), Partial Differential Equations. Proceed-
, 1986. VI, 294 pages. 1988.

1307: T. Murai, A Real Variable Method for the Cauchy Transform,
Analytic Capacity. VIII, 133 pages. 1988.

1308: P. Imkeller, Two-Parameter Martingales and Their Quadra-
ariation. IV, 177 pages. 1988.

1309: B. Fiedler, Global Bifurcation of Periodic Solutions with
metry. VIII, 144 pages. 1988.

1310: O.A. Laudal, G. Pfister, Local Moduli and Singularities. V,
pages. 1988.

1311: A. Holme, R. Speiser (Eds.), Algebraic Geometry, Sun-
ce 1986. Proceedings. VI, 320 pages. 1988.

1312: N.A. Shirokov, Analytic Functions Smooth up to the
undary. III, 213 pages. 1988.

1313: F. Colonius, Optimal Periodic Control. VI, 177 pages.
8.

1314: A. Futaki, Kähler-Einstein Metrics and Integral Invariants. IV,
pages. 1988.

1315: R.A. McCoy, I. Ntantu, Topological Properties of Spaces of
tinuous Functions. IV, 124 pages. 1988.

1316: H. Korezlioglu, A.S. Ustunel (Eds.), Stochastic Analysis and
ted Topics. Proceedings, 1986. V, 371 pages. 1988.

1317: J. Lindenstrauss, V.D. Milman (Eds.), Geometric Aspects of
ctional Analysis. Seminar, 1986–87. VII, 289 pages. 1988.

1318: Y. Felix (Ed.), Algebraic Topology – Rational Homotopy.
ceedings, 1986. VIII, 245 pages. 1988

1319: M. Vuorinen, Conformal Geometry and Quasiregular
pings. XIX, 209 pages. 1988.

Vol. 1320: H. Jürgensen, G. Lallement, H.J. Weinert (Eds.), Semigroups, Theory and Applications. Proceedings, 1986. X, 416 pages.
1988.

Vol. 1321: J. Azéma, P.A. Meyer, M. Yor (Eds.), Séminaire de
Probabilités XXII. Proceedings. IV, 600 pages. 1988.

Vol. 1322: M. Métivier, S. Watanabe (Eds.), Stochastic Analysis.
Proceedings, 1987. VII, 197 pages. 1988.

Vol. 1323: D.R. Anderson, H.J. Munkholm, Boundedly Controlled
Topology. XII, 309 pages. 1988.

Vol. 1324: F. Cardoso, D.G. de Figueiredo, R. Iório, O. Lopes (Eds.),
Partial Differential Equations. Proceedings, 1986. VIII, 433 pages.
1988.

Vol. 1325: A. Truman, I.M. Davies (Eds.), Stochastic Mechanics and
Stochastic Processes. Proceedings, 1986. V, 220 pages. 1988.

Vol. 1326: P.S. Landweber (Ed.), Elliptic Curves and Modular Forms in
Algebraic Topology. Proceedings, 1986. V, 224 pages. 1988.

Vol. 1327: W. Bruns, U. Vetter, Determinantal Rings. VII, 236 pages.
1988.

Vol. 1328: J.L. Bueso, P. Jara, B. Torrecillas (Eds.), Ring Theory.
Proceedings, 1986. IX, 331 pages. 1988.

Vol. 1329: M. Alfaro, J.S. Dehesa, F.J. Marcellan, J.L. Rubio de
Francia, J. Vinuesa (Eds.): Orthogonal Polynomials and their Applications. Proceedings, 1986. XV, 334 pages. 1988.

Vol. 1330: A. Ambrosetti, F. Gori, R. Lucchetti (Eds.), Mathematical
Economics. Montecatini Terme 1986. Seminar. VII, 137 pages. 1988.

Vol. 1331: R. Bamón, R. Labarca, J. Palis Jr. (Eds.), Dynamical
Systems, Valparaiso 1986. Proceedings. VI, 250 pages. 1988.

Vol. 1332: E. Odell, H. Rosenthal (Eds.), Functional Analysis. Proceedings, 1986–87. V, 202 pages. 1988.

Vol. 1333: A.S. Kechris, D.A. Martin, J.R. Steel (Eds.), Cabal Seminar
81–85. Proceedings, 1981–85. V, 224 pages. 1988.

Vol. 1334: Yu.G. Borisovich, Yu. E. Gliklikh (Eds.), Global Analysis
– Studies and Applications III. V, 331 pages. 1988.

Vol. 1335: F. Guillén, V. Navarro Aznar, P. Pascual-Gainza, F. Puerta,
Hyperrésolutions cubiques et descente cohomologique. XII, 192
pages. 1988.

Vol. 1336: B. Helffer, Semi-Classical Analysis for the Schrödinger
Operator and Applications. V, 107 pages. 1988.

Vol. 1337: E. Sernesi (Ed.), Theory of Moduli. Seminar, 1985. VIII, 232
pages. 1988.

Vol. 1338: A.B. Mingarelli, S.G. Halvorsen, Non-Oscillation Domains
of Differential Equations with Two Parameters. XI, 109 pages. 1988.

Vol. 1339: T. Sunada (Ed.), Geometry and Analysis of Manifolds.
Procedings, 1987. IX, 277 pages. 1988.

Vol. 1340: S. Hildebrandt, D.S. Kinderlehrer, M. Miranda (Eds.),
Calculus of Variations and Partial Differential Equations. Proceedings,
1986. IX, 301 pages. 1988.

Vol. 1341: M. Dauge, Elliptic Boundary Value Problems on Corner
Domains. VIII, 259 pages. 1988.

Vol. 1342: J.C. Alexander (Ed.), Dynamical Systems. Proceedings,
1986–87. VIII, 726 pages. 1988.

Vol. 1343: H. Ulrich, Fixed Point Theory of Parametrized Equivariant
Maps. VII, 147 pages. 1988.

Vol. 1344: J. Král, J. Lukeš, J. Netuka, J. Veselý (Eds.), Potential
Theory – Surveys and Problems. Proceedings, 1987. VIII, 271 pages.
1988.

Vol. 1345: X. Gomez-Mont, J. Seade, A. Verjovski (Eds.), Holomorphic
Dynamics. Proceedings, 1986. VII, 321 pages. 1988.

Vol. 1346: O. Ya. Viro (Ed.), Topology and Geometry – Rohlin
Seminar. XI, 581 pages. 1988.

Vol. 1347: C. Preston, Iterates of Piecewise Monotone Mappings on
an Interval. V, 166 pages. 1988.

Vol. 1348: F. Borceux (Ed.), Categorical Algebra and its Applications.
Proceedings, 1987. VIII, 375 pages. 1988.

Vol. 1349: E. Novak, Deterministic and Stochastic Error Bounds in
Numerical Analysis. V, 113 pages. 1988.

Vol. 1350: U. Koschorke (Ed.), Differential Topology. Proceedings, 1987. VI, 269 pages. 1988.

Vol. 1351: I. Laine, S. Rickman, T. Sorvali, (Eds.), Complex Analysis, Joensuu 1987. Proceedings. XV, 378 pages. 1988.

Vol. 1352: L.L. Avramov, K.B. Tchakerian (Eds.), Algebra – Some Current Trends. Proceedings, 1986. IX, 240 Seiten. 1988.

Vol. 1353: R.S. Palais, Ch.-l. Terng, Critical Point Theory and Submanifold Geometry. X, 272 pages. 1988.

Vol. 1354: A. Gómez, F. Guerra, M.A. Jiménez, G. López (Eds.), Approximation and Optimization. Proceedings, 1987. VI, 280 pages. 1988.

Vol. 1355: J. Bokowski, B. Sturmfels, Computational Synthetic Geometry. V, 168 pages. 1989.

Vol. 1356: H. Volkmer, Multiparameter Eigenvalue Problems and Expansion Theorems. VI, 157 pages. 1988.

Vol. 1357: S. Hildebrandt, R. Leis (Eds.), Partial Differential Equations and Calculus of Variations. VI, 423 pages. 1988.

Vol. 1358: D. Mumford, The Red Book of Varieties and Schemes. V, 309 pages. 1988.

Vol. 1359: P. Eymard, J.-P. Pier (Eds.), Harmonic Analysis. Proceedings, 1987. VIII, 287 pages. 1988.

Vol. 1360: G. Anderson, C. Greengard (Eds.), Vortex Methods. Proceedings, 1987. V, 141 pages. 1988.

Vol. 1361: T. tom Dieck (Ed.), Algebraic Topology and Transformation Groups. Proceedings, 1987. VI, 298 pages. 1988.

Vol. 1362: P. Diaconis, D. Elworthy, H. Föllmer, E. Nelson, G.C. Papanicolaou, S.R.S. Varadhan. École d'Été de Probabilités de Saint-Flour XV–XVII, 1985–87. Editor: P.L. Hennequin. V, 459 pages. 1988.

Vol. 1363: P.G. Casazza, T.J. Shura. Tsirelson's Space. VIII, 204 pages. 1988.

Vol. 1364: R.R. Phelps, Convex Functions, Monotone Operators and Differentiability. IX, 115 pages. 1989.

Vol. 1365: M. Giaquinta (Ed.), Topics in Calculus of Variations. Seminar, 1987. X, 196 pages. 1989.

Vol. 1366: N. Levitt, Grassmannians and Gauss Maps in PL-Topology. V, 203 pages. 1989.

Vol. 1367: M. Knebusch, Weakly Semialgebraic Spaces. XX, 376 pages. 1989.

Vol. 1368: R. Hübl, Traces of Differential Forms and Hochschild Homology. III, 111 pages. 1989.

Vol. 1369: B. Jiang, Ch.-K. Peng, Z. Hou (Eds.), Differential Geometry and Topology. Proceedings, 1986–87. VI, 366 pages. 1989.

Vol. 1370: G. Carlsson, R.L. Cohen, H.R. Miller, D.C. Ravenel (Eds.), Algebraic Topology. Proceedings, 1986. IX, 456 pages. 1989.

Vol. 1371: S. Glaz, Commutative Coherent Rings. XI, 347 pages. 1989.

Vol. 1372: J. Azéma, P.A. Meyer, M. Yor (Eds.), Séminaire de Probabilités XXIII. Proceedings. IV, 583 pages. 1989.

Vol. 1373: G. Benkart, J.M. Osborn (Eds.), Lie Algebras, Madison 1987. Proceedings. V, 145 pages. 1989.

Vol. 1374: R.C. Kirby, The Topology of 4-Manifolds. VI, 108 pages. 1989.

Vol. 1375: K. Kawakubo (Ed.), Transformation Groups. Proceedings, 1987. VIII, 394 pages. 1989.

Vol. 1376: J. Lindenstrauss, V.D. Milman (Eds.), Geometric Aspects of Functional Analysis. Seminar (GAFA) 1987–88. VII, 288 pages. 1989.

Vol. 1377: J.F. Pierce, Singularity Theory, Rod Theory, and Symmetry-Breaking Loads. IV, 177 pages. 1989.

Vol. 1378: R.S. Rumely, Capacity Theory on Algebraic Curves. III, 437 pages. 1989.

Vol. 1379: H. Heyer (Ed.), Probability Measures on Groups IX. Proceedings, 1988. VIII, 437 pages. 1989

Vol. 1380: H.P. Schlickewei, E. Wirsing (Eds.), Number Theory, Ulm 1987. Proceedings. V, 266 pages. 1989.

Vol. 1381: J.-O. Strömberg, A. Torchinsky. Weighted Hardy Spaces. V, 193 pages. 1989.

Vol. 1382: H. Reiter, Metaplectic Groups and Segal Algebras. XI, 128 pages. 1989.

Vol. 1383: D.V. Chudnovsky, G.V. Chudnovsky, H. Cohn, M.B. Nathanson (Eds.), Number Theory, New York 1985–88. Seminar. V, 256 pages. 1989.

Vol. 1384: J. Garcia-Cuerva (Ed.), Harmonic Analysis and Partial Differential Equations. Proceedings, 1987. VII, 213 pages. 1989.

Vol. 1385: A.M. Anile, Y. Choquet-Bruhat (Eds.), Relativistic Fluid Dynamics. Seminar, 1987. V, 308 pages. 1989.

Vol. 1386: A. Bellen, C.W. Gear, E. Russo (Eds.), Numerical Methods for Ordinary Differential Equations. Proceedings, 1987. VII, 136 pages. 1989.

Vol. 1387: M. Petković, Iterative Methods for Simultaneous Inclusion of Polynomial Zeros. X, 263 pages. 1989.

Vol. 1388: J. Shinoda, T.A. Slaman, T. Tugué (Eds.), Mathematical Logic and Applications. Proceedings, 1987. V, 223 pages. 1989.

Vol. 1000: Second Edition. H. Hopf, Differential Geometry in the Large. VII, 184 pages. 1989.

Vol. 1389: E. Ballico, C. Ciliberto (Eds.), Algebraic Curves and Projective Geometry. Proceedings, 1988. V, 288 pages. 1989.

Vol. 1390: G. Da Prato, L. Tubaro (Eds.), Stochastic Partial Differential Equations and Applications II. Proceedings, 1988. VI, 258 pages. 1989.

Vol. 1391: S. Cambanis, A. Weron (Eds.), Probability Theory on Vector Spaces IV. Proceedings, 1987. VIII, 424 pages. 1989.

Vol. 1392: R. Silhol, Real Algebraic Surfaces. X, 215 pages. 1989.

Vol. 1393: N. Bouleau, D. Feyel, F. Hirsch, G. Mokobodzki (Eds.), Séminaire de Théorie du Potentiel Paris, No. 9. Proceedings. VI, 265 pages. 1989.

Vol. 1394: T.L. Gill, W.W. Zachary (Eds.), Nonlinear Semigroups, Partial Differential Equations and Attractors. Proceedings, 1987. IX, 233 pages. 1989.

Vol. 1395: K. Alladi (Ed.), Number Theory, Madras 1987. Proceedings. VII, 234 pages. 1989.

Vol. 1396: L. Accardi, W. von Waldenfels (Eds.), Quantum Probability and Applications IV. Proceedings, 1987. VI, 355 pages. 1989.

Vol. 1397: P.R. Turner (Ed.), Numerical Analysis and Parallel Processing. Seminar, 1987. VI, 264 pages. 1989.

Vol. 1398: A.C. Kim, B.H. Neumann (Eds.), Groups – Korea 1988. Proceedings. V, 189 pages. 1989.

Vol. 1399: W.-P. Barth, H. Lange (Eds.), Arithmetic of Complex Manifolds. Proceedings, 1988. V, 171 pages. 1989.

Vol. 1400: U. Jannsen. Mixed Motives and Algebraic K-Theory. XIII, 246 pages. 1990.

Vol. 1401: J. Steprāns, S. Watson (Eds.), Set Theory and its Applications. Proceedings, 1987. V, 227 pages. 1989.

Vol. 1402: C. Carasso, P. Charrier, B. Hanouzet, J.-L. Joly (Eds.), Nonlinear Hyperbolic Problems. Proceedings, 1988. V, 249 pages. 1989.

Vol. 1403: B. Simeone (Ed.), Combinatorial Optimization. Seminar, 1986. V, 314 pages. 1989.

Vol. 1404: M.-P. Malliavin (Ed.), Séminaire d'Algèbre Paul Dubreil et Marie-Paul Malliavin. Proceedings, 1987–1988. IV, 410 pages. 1989.

Vol. 1405: S. Dolecki (Ed.), Optimization. Proceedings, 1988. V, 223 pages. 1989.

Vol. 1406: L. Jacobsen (Ed.), Analytic Theory of Continued Fractions III. Proceedings, 1988. VI, 142 pages. 1989.

Vol. 1407: W. Pohlers, Proof Theory. VI, 213 pages. 1989.

Vol. 1408: W. Lück, Transformation Groups and Algebraic K-Theory. XII, 443 pages. 1989.

Vol. 1409: E. Hairer, Ch. Lubich, M. Roche. The Numerical Solution of Differential-Algebraic Systems by Runge-Kutta Methods. VII, 139 pages. 1989.

Vol. 1410: F.J. Carreras, O. Gil-Medrano, A.M. Naveira (Eds.), Differential Geometry. Proceedings, 1988. V, 308 pages. 1989.